GREETINGS,

CARBON-BASED

BIPEDS!

Fiction

Across the Sea of Stars★

Against the Fall of Night

Childhood's End

The City and the Stars

The Deep Range

Dolphin Island

Earthlight

Expedition to Earth

A Fall of Moondust

The Fountains of Paradise

From the Oceans, from the Stars★

The Ghost from the Grand Banks

Glide Path

The Hammer of God

Imperial Earth

Islands in the Sky

The Lion of Comarre

The Lost Worlds of 2001

A Meeting with Medusa★

More Than One Universe★

The Nine Billion Names of God★

The Other Side of the Sky

Prelude to Mars★

Prelude to Space

Reach for Tomorrow

Rendezvous with Rama

The Sands of Mars

The Sentinel★

The Songs of Distant Earth

Tales from Planet Earth★

Tales from the White Hart

Tales of Ten Worlds

The Wind from the Sun

2001: A Space Odyssey

2010: Odyssey Two

2061: Odyssey Three

3001: The Final Odyssey

WITH GENTRY LEE

Cradle

Rama II

The Garden of Rama

Rama Revealed

WITH MIKE McQUAY

Richter 10

WITH MIKE KUBE-McDOWELL

Trigger

Edited by Arthur C. Clarke

Arthur C. Clarke's July 20, 2019

Arthur C. Clarke's Venus Prime, I–IV, by Paul Preuss

Beyond the Fall of Night by Gregory Benford

The Coming of the Space Age

Project Solar Sail

Science Fiction Hall of Fame, I

Three for Tomorrow

Time Probe

UK Only

An Arthur C. Clarke Omnibus★

An Arthur C. Clarke 2nd Omnibus★

The Best of Arthur C. Clarke★

Four Great S.F. Novels★

Of Time and Stars★

2001, Deep Range, Moondust, Etc.★

★Anthologies

Arthur C. Clarke

Edited by Ian T. Macauley

St. Martin's Press / New York

GREETINGS,

Collected Essays

CARBON-BASED

1934–1998

BIPEDS!

Sources for the previously published essays in this volume appear on pages 541–545 and constitute a continuation of this copyright page.

The lines from poems by Lord Dunsany in the essay "Dunsany, Lord of Fantasy" are reprinted by permission of the Estate of Lord Dunsany.

Title page photograph of Arthur C. Clarke is used courtesy of Namas Bhojani.

Library of Congress Cataloging-in-Publication Data

Clarke, Arthur Charles.
    Greetings, carbon-based bipeds! : collected essays, 1934–1998 /
Arthur C. Clarke : edited by Ian T. Macauley.—1st ed.
        p.   cm.
    Includes bibliographical references and index.
    ISBN 0-312-19893-0
    1. Science. 2. Technology.   I. Macauley, Ian T.   II. Title.
Q113.C54P   1999
500—dc21   99-22074
                                                                                    CIP

Book design by Richard Oriolo

First Edition: August 1999

10   9   8   7   6   5   4   3   2   1

Contents

Part III. The 1960s: Kubrick and Cape Kennedy

Part IV. The 1970s:
Tomorrow's Worlds

Part V. The 1980s:
Stay of Execution

Part VI. The 1990s:
Countdown to 2000

Part VII. Postscript:
2000 and Beyond

A c k n o w l e d g m e n t s

Tracing a literary career spanning almost seven decades has not been easy: for years I resisted invitations from editors and readers alike to undertake this mammoth task, and I still cannot quite believe that I was foolish enough to attempt it. The person who takes much of the blame is Robert Weil, until recently senior editor at St. Martin's. His arm-twisting finally did the trick.

The task was spread over two years and three continents. In the UK, brother Fred, secretary Chris Howse, niece Angela Edwards, and their colleagues at the Rocket Publishing Company and the Arthur C. Clarke Foundation of Great Britain excavated early pieces from long-vanished publications. (David N. Samuelson's *Authur C. Clarke: A Primary and Secondary Bibliography* [Boston: G. K. Hall & Co., 1984] was an invaluable tool in this search.)

The monumental task of getting this material into shape, correcting the proofs, and so on, was undertaken by my longtime friend Ian Macauley—to whom I dedicated one of my first novels, *Islands in the Sky,* back in 1952! His wife, Marnie Winston-Macauley, also lent her talents and time generously, as did their son Simon.

Here in Colombo, my office staff waded through hundreds of manuscripts. Executive Officer Nalaka Gunawardene—who now serves as my "backup memory"—helped me in the difficult task of deciding which pieces were worthy of publication. Technical Assistant Rohan de Silva spent hours scanning early texts with optical character recognition (OCR) software, which my secretaries Tony Thurgood and Rohan Amarasekera then checked diligently for errors.

Above all, my gratitude to my longtime partner, Hector Ekanayake, his wife, Valerie, and their three lovely daughters, Cherene, Tamara, and Melinda, for providing and sustaining the environment in which I have lived and worked since the 1960s.

And last but far from least—my killer Chihuahua, Pepsi, ferocious guardian of my privacy.

Sir Arthur C. Clarke, CBE
Colombo, Sri Lanka
February 3, 1999

P r e f a c e

During the last sixty years, I must have written at least a thousand pieces of nonfiction of every possible length, from a few paragraphs to entire books. Everything from the first half century that seemed worth preserving has appeared in *The Challenge of the Spaceship* (1959), *Voices from the Sky* (1965), *The View from Serendip* (1977), *1984: Spring—a Choice of Futures* (1984), and *By Space Possessed* (1993). In addition, my technical papers were assembled in the Marconi Award volume *Ascent to Orbit* (1984), and in 1996 David Aronowitz, after heroic (some might say misplaced) archaeological research, published my very earliest writings in a deluxe edition, cunningly entitled *Childhood's End*.

For some time, Robert Weil of St. Martin's Press has urged me to assemble a representative selection of my nonfiction writings in a single volume, and the success of my friend Martin Gardner's splendid anthology *The Night Is Large* has demolished the last of my feeble excuses. Gardner's book is divided into sections devoted to his many interests: science, art, philosophy, religion, and so on. But not being such a polymath (who is?), I have put my essays in chronological order, with much help from David Samuelson's bibliography and Neil McAleer's biography.

Each of the six decades from the 1940s through the 1990s is preceded by a brief introduction, which will also serve as a reminder of the profound cultural, political, and scientific revolutions that were taking place while the pieces were being written and that are, of course, being reflected in them. Tempted though I occasionally have been to do a little editing, with an advantage of 20/20 hindsight, each essay is presented much as it first appeared in print. Some repetition is therefore inevitable. When necessary, footnotes cover developments that have confirmed, or refuted, my assertions. Often it has been more interesting to see where (and why) I went wrong, than where I happened to be right.

Sir Arthur C. Clarke, CBE

THE

1930s AND 1940s:

ROCKETS AND RADAR

A teenaged Clarke working on one of his early experiments in communications,
surrounded by his beloved collection of science fiction books and magazines.

Introduction

The very first printed item my indefatigable bibliographer David N. Samuelson (*Arthur C. Clarke: A Primary and Secondary Bibliography*, Boston, G. K. Hall & Co., 1984) has been able to locate is in the autumn 1932 issue of the *Huish Magazine*, published four times a year by the school I attended from 1927 to 1936. Every weekday I would cycle almost six miles from Ballifants Farm, Bishops Lydeard, to the county town of Taunton, attend classes at Huish Grammar School, and then cycle home again. When in April 1993 the old school was transformed into Richard Huish College, I was honored to be offered the title of *president*, but instead settled for *master*, having long been inspired by two famous verses by that most prolific of writers, Anon.:

> *I am the master of this college:*
> *What I don't know isn't knowledge.*

and

> *In all infinity*
> *There is no one so wise*
> *As the master of Trinity.*

The *Huish Magazine* was edited by our English instructor, a fierce little Welshman, Capt. E. B. Mitford, M.C. Many years ago I was happy to dedicate my collection of short stories *The Nine Billion Names of God* to him.

About once a week Mitty would assemble, after school hours, half a dozen students who showed intimations of literacy and cajole them into writing something for the magazine. These struggling authors were rewarded immediately—and not, as all too often happens, on publication or even later. Their royalties came from a large bag of toffees.

It was at Huish Grammar that I had my first encounter with the world of high tech, in connection with an "assignment" from the magazine's editor, Mitty. At the time I was living in hopes that Mitty had forgotten about my existence, but alas I was mistaken. For one day, that potentate asked me when I had last contributed anything to the magazine. I took the hint and went hastily in search of "copy," first at the school and then at the nearby Technical Institute.

Wandering disconsolately around Huish, I saw some students emerge from the school and travel westwardly and swiftly toward the institute.

Joining the group, I walked with them until we arrived at a rather unimposing edifice, which we entered. I found myself in a large room, filled with strange instruments and incomprehensible machinery. The place reverberated to the roar of many voices and the shrill sound of violated dynamos. From various brass knobs, continuous streams of sparks ripped and crackled, while every so often I was dazzled by crazily swinging beams of light from the mirrors of tormented galvanometers. And ever and anon came the sound of splintering glass, mingled with cries of pain and anguish.

Into this maelstrom of general upheaval I was thrust headlong. Recovering from the first shock, I sought a safe refuge from which I could survey the uproar without exposing myself to much physical danger. I took up a position behind a bench parallel to a wall and was able to perceive, at first, a group of youths gathered around a strange apparatus of spinning disks, kept in motion by a technician cranking a handle. This apparatus, which I later used to transmit my voice over a beam of light, became the subject of my article for the magazine, and you can read about it in *How the World Was One.*

I acquired my first science fiction magazine sometime toward the end of 1930, in my thirteenth year—and my life was irrevocably changed. The March 1930 issue of *Astounding Stories*—or to give its full name, *Astounding Stories of Super-Science*—was not the first science fiction magazine I had ever seen, as I wrote in my 1989 autobiography, *Astounding Days.* My initiation had been provided two years earlier by the November 1928 issue of *Amazing Stories.* I digress slightly to explain just how amazing that issue of *Amazing* really was. I had forgotten it for almost five decades, when the realities of the Space Age caught up with it in a manner so uncanny that it still sends a frisson down my spine.

The magnificent cover of this issue was by the most famous of all science fiction artists, Frank R. Paul (1884–1963). It showed Jupiter looming over the improbably tropical landscape of one of its inner moons, and being admired by visitors from Earth as they streamed out of their silo-shaped spaceship. The giant planet is dominated by the oval of the great red spot and by bands of clouds sculpted into loops and swirls by storms the size of Earth. The details of these atmospheric disturbances could not possibly have been observed by telescopes in 1928—yet they are precisely the same as those revealed by the Voyager space probes a half century later!

The 1920–30 period, just before the Depression, was the heyday of the pulp magazines, of which *Amazing* and *Astounding* certainly were. The pulps were the popular fiction of the barely literate masses, who have now defected to TV, and though the science fiction ones were high on my list, the majority of the population were reading westerns, romances, and mysteries. I read that March 1930 *Astounding* from cover to cover, doubtless when I should have been doing algebra or geometry or (ugh) Latin at Huish. Over the next few years I acquired more copies of *Astounding* mostly by chance. It was commonly believed at that time that these "Yank pulps" invaded the United Kingdom as ballast in returning cargo ships. These magazines eventually ended up at Woolworth at three pennies apiece, and I sacrificed part of every lunch hour to sift through the dross in search of a science fiction gem, filling up the gaps in my *Astounding* collection.

The first issue of *Astounding* (January 1930) has a cover neatly combining all the clichés of pulp fiction. In the foreground, an aviator who has crashed near the South Pole is battling with an unusual species of beetle; an entomologist would be more amazed to discover that it has a fine set of needle-sharp teeth, rather than the minor detail that it is man-sized. And you will not be surprised that there is a damsel in a very skimpy fur coat, especially for the Antarctic.

But I would not like anyone to think that my boyhood reading was limited to *Astounding*. I devoured "real" books as well, encountering the masters H. G. Wells and Jules Verne at W. H. Smith booksellers and the Taunton Public Library. At W. H. Smith's, I could not afford the shillings to purchase these books, so I again spent my lunch hours reading swiftly, both at the bookshop and at the library.

Through *Astounding* I learned of the multidimensional world. In the January 1931 issue, Murray Leinster wrote of hyperspace in his "The Fifth Dimension Catapult." We now take these dimensions for granted since Einstein showed us in his *Theory of Relativity* that time and space make up a fourth dimension; quantum physics gives us ultimate dimensions as every probability is recorded, from the fifth to the eleventh and ad infinitum! To those of us who are stuck in two dimensions and cannot conceive of a cube, we are offered the literature of flatlands, worlds without thickness. E. A. Abbott's *Flatland: A Romance of Many Dimensions* (1884) is a case in point.

In 1933, an *Astounding* editorial gave us the concept of the expanding universe, a theory many of us accept today. But it is something of a surprise that it was familiar to some way back then. Even though Hubble had discovered the "redshift" four years earlier, it had pretty well been dismissed as a Doppler effect caused by recession of more distant galaxies. In the thirties, astronomers were content to go along with the Doppler theory until Grote Reber, who helped found the science of radio astronomy, measured the radio waves of distant galaxies and proved that most stars shifted to the red and were moving away from us. *Ascent to Orbit,* my scientific autobiography, reprints a 1949 article, "The Radio Telescope," which reproduced Reber's first map.

To what extent can science fiction be blamed for any of these technical advances or "regressions"? Ronald Reagan usurped the phrase *star wars* after seeing too many movies, and we are now stuck with this concept of technological warfare. This was addressed much earlier in my short story "Superiority" (1951), which was reprinted in *Expedition to Earth* (1953). This story describes the inevitable fate of those who become intoxicated with technological obscenities.

The short story "Twilight," written under a John W. Campbell pseudonym, appeared in the December 1934 *Astounding*. Campbell, then its new editor, wrote a haunting mood piece that described an age 7 million years hence when the human race has lost all curiosity and merely exists idly in a world run by the perfect machines its ancestors had created. "Twilight" had a major impact on my own career and certainly influenced "Against the Fall of Night," which I started writing in 1937, completed in 1946, and offered to Campbell several times and was rejected, before it finally appeared in 1948 in *Startling Stories*.

As I approached the critical age of nineteen, I do not recall that I had any particular ambition, except to read science fiction magazines, look at the Moon through my homemade telescopes, and play with Meccano—the most wonderful toy ever invented up to that time. With its perforated girders and strips, its angle brackets and flanges, its rods and cranks, its threaded pins and axle rods, and its vast array of gleaming brass gearwheels, a Meccano set permitted the construction of almost any conceivable piece of engineering. It came in ten sizes, and one of my boyhood dreams was to own the fabulous No. 10 set, but its cost was so astronomical (twenty pounds, I believe!) that such an ambition was pure fantasy. Well, I now have a cure for the complete Meccano syndrome: it's called a personal computer!

In the late 1930s, when I was an officer of the British Interplanetary Society as well, of course, as a dyed-in-the-wool science fiction fan, I began to regularly correspond with John W. Campbell of *Astounding Stories,* writing in 1939 of the need to keep spaceship crews down to the minimum. In 1946, I debated the advantage of nuclear fuel over chemicals for rocket propulsion. In 1948, while at King's College, London, I wrote Campbell about the the advantages of cavity magnetrons, which had already been developed by the Russians.

Dunsany, Lord of Fantasy

In the 1920s and 1930s, fantasy and science fiction writers such as Lord Dunsany,
Clark Ashton Smith, and C. S. Lewis beguiled readers with prose that
set fire to their souls. One such soul was a young Arthur C. Clarke,
who in the early 1940s began corresponding with Dunsany.
Here is Clarke's tribute to this genius.

As the first half of the decade was dominated by World War II, it is not surprising that during 1940–44 I wrote only eight articles—and a mere six very short pieces of fiction (1943 was a complete blank!).

The essay that follows was actually written in 1942, but this version was published two years later. Around that time I began corresponding with the famed Irish dramatist Lord Dunsany, and when my own *The Exploration of Space* appeared in 1951, he invited me to visit him at Dunstall Castle, a few miles southeast of London.

I took with me a copy of his fantasy *The Charwoman's Shadow* (Putnam's, 1926), which he duly autographed with a sweeping *DUNSANY* running right across the page; it was the only time I ever saw anyone use a quill pen and then sprinkle the result with fine sand to dry the ink. But he did more than this: in the quarter century since it was written, he had noticed a slight error in geography, which he corrected in two places with the same quill pen. "The Country Towards Moon's Rising" was transformed into "the Country Beyond Moon's Rising." What a difference!

The ending of *The Charwoman's Shadow* is the finest piece of pure magic I know in the whole of literature, and now that I can personally identify with the first sentence, I find it almost unbearably moving:

> And there came upon him at last those mortal tremors that are about the end of all earthly journeys. He hastened then. And before the human destiny overtook him he saw one morning, clear where the dawn had been, the luminous rock of the bastions and glittering rampart that rose up sheer from the frontier of "the Country Beyond Moon's Rising." This he saw though his eyes were dimming now with fatigue and his long sojourn on earth; yet if he saw dimly he heard with no degree of uncertainty the trumpets that rang out from those battlements to welcome him after his sojourn, and all that followed him gave back the greeting with such cries as once haunted certain valleys at certain times of the moon. Upon those battlements and by the opening gates were gathered the robed Masters that had trafficked with time and dwelt awhile on Earth, and handed the mysteries on, and had walked round the back of the grave by the way that they knew, and were even beyond damnation. They raised their hands and blessed him.
>
> And now for him, and the creatures that followed after, the gates were wide that led through the earthward rampart of "the Country Beyond Moon's Rising." He limped towards it with all his magical following. He went therein, and the Golden Age was over.

So ends *The Charwoman's Shadow;* now for the essay I wrote almost ten years before I had the privilege of meeting the author himself.

Dunsany is a poet in the truest sense, but it is prose rather than in verse that his finest work has been done. No one has ever approached his skill in suggesting, so flawlessly and with such economy of means, that the world is not exactly as we suppose. No one can make the blood run cold with a simple phrase, no one can suggest so much while saying so little. His stories sparkle with ideas, often single sentences that challenge the mind with vertiginous implications. Under the magic of his art, the commonest things become enchanted, and when his imagination soars away from earth, we enter realms of fantasy indeed.

> *By walls of cities not of Earth*
> *All my winged dreams have run,*
> *And known the demons that had birth*
> *In planets of another sun.*

Let me quote a few passages that may give some idea of the flavor of his finest stories, little though I like to wrench these jewels from their settings:

> Something so huge that it seemed unfair to man that it should move so softly stalked splendidly by them. . . .
>
> There the Gibbelins lived and discreditably fed.
>
> There watched him ceaselessly from the Under Pits those eyes whose duty it is.
>
> . . . then he began to fall. It was long before he believed and truly knew that this was really he that fell from this mountain, for we do not associate such dooms with ourselves; but when he had fallen for some while through the evening and saw below him, where there had been nothing to see, the glimmer of tiny fields, then his optimism departed.

And one more passage, the conclusion of "The Probable Adventures of the Three Literary Men," who were so rash as to try to steal the golden box of poems from the House at the End of the World.

> . . . and then it befell that as they drew near safety, in the night's most secret hour, some hand in an upper chamber lit a shocking light, lit it and made no sound.
>
> For a moment it might have been an ordinary light, fatal as even that could very well be at such a moment as this; but when it began to follow them like an eye and grow redder and redder as it watched them, then even optimism despaired.
>
> And Sippy very unwisely attempted flight, and Slorg even as unwisely tried to hide; but Slith, knowing well why that light was lit in that secret upper chamber and who it was that lit it, leaped over the edge of the World and is falling from us still through the unreverberate blackness of the abyss.

The critics have not always been kind to Dunsany. This lack of appreciation may in part be due to Dunsany's unfortunate choice of parents, for an artist with a title is regarded as a dilettante and is not taken seriously.

> *From little fountain pens they wring*
> *The last wee drop of inky spite;*
> *"We do not like the kind of thing*
> *That lords," they say, "most likely write."*

But who can deny that the man who penned these lines is a true artist:

> *So much there is to catch,*
> *And the years so short,*
> *That there is scarce time to snatch*
> *Pen, palette, or aught,*
> *And to seize some shape we can see,*
> *That others may keep*
> *Its moment of mystery,*
> *Then go to our sleep.*

The radio has done much to make Dunsany known to a wider public, for he has written plays for broadcasting as well as the stage. *Golden Dragon City* and *The Use of Man* have been broadcast several times; those who are familiar with the latter and its implied denunciation of hunting may be surprised to know that its author is a master of foxhounds.

Joseph Jorkens, the well-known club raconteur who can always be relied upon to produce a good story in return for a drink, is Dunsany's best-known character and his adventures seem to be popular with the general public. At any rate such statistical studies as I have made in libraries appear to show this. Jorkens thinks nothing of finding icebergs in the Red Sea in midsummer, or a diamond that fully justifies his description of it as "a large one"—he walked across it for many hours under the impression that it was a frozen lake.

Nevertheless, much as I like Jorkens, the far rarer stories of World's End and other places, related in *Tales of Wonder, A Dreamer's Tales,* and *The Book of Wonder,* appeal to me more strongly. And of them all this is the one that at present moves me most. It is called "The Field" and was written more than thirty years ago.

Not far from London is a field, beautiful and peaceful, where the poet loved to rest. Yet as he grew to know it better, there seemed something ominous about the place and the feeling grew with each successive visit.

He made inquiries and found that nothing had happened there, so that it was from the future that the field's troubles came.

> Once to distract my thoughts I tried to gauge how fast the system was trickling, but I found myself wondering if it flowed faster than blood. . . . And then the fancy came to me that it would be a terribly cold place to be in the starlight, if for some reason one was hurt and could not get away.

So at last he took to the field a friend who would be able to tell him what evil thing was going to happen there.

> By the side of the stream he stood and seemed very sad. Once or twice he looked up and down it mournfully, then he bent and looked at the kingcups, first one and then the other, very closely, and shaking his head.
> For a long while he stood in silence, and all my old uneasiness returned, and my bodings for the future.
> And then I said, "What manner of field is it?"
> And he shook his head sorrowfully.
> "It is a battlefield," he said.

Is this only a story? Or is there such a field? There is foreboding there, matched equally by another line of Dunsany's:

> Over mossy girders the old folk come back.

I cannot leave Dunsany without making some mention of the incomparable artist, S. H. Sime, who has illustrated so many of his stories. No one has ever captured the spirit of fantasy more perfectly than Sime, though sometimes Finlay (whose style is similar) has approached him.

Let us gather one more quotation. These words prefaced the first volume of Georgian poetry when it appeared in 1912:

> Of all materials for labor, dreams are the hardest; and the artificer in ideas is the chief of workers, who out of nothing will make a piece work that will stop a child from crying or lead nations to higher things. For what is it to be a poet? It is to see at a glance the glory of the World, to see beauty in all its forms and manifestations, to feel ugliness like a pain, to resent the wrongs of others as bitterly as one's own, to know mankind as others know single men, to know Nature as botanists know a flower, to be thought a fool, to hear at moments the clear voice of God.

———

As I take farewell of Lord Dunsany, I am happy to report that our correspondence has recently been located and will soon be published, together with my much more extensive exchange of letters with C. S. Lewis. I have not seen it for decades and fear that some embarrassments await.

Rockets

Fifty years ago, long before rockets took astronauts to the moon, there were few believers in the scientific community. One such believer in the future of space travel was Willy Ley, who wrote about the practical applications of rocketry. Here Clarke reviews and discusses Mr. Ley's visions.

Early in 1944 I must have decided that writing might play a significant part in my life, and therefore I'd better keep track of what I was doing. So I purchased a small (10 × 16 cm) notebook, whose disintegrating remains are spread out on my desk at this moment. The full entry on page one is

> What I believe   1,000 words
> Statement of philosophy   Written Jan. 1944
>   Circulated with Probe Newsletter

I have no idea what profundities this long-lost piece contained. It would be interesting to compare it with the—I hope—more mature

Credo, written almost half a century later. And I should explain that the *Probe Newsletter* consisted of a few sheets put out from time to time by Harold Chibbett, now deceased, a leading member of the London Science Fiction Circle, who was fascinated by the paranormal. I am sure he would have been delighted to have known that he was a major component of Harry Purvis in *Tales from the White Hart*—and thus a close relation of Dunsany's Mr. Jorkens.

The year 1944 saw the first comprehensive book in English on rockets, written by Willy Ley, who had been sensible enough to get out of Germany while there was still time. I shall always be grateful to Willy for keeping me supplied with American science fiction magazines in the dark days of World War II, and I was happy to review many of his most important books.

Rockets: The Future of Travel Beyond the Stratosphere

by Willy Ley (Viking Press, 1944)

Willy Ley's long-awaited book need not be recommended to all who are interested in astronautics—and presumably all science fiction fans are included in that category. It supersedes at once all earlier books on the subject in English, and it comes at a time when the rocket has done all and more than predicted by its most enthusiastic supporters. In one way, Willy Ley has been unlucky in the date of publication—early 1944. Developments since the beginning of the year have already made some of his conclusions out-of-date, and two of the most important events in the history of the rocket—the first pure rocket airplane and the giant long-range rocket shell—were not revealed to the world until after the book went to press. Let us hope that the first edition will sell quickly so that subsequent printings can be brought up-to-date.

The first fifty pages of the book contain a detailed history of interplanetary fiction since the first imaginary voyage to the Moon, written in A.D. 120. It is a little odd to think that these stories, which have been popular for eighteen centuries, will soon be written no more.

The history of the rocket in warfare is much more extensive than generally realized and Ley devotes a chapter to it. How many people know that in 1807 Copenhagen was largely destroyed by a bombardment of twenty-five thousand rockets? But the development of a rifle artillery closed the history of war rockets for over a century, and only now are they coming back into their own. Ley shows a curious conservatism in his remarks about rocket weapons, and he makes no mention of the extremely important rocket boosters now used in many aircraft.

Under the appropriate title "Prophets With Some Honor," Ley traces the history of the spaceship on the scientific rather than the fictional side. Most of the credit for the serious conceptions appears to be due to the German Hermann Ganswindt and the Russian Konstantin Eduardovitch Tsiolkovsky. It is pleasant to know that for the last ten years of his life Tsiolkovsky was a famed and respected man in the Soviet Union, certainly the first expert in astronautics ever to win national renown for his work.

However, the real foundations of astronautics were well and truly laid by Hermann Julius Oberth in 1923 with his mathematical study *Die Rakete zu dem Planeternraumen*. In this book, for the first time, the fundamental equations of astronautics were set out and the problem was analyzed as an engineering undertaking that might be achieved within the next few decades.

Oberth, a Hungarian-born German, is just over fifty, and a lot of rubbish has recently appeared about him in the British press. He may or may not be associated with the V-2 (that would help to explain the delay in its appearance—as a practical experimenter, Oberth left everything to be desired), but he will certainly be honored in the years to come as the founder of modern astronautics.

Ley devotes two fascinating chapters (reproduced with some alterations from the *Astounding* article "The End of the Rocket Society") to the history of astronautics in Germany. It contains many lessons that all interplanetary societies would do well to bear in mind.

The German rocket society, the Verein für Raumschiffart, had succeeded in virtually perfecting the small liquid-fuel rocket motor when Hitler came into power. It seems likely that the Nazis took over the society on principle rather than through any realization of the rocket's ultimate possibilities. That came later, and the purely scientific work of the German society is undoubtedly responsible for the enormous interest the Germans have shown over reaction propulsion during this war.

Only the last seventy pages of the book deal with the spaceship and the problems associated with it. Ley points out that it is not difficult to send an unmanned rocket outside the Earth's atmosphere, though when he wrote, he certainly had no idea just how quickly it was going to be done. It is, however, immensely more difficult to effect a landing on even the nearest body, still less return to Earth. With the fuels we possess today, interplanetary travel is barely possible using complicated engineering principles and tolerating shocking overall inefficiences. For it to be practical, we must have higher exhaust velocities and/or refueling stations in space. Both of these will come in time, though when, Ley does not venture to predict.

Ley makes no mention of the British Interplanetary Society cellular design, which overcomes some of the limitations of present-day fuels, in particular the impossibly large rates of combustion at takeoff. The American Rocket Society will probably take a poor view of the paragraph he devotes to their work. However, as they have officially deserted astronautics and are now purely a rocket society, they deserve it.

One of the most interesting parts of the book is the appendix, which contains among other things a very extensive bibliography. Ley points out that "a complete literary history of that variety of imaginative literature which is based on facts and theories of a scientific nature—so-called 'science fiction'—still remains to be written." When the extent of this class of literature is considered, this is certainly a remarkable oversight.

Ley's book is written for the popular and semitechnical press, but nevertheless contains a good deal of "meat." It is extremely well-written and readable, with occasional flashes of wit. There is no doubt that it will admirably fulfill the author's intentions.

The Coming Age

of Rocket Power

Rocketry in the 1940s was far more theoretical than practical and only began to emerge as a true science as the 1950s arrived. In the following book review, Clarke discusses the shortcomings and potentials of the forces that would take man into space.

Things were now happening rapidly in the rocket field, and it would soon be impossible to keep up with the literature. But I was doing my best.

The Coming Age of Rocket Power

by G. Edward Pendray (Harper Bros., 1945)

This volume will inevitably be compared with Ley's *Rockets,* and the two books present an interesting contrast in viewpoints. Pendray has the advantage of writing in the post V-2 era, but he has not made the most of this, possibly owing to lack of time (the preface is dated December 1, 1994). Throughout the book the emphasis is on the immediate, practical implications of the rocket, and those whose interests lie primarily in astronautics will be disappointed. However, by the very moderation of its claims, the book will make quaint reading in fifty years' time.

One of the most interesting chapters in the book is the description of the work of Prof. Robert Hutchins Goddard, whom Pendray seems to have coaxed some way out of his shell. The American society is also well featured and some enlightened references are made to the current work of Reaction Motors Inc., which appears to have absorbed the leading lights of the American Rocket Society. But Pendray does less than justice to the Germans, whose achievements have completely overshadowed those of all other nations. He does not seem to appreciate the importance of the V-2, which has not unjustly been called the first spaceship. There are some misleading statements about its performance, but considering the time of publication this is not the author's fault.

After an interesting and rather depressing analysis of the economics of long-range rocket transport, Pendray devotes a brief chapter to the lunar voyage (with a passing crack at the "intellectual escapists of Europe.") Throughout this chapter one gets the impression that the author is trying hard to be respectable. Witness: "as a practical matter we do not seem to be much nearer the achievement of space flight now than were Bishop Godwin and his swans."

Alas for the flying years! It is a long way from Gawain Edwards of *Wonder Stories* to the important executive of Westinghouse. The shades of the prison house have begun to close upon the growing boy.

But in spite of all this, Pendray expresses well enough the feeling that most rocket enthusiasts have, however matter-of-fact they pretend to be:

"We feel somewhat privileged, as though we had stood in these years at some obscure crossroads in history, and seen the world change. We do not know exactly what we have loosed into the earth . . . but we feel in our souls that it is magnificent and wonderful, and that the human race will be richer for it in time to come."

This is an excellent little book, within its self-imposed limitations, perhaps as good as can be written at the present time.

Extraterrestrial

Relays

In this 1945 technical paper, Clarke details his concept of geosynchronous
communications satellites—that is, those that stay in a fixed position
above the Earth. Today mankind, thanks to this "discovery," takes
for granted instant and continuous telephone, fax, television,
E-mail, and computer services.

There were seventy-four items in my first flimsy notebook before I switched to a more impressive loose-leaf volume in 1950. Almost all the fiction, I am pleased to see, is still in print, though sometimes in greatly enlarged and revised from. It includes, as Item 21, the much rejected *Against the Fall of Night*—the precursor of *The City and the Stars*—as well as "Guardian Angel," which later became part of *Childhood's End.* And Item 44 is *Prelude to Space,* written in just twenty days in the summer of 1947 (though it had been preceded by months of note taking).

The next year, in November 1948, I started to write a story that was clearly a blatant exercise in wish fulfillment. Its theme was a science fiction writer's first trip to Mars, where he would see how well the reality lived up to his early fantasies. The project was quickly abandoned—but three years later became *The Sands of Mars*. And the novel itself, on CD-ROM with a video message from myself, would have been on the Red Planet by now, for the amusement of future colonists—if, alas, the Russian Mars Lander hadn't ended up at the bottom of the Pacific.

Item 62 was written in a white heat over Christmas 1948, as an entry for a British Broadcasting Corporation short-story competition. "The Sentinel" was promptly rejected, though I sold it later and published it in my first collection, *Expedition to Earth*. The basic idea was simple: the discovery on the Moon of an alien artifact placed there millions of years before, to warn the rest of the universe that a space-faring species had arisen on the third planet of an obscure sun. Although the BBC gave it thumbs-down, I still think it would be a good idea for a movie.

Quite apart from the pressures of service life—though I have only happy memories of the years during which I climbed from Aircraftsman II (under training) to Flight Lieutenant—I was discouraged from writing by the thought that any and all literary efforts would have to be scrutinized by the Royal Air Force censor. And, perhaps even worse, my immediate superior officers.

I am not sure, however, if either they or the censor saw the four-page memo "The Space Station: Its Radio Applications," which I typed out on my Remington noiseless portable in May 1945. The top copy is now in the Smithsonian, and the full text will be found in *Ascent to Orbit*.

The history of my 1945 communication-satellite paper has been told countless times (you will find my final—I hope—version below), but as it is the most important thing I ever wrote, it can hardly be excluded. However, remembering how Stephen Hawking's editor told him that every equation would halve his sales, I have stripped out all the technicalities. So here are the words without the music.

The postscript, written in the euphoria of the Atomic Dawn, now makes embarrassing reading. Nuclear rockets never lived up to their early promise, and though they may one day have deep-space applications, their use will never be permitted in heavily built-up areas such as planet Earth.

Thus from *Wireless World* of October 1945 we have my paper "Can Rocket Stations Give Worldwide Radio Coverage?"

Although it is possible, by a suitable choice of frequencies and routes, to provide telephony circuits between any two points or regions of the Earth for a large part of the time, long-distance communication is greatly hampered by the peculiarities of the ionosphere, and there are even occasions when it may be impossible. A true broadcast service, giving constant field strength at all times over the whole globe, would be invaluable, not to say indispensable, in a world society.

Unsatisfactory though the telephony and telegraph position is, that of television is far worse, since ionospheric transmission cannot be employed at all. The service area of a television station, even on a very good site, is only about one hundred miles across. To cover a small country like Great Britain would require a network of transmitters, connected by coaxial lines, waveguides, or VHF relay links. A recent theoretical study has shown that such a system would require repeaters at intervals of fifty miles or less. A system of this kind could provide television coverage, at a very considerable cost, over the whole of a small country. It would be out of the question to provide a large continent with such a service, and only the main centers of population could be included in the network.

The problem is equally serious when an attempt is made to link television services in different parts of the globe. A relay chain several thousand miles long would cost millions, and transoceanic services would still be impossible. Similar considerations apply to the provision of wideband frequency modulation and other services, such as high-speed facsimile, which are by their nature restricted to the ultrahigh frequencies.

Many may consider the solution proposed in this discussion too far-fetched to be taken very seriously. Such an attitude is unreasonable, as everything envisaged here is a logical extension of developments in the last ten years—in particular the perfection of the long-range rocket of which the V-2 was the prototype. While this article was being written, it was announced that the Germans were considering a similar project, which they believed possible within fifty to a hundred years.

Before proceeding further, it is necessary to discuss briefly certain fundamental laws of rocket propulsion and astronautics. A rocket that achieved a sufficiently great speed in flight outside the Earth's atmosphere would never return. This orbital velocity is five miles per second, and a rocket that obtained it would become an artificial satellite, circling the world forever with no expenditure of power—a second moon, in fact. The German transatlantic rocket A-10 would have reached more than half this velocity.

It will be possible in a few more years to build radio-controlled rockets that can be steered into such orbits beyond the limits of the atmosphere and left to broadcast scientific information back to Earth. A little later, manned rockets will be able to make similar flights with sufficient excess power to break the orbit and return to Earth.

There are an infinite number of possible stable orbits, circular and elliptical, in which a rocket would remain if the initial conditions were correct. The velocity of five miles per second applies only to the closest possible orbit, one just outside the atmosphere, and the period of revolution would be about ninety minutes. As the radius of the orbit increases, the velocity decreases, since gravity is diminishing and less centrifugal force is needed to balance it. The Moon, of course, is a particular case.

One orbit, with a radius of twenty-six thousand miles, has a period of exactly twenty-four hours. A body in such an orbit, if its plane coincided with that of the Earth's equator, would revolve with the Earth and would thus be stationary about the same spot above the planet. It would remain fixed in the sky of a whole hemisphere and unlike all other heavenly bodies would neither rise nor set. A body in a smaller orbit would revolve more quickly than the Earth and so would rise in the west, as indeed happens with the inner moon of Mars.

Using material ferried up by rockets, it would be possible to construct a space station in such an orbit. The station could be provided with living quarters, laboratories, and everything needed for the comfort of its crew, who would be relieved and provisioned by regular rocket service. This project might be undertaken for purely scientific reasons as it would contribute enormously to our knowledge of astronomy, physics, and meteorology. A good deal of literature has already been written on the subject.

Although such an undertaking may seem fantastic, it requires for its fulfillment rockets only twice as fast as those already in the design stage. Since the gravitational stresses involved in the structure are negligible, only the very lightest materials would be necessary and the station could be as large as required.

Let us now suppose that such a station were built in this orbit. It could be provided with receiving and transmitting equipment (the problem of power will be discussed later) and could act as a repeater to relay transmissions between any two points on the hemisphere beneath, using any frequency that will penetrate the ionosphere. If directive arrays were used, the power requirements would be very small, as direct line-of-sight transmission would be used. There is the further important point that arrays on Earth, once set up, could remain fixed indefinitely.

Moreover, a transmission received from any point on the hemisphere could be broadcast to the whole of the visible face of the globe, and thus the requirements of all possible services would be met.

It may be argued that we have as yet no direct evidence of radio waves passing between the surface of Earth and outer space; all we can say with certainty is that the shorter wavelengths are not reflected back to Earth. Direct evidence of field strength above the Earth's atmosphere could be obtained by V-2 rocket technique, and it is hoped that someone will do something about this soon as there must be quite a surplus stock somewhere! Alternatively, given sufficient transmitting power, we might obtain the necessary evidence by exploring for echoes from the Moon.

A single station could only provide coverage to half the globe, and for world service three would be required, though more could be readily utilized. The stations would be arranged equidistantly around the Earth, and the following longitudes appear to be suitable:

30E—Africa and Europe

150E—China and Oceana

90W—The Americas

The stations in the chain would be linked by radio or optical beams, and thus any conceivable beam or broadcast service could be provided.

The technical problems involved in the design of such stations are extremely interesting, but only a few can be gone into here. Batteries of parabolic reflectors would be provided, of apertures depending on the frequencies employed. Assuming the use of waves of three thousand megacycles per second, mirrors about a yard across would beam almost all the power onto the Earth. Larger reflectors could be used to illuminate single countries or regions for the more restricted services, with consequent economy of power. On the higher frequencies it is not difficult to produce beams less than a degree in width, and as mentioned before, there would be no physical limitations on the size of the mirrors. (From the space station, the disk of the Earth would be a little over seventeen degrees across.) The same mirrors could be used for many different transmissions if precautions were taken to avoid cross-modulation.

It is clear from the nature of the system that the power needed will be much less than that required for any other arrangement, since all the energy radiated can be uniformly distributed over the service area, and none is wasted. The power required for the broadcast service would be about 1.2 kilowatts.

Ridiculously small though it is, this figure is probably much too generous. Small parabolas about a foot in diameter would be used for receiving at the Earth end and would give a very good signal/noise ratio. There would be very little interference, partly because of the frequency used and partly because the mirrors would be pointing toward the sky, which could contain no other source of signal. A field strength of ten microvolts/meter might well be ample, and this would require a transmitter output of only fifty watts.

When it is remembered that these figures relate to the broadcast service, the efficiency of the system will be realized. The point-to-point beam transmissions might need powers of only ten watts or so. The slight falling off in field strength due to this cause toward the edge of the service area could be readily corrected by a nonuniform radiator.

The efficiency of the system is strikingly revealed when we consider that the London Television service required about three kilowatts average power for an area less than fifty miles in radius.

A second fundamental problem is the provision of electrical energy to run the large number of transmitters required for the different services. In space beyond the atmosphere, a square meter normal to the solar radiation intercepts 1.35 kilowatts of energy. Solar engines have already been devised for terrestrial use and are an economic proposition in tropical countries. They employ mirrors to concentrate sunlight on the boiler of a low-pressure steam engine. Although this arrangement is not very efficient, it could be made much more so in space, where the operating components are in a vacuum, the radiation is intense and continuous, and the low-temperature end of the cycle could be not far from absolute zero. Thermoelectric and photoelectric developments may make it possible to utilize the solar energy more directly.

Though there is no limit to the size of the mirrors that could be built, one 50 meters in radius would intercept over ten thousand kilowatts, and at least a quarter of this energy should be available for use.

The station would be in continuous sunlight except for some weeks around the equinoxes, when it would enter the Earth's shadow for a few minutes every day during the eclipse period. For this calculation, it is legitimate to consider the Earth as fixed and the Sun as moving around it. The station would graze the Earth's shadow on the last day in February. Every day, as it made its diurnal revolution, it would cut more deeply into the shadow, undergoing its period of maximum eclipse on March 21. On that day it would only be in darkness for one hour and nine minutes. From then onward the period of eclipse would shorten, and after April 11 the station would be in continuous sunlight again until the same thing happened six months later at the autumnal equinox, between September 12 and October 14. The total period of darkness would be about two days per year, and as the longest period of eclipse would be little more than an hour, there should be no difficulty in storing enough power for an uninterrupted service.

CONCLUSION

Briefly summarized, the advantages of the space station are as follows:

1. It is the only way in which true world coverage can be achieved for all possible types of service.

2. It permits unrestricted use of a band at least one hundred thousand megacycles per second wide, and with the use of beams an almost unlimited number of channels would be available.

3. The power requirements are extremely small, since the efficiency of "illumination" would be almost 100 percent. Moreover, the cost of power would be very low.

4. However great the initial expense, it would only be a fraction of that required for the world networks replaced, and the running costs would be incomparably less.

EPILOGUE — ATOMIC POWER

The advent of atomic power has at one bound brought space travel half a century nearer. It seems unlikely that we will have to wait as much as twenty years before atomic-powered rockets are developed, and such rockets could reach even the remoter planets with a fantastically small fuel/mass ratio— only a few percent.

In view of these facts, it appears hardly worthwhile to expend much effort on the building of long-distance relay chains. Even the local networks that will soon be under construction may have a working life of only twenty to thirty years.

The Moon and

Mr. Farnsworth

*Patriotic zealotry of any stripe is anathema to a science writer such as Clarke.
Although Americans stirred with national pride when their nation was the
first to land men on the Moon, Clarke takes apart R. L. Farnsworth,
a head of the United States Rocket Society, who would keep
the Moon all to his nation.*

Here is a (very) minor mystery: I cannot discover where the following piece of invective was printed, as it is neither listed in my own records, or in Professor Samuelson's extensive bibliography. But it was probably written in the 1945–46 period, as the *Bulletin of the British Interplanetary Society* for January 1947 contains an even more scathing review by P. E. Cleator, the founder of the society. See also the chapter "The Lackeys of Wall Street" for another treatment of this imflammatory subject.

To look a gift horse in the mouth is generally considered bad manners. To kick one in the teeth, as I now propose doing, might appear inexcusable;

but the provocation is, I think, sufficient. It consists of a tastefully printed pamphlet entitled "Rockets, New Trail to Empire," which has been sent to this country by its author, R. L. Farnsworth, president of the United States Rocket Society Inc. (Box 29, Glen Ellyn, Illinois).

It is very difficult to criticize this disturbing document without quoting almost the whole of it. Mr. Farnsworth, with that boundless enthusiasm typical of his nation, has used all the resources of high-pressure salesmanship to "put over" astronautics. Most of his facts (apart from some bad technical errors) are correct; so, I fear, are many of his conclusions. It is the attitude, rather than the factual content, of the leaflet which I wish to condemn.

Under the misleading title "The Renaissance," Mr. Farnsworth devotes four pages to listing the effects of astronautics on various human activities, from athletics to religion. This section, as indeed much of the pamphlet, reads like a parody of all the aspirations we in this country have cherished for nearly a decade. The emphasis throughout is on big business and territorial aggrandizement: it seems incredible that anyone who can see so clearly the possibilities of the future could still be mentally rooted so firmly in the past. This typical passage represents the spirit Mr. Farnsworth wishes to take into space:

"We can no longer mislead ourselves that commissions and conventions can secure peace. . . . Therefore it becomes imperative that the Moon be held in strong hands and be under the control of one country. Then let us strive for the day when the American flag is planted firmly upon the volcanic ash of the Moon, and that thenceforth, in its entirety, it will become a possession of the United States!"

We are told that the world is a very small place today, but it seems that it is still a long way from Illinois to San Francisco. This war, many of us believed, was fought to destroy the attitude implied in the quotation above—the idea that one country, be it America or Germany, is alone fitted to rule the world. Fewer things are more certain than the fact that if a single country were to seize the Moon, that country would have all the nations of the world at its throat. The Pax Romana, Mr. Farnsworth, ceased to work some centuries ago. The Pax Brittanica had an even shorter run, and the Pax Germanica never came off at all. I don't think we shall see any further attempts in this direction, though I may be wrong. Empire building is now out of favor: it is strange to hear it advocated from the neighborhood of Chicago. Someday, Mr. Farnsworth, you might ask Colonel McCormick of *The Tribune* for his views on empires.

In his booklet, Mr. Farnsworth is playing right into the hands of the most dangerous enemies of astronautics. Here is a passage from a recent novel by the well-known writer C. S. Lewis, Fellow of Magdalen College, Oxford:

He was a man obsessed with the idea which is at this moment circulating all over our planet in obscure works of "scientifiction," in
little Interplanetary Societies and Rocketry Clubs . . . ready, if ever
the power is put into its hands, to open a new chapter of misery
for the universe. It is the idea that humanity, having sufficiently cor-
rupted the planet where it arose, must at all costs contrive to seed
itself over a larger area: that the vast astronomical distances which are
God's quarantine regulations must somehow be overcome. This is for
a start. But beyond this lies the sweet poison of the false infinite—the
wild dream that planet after planet, system after system, in the end
galaxy after galaxy, can be forced to sustain, everywhere and forever,
the sort of life which is contained in the loins of our species. The
destruction of enslavement of other species in the universe, if such
there are, is to these minds a welcome corollary. In Professor Weston
the power had at last met the dream.

When admonished for this, Mr. Lewis replied: "I don't of course think
that at the moment many scientists are budding Westons; but I do think
that a point of view not unlike Weston's is on the way." After reading this
passage of Mr. Farnsworth's I am forced to agree:

"Man, in his present animalistic frame of mind, will be let loose upon
the galaxy! This is a young, hot, lusty, brutal planet, and from its green
continents will stream a lurid burst of life throughout the Solar System and
ultimately beyond."

This may very well be true: if so, it is our clear duty to do all in our
power to hamper and delay the achievement of interplanetary travel, whatever our personal hopes and aspirations may be. But Mr. Farnsworth, it
would seem, regards this prospect with approval.

I feel very much in sympathy with all that he has to say about the
pioneering spirit and its importance. All of us wish to see the Moon colo-
nized and busy with man's enterprises, but do we wish to see our unfortunate satellite become "the Manhattan of the Solar System"? Do we want
to see it covered with entrepreneurs "selling precision tools . . . hotel fix-
tures, insurance . . . and—deadly phrase!—"making everyone from the
States feel at home"?

Mr. Farnsworth realizes, without understanding, the effects interplanetary travel will have on culture and the arts. Perhaps it is a little unfair (I
feel rather like Jack Benny stealing the kiddy's lollipop) but I cannot resist
quoting from his section on music: "With the vast wealth available on Earth,
it should be possible to sponsor a few thousand musical envoys." I wonder
how much it would have cost to "sponsor" the Ninth Symphony? No, Mr.
Farnsworth, it isn't quite as easy as that!

After extolling "the advertising boys, filling the Earth-Moon rockets with freight," Mr. Farnsworth deals profoundly with religion. "Space travel," he remarks, with the air of one delivering a world-shaking truth, "is what religion wants." Well, it's a novel idea, but I doubt if even the Holy Rollers or the Rosicrucians want Mr. Farnsworth's brand of salvation.

The brochure will, I am afraid, have a deplorable effect on any intelligent layman and will attract the most undesirable type of member, if indeed it attracts any at all. It seems, in fact, nicely calculated to reduce astronautics to a laughingstock. This is largely due to the style, which employs all the most blatant tricks of journalistic advertising and an absolute plethora of exclamation marks.

Little serious thought has been given to the social and moral effects of interplanetary travel, and many people are becoming increasingly critical of those who wish to enlarge man's powers before he is fit to use those he already possesses. The time is coming when—with the lesson of the V-2 behind us—we may have to justify our activities to the world and to prove that the conquest of space will indeed benefit mankind as we have so often proclaimed. If we intend to inflict on other worlds the worst excesses of a materialistic and spiritually barren civilization (and here I am not thinking specifically of modern American culture, but of certain aspects of all Western societies), our case is lost before we begin to plead it.

The "quarantine" will have to remain in force for a few more centuries yet if many advocates of interplanetary travel think as Mr. Farnsworth appears to do.

The Challenge

of the Spaceship

The Cold War had not quite started when the following piece was written.
It reflects Clarke's hopes not only for a war-free world but also for
the beneficent uses of space travel.

My wartime exposure to high technology had convinced me that I was not really suited for a career in His Majesty's Exchequer and Audit Department, even though it promised a secure and leisurely lifestyle, with a good retirement pension at the end. So—after much arm-twisting by my two most closest friends, De Haviland rocket engineer Val Cleaver and Flight Sgt. Johnny Maxwell of the Royal Canadian Air Force—I astonished my colleagues in the Civil Service by handing in my resignation even before being demobilized from the RAF in June 1946.

I was encouraged to make this hazardous move by winning the first prize in an essay contest organized by the *Royal Air Force Quarterly* on the future of the rocket in warfare. It will be found in my collection of technical papers, *Ascent to Orbit* (John Wiley, 1984). Even in 1946 I found the subject extremely depressing, as demonstrated by the opening quotation from Shelley:

Cease! Drain not to its dregs the urn
Of bitter prophecy.
The world is weary of the past,
Oh, might it die or rest at last!

However, "The Rocket and the Future of Warfare" played a very important role in my career. It introduced me to an energetic young member of Parliament, Capt. Raymond Blackburn, to whom I shall always be grateful for pulling strings on my behalf and getting me a university scholarship. (I had been turned down on the not unreasonable grounds that, as a defector from the Civil Service, no more public funds should be wasted on me.)

Many years later, I was surprised—and not particularly flattered—to learn that in certain military circles I was regarded as the inventor of the MAD (mutual assured destruction) doctrine, on the strength of this essay's closing paragraphs.

Judge for yourself:

One returns again to the conclusion that the only defense against the weapons of the future is to prevent them from ever being used. In other words, the problem is political and not military at all. A country's armed forces can no longer defend it; the most they can promise is the destruction of the attacker.

In such circumstances, the statement that the United Nations is the last hope of mankind is literally and terribly true. It is therefore necessary to consider in what way the rocket can be used as an instrument of world, rather than regional, security.

Now that the world possesses rockets and atomic bombs, mankind has a tendency to discount the weapons of which it was so terrified a few years ago. Therefore even if there is no intention of using them except as a last resort, the Security Council should for psychological reasons posses long-range atomic rockets. However, the weapons that it would use if force proved necessary would be the air contingents of its members, employing ordinary explosives and machines of the type that exist today. Behind these would be the threat, never materializing save in dire emergency, of the mightier forces against which there could be no defense.

Not more than twenty launching sites with interlocking circles of fire should be sufficient to give world coverage. The sites would be staffed by men drawn from every nation; it would be the aim to inculcate in these men a supranational outlook. This is by no means impossible even today, as the International Red Cross has shown, and this viewpoint is becoming more widespread in spite of superficial appearances to the contrary. The fact that the personnel required would be largely scientific would assist the realization of this aim.

These launching sites would have to be supported by a research organization of such a caliber that no individual nation could hope to match it. This body might in time act as the nucleus around which the scientific service of the world state would form, perhaps many years in advance of its political realization.

The necessity for these measures should be kept under continual review, for the world would never feel completely at ease while they existed. A heavily armed police force decreases rather than increases the citizen's sense of security—but, on the other hand, we must recognize that the world today resembles the lawless Wild West of the last century, in which an unarmed sheriff would have had little chance of enforcing order. When a world economic system is functioning smoothly, when all standards of living are approaching the same level, when no national armaments are left—then the launching sites could be dismantled.

Only along these or similar lines of international collaboration can security be found; any attempt by great powers to seek safety in their own strength will ultimately end in a disaster which may be measureless.

Upon us, the heirs to all the past and the trustees of a future that our folly can slay before its birth, lies a responsibility no other age has ever known. If we fail in our generation, those who come after us may be too few to rebuild the world, when the dust of the cities has descended and the radiation of the rocks has died away.

My resignation from the Civil Service was also prompted by the fact that I had already sold quite a number of articles and short stories and had also broken into the American science fiction market. (My first major sale, the twelve-thousand-word novella "Rescue Party," was made to John W. Campbell's *Astounding Stories* in September 1945.) Equally important, I had learned that government grants were now available to send ex-servicemen to college; if I could get my hands on one of these, it would support me for a few years and help fill the gaping holes in my science education. After that—well, I would look around for gainful employment, hopefully connected in some way with space.

The British Interplanetary Society had now emerged from its enforced period of suspended animation and had resumed a series of public lectures. As its first postwar chairman, I delivered one of these at St. Martin's Technical School on Charing Cross Road on October 5, 1946. "The Challenge of the Spaceship," with its subtitle, "Astronautics and Its Impact Upon Human Society," was widely reprinted, and I am particularly proud of the fact that when I sent a copy to George Bernard Shaw, I received one of his famous pink postcards with a request: "How can I join the society?" I only hope that I am equally interested in new ideas in my own ninety-first year; GBS promptly became a member and renewed his subscription for the remaining three years of his life.

After its initial 1946 publication in the *Journal of the British Interplanetary Society*, I revised the essay when the Space Age actually opened in 1957, and here is the final, 1959 version. It differs from the 1946 one only in removal of dated material—and, of course, by including references to the truly astonishing events of the previous dozen years, which I had never dreamed would occur so quickly when I first delivered the paper.

A historian of the twenty-first century looking past our own age to the beginnings of human civilization will be conscious of four great turning points that mark the end of one era and the dawn of a new and a totally different mode of life. Two of these events are lost, probably forever, in the primeval night before history began. The invention of agriculture led to the founding of settled communities and gave man the leisure and social intercourse without which progress is impossible. The taming of fire made him virtually independent of climate and, most important of all, led to the working of metals and so put him upon the road of technological development—that road was to lead, centuries later, to the steam engine, the Industrial Revolution, and the age of steel and gasoline and surface transportation through which we are now passing.

The third revolution began, as all the world knows, in a squash court in Chicago on December 2, 1942, when the first self-sustaining nuclear reaction was started by man. We are still too close to that cataclysmic event to see it in its true perspective, but we know that it will change our world, for better or worse, almost beyond recognition. And we know, too, that it is linked with the fourth and in some ways greatest change of all—the crossing of space and the exploration of the other planets. For though the first space vehicles were chemically fueled, only atomic energy is adequate to lift really large payloads out of the Earth's gravitational field—that invisible maelstrom whose tug can still be felt a million miles away.

Prophecy is a dangerous and thankless business, frequently fatal to those who practice it. We have, however, learned from past experience that even the most extravagant forecast seldom overtakes the truth. H. G.

Wells once wrote—and was no doubt laughed to scorn for his folly—that the airplane might have some influence upon warfare by the year 1950. He never dared to imagine that by that date aircraft would not only have become of supreme importance but would have been challenged by still newer weapons.

It is certainly not being rash—it may indeed be conservative—to assume that by the last quarter of this century an efficient method of nuclear propulsion for space vehicles would have been perfected. Atomic power is hardly likely to advance the conquest of space by more than ten years, but it may make it practical almost from the beginning. It would mean that the whole Solar System, and not merely the Moon, would be immediately accessible to man. As our first space probes have demonstrated, it requires very little more power to reach the planets than it does to go to the Moon, but the most economical voyages involve months or even years of free coasting along orbits curving halfway around the Sun. With atomic power these journeys could be cut to a fraction of the time. For example, the "cheapest" journey to Mars—as far as fuel is concerned—lasts 285 days. With a nuclear-propelled ship, traveling by a more direct route at quite a moderate speed, it need take only a few weeks.

There are still some scientists who consider that there is no point in sending men into space, even when it becomes technically possible; machines, they argue, can do all that is necessary. Such an outlook is incredibly shortsighted; worse than that, it is stupid, for it completely ignores human nature.

Though the specific ideals of astronautics are new, the motives and impulses underlying them are old as the race—and in the ultimate analysis, they owe as much to emotion as to reason. Even if we could learn nothing in space that our instruments would not already tell us, we should go there just the same.

Some men compose music or spend their lives trying to catch and hold forever the last colors of the dying day, or a pattern of clouds that, through all eternity, will never come again. Others make voyages of exploration across the world, while some make equally momentous journeys in quiet studies with no more equipment than pencil and paper. If you ask these men the purpose of their music, their painting, their exploring, or their mathematics, they would probably say that they hoped to increase the beauty or the knowledge of the world. That answer would be true, and yet misleading. Very few indeed would give the simpler, more fundamental reason that they had no choice in the matter—that what they did, they did because they had to.

The urge to explore, to discover, to "follow knowledge like a sinking star," is a primary human impulse that needs, and can receive, no further justification than its own existence. The search for knowledge, said a modern Chinese philosopher, is a form of play. If this be true, then the spaceship, when it comes, will be the ultimate toy that may lead mankind from its cloistered nursery out into the playground of the stars.

However, it is not hard to think of endless and entirely valid "practical" reasons why one should wish to cross space. There is no doubt that eventually sheer necessity would bring about the conquest of the other planets. It may well be impossible to have a virile, steadily advancing culture limited to a single world, and taking the long-term view, we know that our Earth will one day become uninhabitable. The Sun is still evolving, growing steadily hotter as its central fires become banked up beneath their accumulated "ash" of helium. In the far future, the oceans will boil back into the skies from which they once condensed, and life must pass from the planet Earth.

But the human race will not wait until it is kicked out. Long before the Sun's radiation has shown any measurable increase, Man will have explored all the Solar System and, like a cautious bather testing the temperature of the sea, will be making breathless little forays into the abyss that separates him from the stars.

The last quarter of this century will be an age of exploration such as Man has never before known. By the year 2000, most of the major bodies in the Solar System will probably have been reached, but it will take centuries to examine them all in any detail. Those who seem to think that the Moon is the goal of interplanetary travel should remember that the Solar System contains eight other planets, more than thirty moons, and some thousands of asteroids. The total area of the major bodies is about 250 times that of Earth, though the four giant planets probably do not possess stable surfaces on which landings could be made. Nevertheless, that still leaves an area ten times as great as all the continents of Earth.

This then is the future that lies before us, if our civilization survives the diseases of its childhood. It is a future that some may find terrifying, as no doubt our ancestors found the hostile emptiness of the great oceans. But the men who built our world crossed those oceans and overcame those fears. If we fail before the same test, our race will have begun its slide into decadence. Remember, too, that when the great explorers of the past set sail into the unknown, they said good-bye for years to their homes and everything they knew. Our children will face no such loneliness. When they are among the outermost planets, when the Earth is lost in the glare of the Sun, and the Sun itself is no more than the brightest of the stars, they will still be able to hear its voice and to send their own words in a few hours back to the world of men.

Let us now consider the effects that interplanetary travel must have upon human institutions and ideas. The most obvious and direct result of the crossing of space will be a revolution in almost all branches of science.

I will not attempt to list more than a few of the discoveries we may make when we can set up research stations and observatories upon the other planets or in satellite orbits. One can never predict the outcome of any science investigation, and the greatest discoveries of all—the ones that will most influence human life—may come from sciences as yet unborn.

Astronomy and physics will, of course, be the fields of knowledge most immediately affected. In both these sciences there are whole areas where research has come to a dead end, or has never been started, because our terrestrial environment makes it impossible.

The atmosphere, which on a clear night looks so transparent, is in reality a colored filter blocking all rays beyond the ultraviolet. Even in the visible spectrum the light that finally struggles through the shifting strata above our head is so distorted that the images it carries dance and tremble in the field of the telescope.

An observatory on the Moon, working with quite small instruments, would be many times as effective as one on Earth. Far greater magnifications could be employed, and far longer exposures used. In addition, the low gravity would make relatively simple the building of larger telescopes than have ever been constructed on this planet.

In physics and chemistry, access to vacuums of unlimited extent will open up quite new fields of investigation. The electronic scientist may well look forward to the day when he can build radio tubes miles long, if he wishes, merely by setting up his electrodes in the open! It is also interesting to speculate whether we may not learn more about gravity when we can escape partly or wholly from its influence.

Artificial satellites have already given dramatic notice of what may be achieved when we can establish permanent, manned stations in space for observation and research. Accurate weather forecasting—which, it has been estimated, would be worth $5 billion a year to the United States alone—will probably be impossible until we can hoist the meteorologists out into space, however reluctant they may be to go there. Only from a height of several thousand miles is it possible to observe the Earth's weather pattern as a whole, and to see literally at a glance the movement of storms and rain areas.

One application of space stations and satellites, whose importance it is impossible to overestimate, is their use for communications and television relaying. Should any one nation establish a satellite relay chain, it would do more than dominate the world's communications. The cultural and political impact of television news and entertainment broadcast directly to every home on Earth would be immeasurable. When one considers the direct effect of television upon ostensibly educated populations, the impact upon the semiliterate peoples of Africa and Asia may be decisive. It may well determine whether English or Russian becomes the leading world language by the end of this century.

Yet these first direct results of astronautics may be less important, in the long run, than its indirect consequences. This has proved true in the past of most great scientific achievements. Copernican astronomy, Darwin's theory of evolution, Freudian psychology—these had few immediate practical results, but their effect on human thought was tremendous.

We may expect the same of astronautics. With the expansion of the world's mental horizons may come one of the greatest outbursts of creative activity ever known. The parallel with the Renaissance, with its great flowering of the arts and sciences, is very suggestive. "In human records," wrote the anthropologist J. D. Unwin, "there is no trace of any display of productive energy which has not been preceded by a display of expansive energy. Although the two kinds of energy must be carefully distinguished, in the past they have been . . . united in the sense that one has developed out of the other." Unwin continues with this quotation from Sir James Frazer: "Intellectual progress, which reveals itself in the growth of art and science . . . receives an immense impetus from conquest and empire." Interplanetary travel is now the only form of "conquest and empire" compatible with civilization. Without it, the human mind, compelled to circle forever in its planetary goldfish bowl, must eventually stagnate.

We all know the narrow, limited type of mind that is interested in nothing beyond its town or village and bases its judgments on those parochial standards. We are slowly—perhaps too slowly—evolving from that mentality toward a world outlook. Few things will do more to accelerate that evolution than the conquest of space. It is not easy to see how the more extreme forms of nationalism can long survive when men begin to see the Earth in its true perspective as a single small globe among the stars.

There is, of course, the possibility that as soon as space is crossed, all the great powers will join in a race to claim as much territory as their ships can reach. Some American writers have even suggested that for its own protection the United States must occupy the Moon to prevent its being used as a launching site for atomic rockets. Fantastic though such remarks may seem today, they represent a danger that would be unwise to ignore. The menace of interplanetary imperialism can be overcome only by worldwide technical and political agreements well in advance of the actual event, and these will require continual pressure and guidance from the organizations that have studied the subject.

The Solar System is rather a large place, though whether it will be large enough for so quarrelsome an animal as Homo sapiens remains to be seen. But it is surely reasonable to hope that the crossing of space will have a considerable effect in reducing the psychological pressures and tensions of our present world. Much depends, of course, on the habitability of the other planets. It is not likely that very large populations will, at least for many centuries, be able to subsist outside the Earth. There may be no worlds in the Solar System upon which men can live without mechanical aids, and some of the greatest achievements of future engineering will be concerned with shaping hostile environments to human needs.

We must not, however, commit the only too common mistake of equating mere physical expansion, or even increasing scientific knowledge, with "progress"—however that may be defined. Only little minds are impressed by sheer size and number. There would be no virtue in possessing the universe if it brought neither wisdom nor happiness. Yet possess it we must, at least in spirit, if we are ever to answer the questions that men have asked in vain since history began.

Perhaps this analogy will make my meaning clearer. Picture a small island inhabited by a race that has not yet learned the art of making ships. Looking out across the ocean, this people can see many other islands, some of them much the same as its own but most of them clearly very different. From some of these islands, it is rumored, the smoke of fires has been seen ascending—though whether these fires are the work of man, no one can say.

Now these islanders are very thoughtful people, and writers of many books with such resounding titles as *The Nature of the Universe, The Meaning of Life, Mind and Reality,* and so on. While admiring their enterprise, I do not think we should take their conclusions very seriously—at least until they have gone a little further afield than their own coral reef. In his long philosophical poem *The Testament of Beauty*, published in 1929, a year before his death at age eighty-five, Robert Bridges remarked that true wisdom lies in first asking what things are, not why.

That task the human race can scarcely begin to undertake while it is still earthbound.

Every thoughtful man has often asked himself: Is our race the only intelligence in the universe, or are there other, perhaps far higher, forms of life elsewhere? There can be few questions more important than this, for upon its outcome may depend all philosophy—yes, and all religion, too.

The first discovery of planets revolving around other suns, which was made in the United States in 1942, has changed all ideas of the plurality of worlds. Planets are far commoner than we had ever believed: there may be thousands of millions in this galaxy alone. Few men today would care to argue that the Earth must be the only abode of life in the whole of space.

It is true—it is even likely—that we may encounter no other intelligence in the Solar System. That contact may have to wait for the day, perhaps ages hence, when we can reach the stars. But sooner or later it must come.

There have been many portrayals in literature of these fateful meetings. Most science fiction writers, with characteristic lack of imagination, have used them as an excuse for stories of conflict and violence indistinguishable from those that stain the pages of our own history. Such an attitude shows a complete misunderstanding of the factors involved.

Remember the penny and the postage stamp that Sir James Jeans in *The Mysterious Universe* balanced on Cleopatra's Needle. The obelisk represented the age of the world, the penny the whole duration of man's existence, and the stamp the length of time in which he had been slightly civilized. The period during which life will be possible on Earth corresponds to a further column of stamps hundreds of yards—perhaps a mile—in length.

Thinking of this picture, we see how infinitely improbable it is that the question of interplanetary warfare can ever arise. Any races we encounter will almost certainly be superhuman or subhuman—more likely the former, since ours must surely be one of the youngest cultures of the universe. Only if we score a bull's-eye on that one stamp in the mile-high column will we meet a race at a level of technical development sufficiently near our own for warfare to be possible. If ships from Earth ever set out to conquer other worlds, they may find themselves, at the end of their journeys, in the position of painted war canoes drawing slowly into New York Harbor.

But if the universe does hold species so greatly in advance of our own, then why have they never visited Earth? There is one very simple answer to this question. Let us suppose that such races exist; let us even suppose that, never having heard of Einstein, they can pass from one end of the galaxy to the other as quickly as they wish.

That will help them less than one might think. In ten minutes, a man may walk along a beach—but in his whole lifetime he could not examine every grain of sand upon it. For all that we know, there may be fleets of survey ships diligently charting and recharting the universe. Even making the most optimistic assumptions, they could scarcely have visited our world in the few thousand years of recorded history.

Perhaps, even at this moment, there lies in some rather extensive filing system a complete report on this planet, with maps that to us would look distorted but still recognizable. That report would show that though Earth was teeming with life, it had no dominant species. However, certain social insects showed considerable promise, and the file might end with the note: "Intelligence may be emerging on this planet. Suggest that intervals between surveys be reduced to a million years."

Very well, you may ask—suppose we encounter beings who judge, condemn, and execute us as dispassionately, and with as little effort, as we spray a pool of mosquito larvae with DDT? I must admit that the possibility exists, and the logical answer—that their reasons will no doubt be excellent—is somewhat lacking in appeal. However, this prospect appears remote. I do not believe that any culture can advance for more than a few centuries at a time on a technological front alone. Morals and ethics must not lag behind science, otherwise the social system will breed poisons that will cause its certain destruction. I believe therefore that with superhuman knowledge must go equally great compassion and tolerance. In this I may be utterly wrong; the future may belong to forces that we should call cruel and evil. Whatever we may hope, we cannot be certain that human aspirations and ideals have universal validity. This we can discover in one way only, and the philosophical mind will be willing to pay the price of knowledge.

I have mentioned before how limited our picture of the universe must be so long as we are confined to this Earth alone. But the story does not end there. Our impressions of reality are determined, perhaps more than we imagine, by the senses through which we make contact with the external world. How utterly different our cosmologies would have been had nature economized with us, as she has done with other creatures, and given us eyes incapable of seeing the stars! Yet how pitiably limited are the eyes we do possess, tuned as they are to a single octave in an endless spectrum! The world in which we live is drenched with invisible radiations, from the microwaves that we have just discovered from the Sun and stars, to the cosmic rays, whose origin is still one of the prime mysteries of modern physics. These things we have discovered in the last generation, and we cannot guess what still lies beneath the threshold of the senses—though recent discoveries in paranormal psychology may hint that the search may only be beginning.

The races of other worlds will have sense and philosophies very different from our own. To recall Plato's famous analogy, we are prisoners in a cave, gathering our impressions of the outside world from shadows thrown upon the walls. We may never escape to reach that outer reality, but one day we may hope to meet other prisoners in adjoining caves, where the shadows may be very different and where we may learn far more than we could ever do by our own unaided efforts.

These are deep waters, and it is time to turn back to the shore, to leave the distant dream for the present reality of fuels and motors, of combustion-chamber pressures and servomechanisms. Yet I make no apology for discussing these remote vistas at some length, if only to show the triviality of the viewpoint that regards interplanetary travel as a schoolboy adventure of no more real value than the scaling of some hitherto inaccessible mountain. The adventure is there, it is true, and that is good in itself—but it is only a small part of a much greater whole.

Not so shortsighted, but equally false, is the view expressed by Prof. C. S. Lewis, who has written of would-be astronauts in this unflattering fashion: "The destruction or enslavement of other species of the universe, if such there are, is to these minds a welcome corollary." In case there are any to whom this prospect still appeals, I would point out that empires, like atomic bombs, are self-liquidating assets. Dominance by force leads to revolution, which in the long run, even if indirectly, must be successful. Humane government leads eventually to self-determination and equality, as the classic case of the British Empire has shown. Commonwealths alone can be stable and enduring, but empires must always contain the seeds of their own dissolution.

The desire to give a comprehensive picture of the outcome of astronautics has compelled me to range—not unwillingly—over an enormous field. However, I do not wish anyone to think that the possibilities we have been discussing need come in this century, or the next. Yet any of them may arise, at any time, as soon as the first ships begin to leave the Earth. Man's first contact with other intelligent races may lie as far away in time as the building of the pyramids—or it may be as near as the discovery of X rays.

Of this, at least, we may be fairly certain. Barring accidents—the most obvious of which I need not specify—the exploration of the planets will be in full swing as this century draws to a close. To examine them in any detail, and to exploit their possibilities fully, will take hundreds of years. But man being what he is, when his first ship circles down into the frozen wastes of Pluto, his mind will already be bridging the gulf still lying between him and the stars.

Interplanetary distances are a million times as great as those to which we are accustomed in everyday life, but interstellar distances are a million-fold greater still. Before them even light is a hopeless laggard, taking years to pass from one star to its neighbor. How man will face this stupendous challenge I do not know; but face it one day he will. Prof. J. D. Bernal was, I believe, the first to suggest that one solution might lie in the use of artificial planets, little self-contained worlds embarking upon journeys that would last for generations, but the thought of these tiny bubbles of life, creeping from star to star on their age-long journeys, carrying whole populations doomed never to set foot upon any planet, never to know the passage of seasons or even the interchange of night and day, is one from which we might well recoil in horror. However, those who would make such journeys would have outlooks very different from our own, and we cannot judge their minds by our standards.

These speculations, intriguing though they are, will hardly concern mankind in this century. We may, I think, confidently expect that it will be a hundred years at least before confinement to the Solar System produces very marked signs of claustrophobia.

Our survey is now finished. We have gone as far as it is possible, at this moment of time, in trying to assess the impact of astronautics upon human affairs. I am not unmindful of the fact that fifty years from now, instead of preparing for the conquest of the outer planets, our grandchildren may be dispossessed savages clinging to the fertile oases in a radioactive wilderness. Yet we must keep the problems of today in their true proportions. They are of vital—indeed of supreme—importance, since they can destroy our civilization and slay the future before its birth. But if we survive them, they will pass into history and the time will come when they will be as little remembered as the causes of the Punic Wars. The crossing of space—even the sense of its imminent achievement in the years before it comes—may do much to turn men's minds outward and away from their present tribal squabbles. In this sense the rocket, far from being one of the destroyers of civilization, may provide the safety valve that is needed to preserve it.

This point may be of the utmost importance. By providing an outlet for man's exuberant and adolescent energies, astronautics may make a truly vital contribution to the problems of the present world. Space flight does not even have to be achieved for this to happen. As soon as there is a general belief in its possibility, that belief will begin to color Man's psychological outlook. In many ways, the very dynamic qualities of astronautics are in tune with the restless, expansive spirit of our age.

In this essay I have tried to show that the future development of mankind, on the spiritual no less than the material plane, is bound up with the conquest of space. To what may be called—using the words in the widest possible sense—the liberal scientific mind, I believe these arguments to be unanswerable. The only real criticism that may be raised against them is the quantitative one that the world is not yet ready for such changes. It is hard not to sympathize with this view, which may be correct, but I have given my reasons for thinking otherwise.

The future of which I have spoken is now being shaped by men working with slide rules in quiet offices, and by men taking instrument readings amid the savage roar of harnessed jets. Some are engineers, some are dreamers—but many are both. The time will come when they can say with T. E. Lawrence: "All men dream; but not equally. Those who dream by night in the dusty recesses of their minds wake in the day to find that it was vanity; but the dreamers of the day are dangerous men, for they may act their dream with open eyes, to make it possible."

Thus it has always been in the past, for our civilization is no more than the sum of all the dreams that earlier ages have brought to fulfillment. And so it must always be, for if men cease to dream, if they turn their backs upon the wonder of the universe, the story of our race will be coming to an end.

First Men

in the Moon

The Moon and dreams of colonization have always held an attraction to mankind.
Now, as man is moving ever closer to the space frontier, perhaps all the
comforts of home will be provided under sheltered domes,
as the following essay projects.

now find it rather surprising that even while I was taking my bachelor of science degree in physics, pure math, and applied maths at King's College from 1946 to 1948, I still found time for a considerable amount of writing. Indeed, my first published novel, *Prelude to Space,* was written during the 1947 summer vacation.

And I have just learned, to my amusement, that a rumor is now circulating that I paid my way through college with the income from pornographic short stories! I can only assume that this tale was started by some literary snob who judged science fiction by the lurid—and utterly misleading—covers of certain 1940s pulp magazines. In fact, one of the valid criticisms of pop science fiction during the first half of the century is that is was not pornographic enough—a situation now amply rectified by contemporary writers.

"The Challenge of the Spaceship" attracted a good deal of interest and was even printed in a prestigious volume of essays entitled *British Thought, 1947*. It triggered many commissions, such as the following item from the now defunct London evening newspaper *The Star,* which paid me the princely sum of ten guineas, or ten pounds and ten shillings.

Though it was written more than half a century ago, it has dated fairly well, and the only serious error is the—then universal—overoptimism about the future role of atomic energy. I reproduce it for two main reasons. First of all, it conveys the general feeling about space flight in those days. And, incredibly, a mere twenty-two years later, it was my privilege to write the epilogue to a book with a very similar title, *First on the Moon,* the official account of the Apollo 11 mission.

A few months ago, American scientists installed movie cameras in a V-2 rocket and so gave the world one of the most remarkable newsreels ever made. No one who has seen it is likely to forget the picture of the receding Earth, with the blackness of space visible beyond its rim, spinning seventy miles below the rocket.

Since then a V-2 has risen over 110 miles into space—almost to the limits of the atmosphere, and far beyond the Heaviside layer, which reflects radio waves back to Earth.

At this altitude, much less than a millionth of the atmosphere remains, and air resistance, even at speeds of thousands of miles an hour, is too small to be measured. This height record will be broken over and over again in the coming years as more powerful rockets are constructed. Already we know, in principle, how to build rockets that can escape from the Earth.

Not long ago the technical director of the largest American rocket research laboratory, in a talk to the British Interplanetary Society, outlined plans for a rocket that could carry a payload of ten pounds to the Moon. Such a machine, he was confident, could be built in a relatively few years.

The use of high-altitude rockets for scientific research will raise some interesting poltical problems in the very near future.

A rocket on its way into space may circle half the world, at a steadily increasing height, with a sublime indifference to frontiers—and territorial gains.

Before long a vertical equivalent of the "three-mile limit" will have to be agreed internationally.

No country can own the space above it for an indefinite distance— otherwise during the Earth's daily rotation every government could lay claim to a large portion of the universe. The United Nations will be faced with a major problem when the first radio-controlled rocket reaches the Moon.

It is to be hoped that before this happens some international agreement can be reached, lest the first stages in the conquest of space be marred by a dangerous and degrading scramble for as much as possible of our satellite's 15 million square miles of territory.

In a few years, we will be seeing the first photographs of the Earth as a single small globe among the stars. Is it too much to hope that this may make the more extreme forms of nationalism look as ridiculous as they are?

The date of the first lunar landing depends very largely on the speed with which atomic energy is harnessed for propulsion. It seems likely that, well before the end of this century, an attempt will be made to form some permanent colony on the Moon.

At first, it will consist almost entirely of scientists—particularly astronomers.

From the Moon visibility is always perfect, for there are no clouds— indeed, there is no detectable atmosphere at all. Moreover, it is two weeks between sunrise and sunset, so observations can be carried out that would be quite impossible from the Earth.

Conditions on the Moon are so severe that there will be no question of human beings "enduring" them—they will have to be completely insulated from their surroundings.

Scientists will live and work in air-conditioned buildings, partly underground, to avoid the temperature changes at the surface.

During the day the exposed lunar rocket may reach the boiling point of water, and at midnight the temperature falls to 250 degrees below freezing.

However, the absense of atmosphere greatly simplifies the problem of temperature control, for buildings can be constructed on thermos-flask principles, with the external walls lightly colored or even silvered.

An air supply will also have to be provided, but fortunately, about 50 percent by weight of ordinary rock consists of oxygen. It would not be too difficult to develop equipment to release this from the lunar rocks, if no more convenient sources can be found.

For work in the open, and for the exploration of the Moon's surface, space suits will be employed. These will resemble diving suits but will be heat insulated and provided with radio so that the occupant can communicate with his colleagues or his base.

Fairly complete designs for equipment of this sort have been worked out by the BIS as a matter of technical interest.

The production of food presents some interesting agricultural problems. During the war, techniques were evolved for growing crops on barren

Pacific islands by hydroponic methods, and these would be very suitable for lunar conditions.

It would be necessary to build pressurized greenhouses, and to obtain the chemicals required from the lunar rocks, but this should be relatively easy.

The Moon shows sign of great volcanic activity from past ages, and no doubt very extensive mineral deposits exist on or near the surface.

Building and mining operations will be greatly facilitated by the low gravity, for objects on the Moon will have only a sixth of their terrestrial weight.

Above all, the lunar agriculturalists will have the satisfaction of knowing that, between dawn and nightfall, there will be fourteen Earth days of intense, unbroken sunlight.

Since life on the Moon presents such complications, it may well be asked why one should bother to go there in the first place. One answer is given by history. From the beginning of time, men have felt the urge for exploration and discovery, and our civilization is nothing more than the sum of their achievements.

The conquest of the Moon will immensely expand our knowledge and understanding of the universe—and in the long run, even on a cash basis, nothing is more precious than knowledge.

We cannot imagine today what our descendants will make of the Moon a few hundred years from now. Though it may be a barren, lifeless world, atomic power will give us both the means of reaching it, and of shaping it to our needs.

Perhaps one day it may be dotted with great transparent domes, each covering a city of millions of people. If this seems fantastic, think how the towers of Manhattan must dwarf any dream that Columbus ever knew.

The Problem

of Dr. Campbell

Mathematics, the language of science, has over the ages been used to "prove"
almost every theory. Should it, therefore, be surprising, as Clarke writes,
that before the concepts of interplanetary flight and astronautics
became popular, a lot of effort went into calculations
disputing the necessity and practicality of
traveling to the Moon?

After graduating from King's College I was then faced with the unpleasant necessity of looking for an honest living, since though I assumed
that my typewriter might provide me with occasional jam, I needed a
more reliable source for my bread and butter. Luckily the dean of
science found a job that might have been designed for me—assistant
editor of *Physics Abstracts,* published by the Institute of Electrical Engineers.

During the brief but happy period I spent at the IEE, all of the
world's leading scientific journals passed over my desk, and I had to mark
the articles that appeared important and farm them out to a small army of
multilingual abstractors. Then I had to edit and index the results—no

easy task, since English was seldom the first language of my assistants, and under what heading do you index some totally new discovery? I fear that the issues of *Physics Abstracts* during the period of my editorial reign (circa 1949) must contain many curiosities.

One of the burdens that we premature space cadets had to bear—at least until the 1955 announcement of the Earth satellite program made astronautics respectable—was persistent attacks from the scientific community. My favorite example is from a book published by one Prof. A. W. Bickerton in 1926—the very year in which Robert Hutchins Goddard flew the world's first liquid-propellant rocket. The learned professor wrote:

"This foolish idea of shooting at the Moon is an example of the absurd length to which vicious specialization will carry scientists working in thought-tight compartments. . . . Our most violent explosive has only one-tenth of the energy necessary to escape from Earth. . . . Hence the proposition appears to be basically impossible."

It seems incredible that anyone could make such a howler in elementary dynamics. Any intelligent schoolboy should have been able to tell the professor that there wasn't the slightest need for the "explosive" to escape from Earth. Although Verne's space gun would not have been practical—at least for live passengers!—it serves as a good thought experiment to show that all the propellants can remain at sea level; only the payload has to be given escape velocity. Even an outward-bound rocket burns its fuel within a few hundred kilometers of the Earth; it doesn't have to carry it all the way to the Moon, as Bickerton seems to have imagined.

A later and much more comprehensive attack on "this foolish idea of shooting at the Moon" was made in the 1940s by Dr. J. W. Campbell, president of the Royal Astronomical Society of Canada. I am still quite proud of the demolition job I did on the good doctor, though it does contain one slightly embarrassing paragraph.

When I agreed with Dr. Campbell in pouring scorn on the idea of a spaceship entering a spinning station under its own power, I never imagined that twenty years later Stanley Kubrick would show exactly this. However, in view of the problems that the (nonrotating!) *Mir* has had with docking modules, this may be the only point that Dr. Campbell got right.

Dr. J. W. Campbell is the president of the Royal Astronomical Society of Canada and for some years has taken a skeptical interest in spaceflight. In 1941 he published a mathematical paper in which he calculated that a spaceship would need an initial weight of about 1,000,000,000,000 tons for the lunar return journey. This interesting and original result was obtained— apparently in complete ignorance of all the work that had been done on the subject over the previous twenty years—by assuming a "final" mass for the rocket of 500 tons (!) and an exhaust velocity of only one mile a second. That practical spaceflight cannot be achieved under these conditions can easily be shown in a few lines and does not require the several pages of mathematics employed by Dr. Campbell. However, his paper must have convinced many at the time that interplanetary travel was quite impossible.

Actually it proves nothing of the sort, for if one assumes sufficiently unfavorable initial conditions, one can obtain even more fantastic figures. A spectacular example of this occurs in Goddard's first Smithsonian paper, where it is shown that with the best fuels available a rocket would have to weigh ten pounds to give a certain performance—whereas with the old-fashioned fuels and construction used in a type of marine rocket which Goddard was unkind enough to name, the initial mass would have to be six times that of Earth!

Dr. Campbell has now returned to the subject in an address entitled "The Problem of Space Travel," given to the Royal Astronomical Society at Toronto. This is a nonmathematical talk, much of which is devoted to criticisms of newspaper and magazine articles on astronautics. There is still no indication that Dr. Campbell has read any technical studies on the subject, and one wonders what he would think of a critique of modern astronomy that was based largely on reports in the daily press.

The lecture opens with quotations pro and con astronautics by various writers, and it is a little unkind of Dr. Campbell to reproduce a polemic by F. R. Moulton, every single item of which was demonstrably incorrect even when it was published in 1935. (With what authority Dr. Moulton speaks on the subject may be judged from his calculation that it would need the energy of sixteen thousand tons of coal at 100 percent efficiency to lift a ten-ton rocket one hundred miles. The Germans are therefore greatly to be congratulated on producing rockets with an efficiency of several hundred thousand percent!)

After a short and lucid summary of the fundamental problems of astronautics, Dr. Campbell deals with various suggestions that have been put forward to overcome them. In particular he is critical of the idea (proposed by Ley and others) that the orientation of a spaceship could be altered by the rotation of internal masses. It is perfectly true, as Dr. Campbell points out, that the response of a spaceship to such momentum changes would be very sluggish, but he seems unaware of the fact that this does not matter. With a rather small, electrically driven flywheel, a rocket the size of the V-2 could be rolled completely over in less than ten minutes—and there is no particular urgency about a maneuver that can be carried out over a period of days or months! The alternative solution (much more usually mentioned and actually employed in the United States Navy's *Neptune*) is not considered by Dr. Campbell. Tangential steering jets are quicker and the momentum change is permanent. A flywheel gradually restores the status quo as it runs down—unless one carefully edges it out through a convenient hatch, which is rather an expensive way to jettison angular momentum!

Dr. Campbell devotes some time to a critical discussion of space stations, but unfortunately much of his information appears to be culled from a well-written but hardly technical article in the American magazine *Look*. Most of his comments are, therefore, somewhat trivial, as they deal with matters that were discussed in great detail by Oberth, Noordung, and von Pirquet twenty years ago. Dr. Campbell pours scorn—as well he might—on *Look*'s optimistic suggestion that anything as unwieldy as a spaceship could nose its way under its own power into "mooring sockets" on a probably spinning space station. However, in calling the approach to the station a "trapeze act," Dr. Campbell should have pointed out that it can be done in slow motion, and the performers have a chance of repeating it if the first attempt fails!

The problem of making contact between two bodies in space seems to us considerably less difficult than refueling a plane in midair—which looks almost impossible on paper, yet which has turned out surprisingly simple in practice. Having watched films of aircraft snatching grounded gliders, we consider that a few nylon cables should be able to neutralize any reasonable relative velocity, after which it would be merely a matter of manning the capstans!

It is difficult to see the grounds for Dr. Campbell's statement that "space suits would have to be worn at all times," unless the station was manned by meteorphobes with a profound mistrust of statistics. (They might get that way, of course, after a long tour of duty!) Perhaps the highest coefficient of irrelevancy, however, should go to the gloomy remark that, when it was eclipsed by the Earth, the station would be "very difficult to see"—and hence, one presumes, to find. Dr. Campbell might be reassured on this point if he bothers to work out the visual range of a hundred-watt bulb in free space. But why look for a space station, even in full sunlight, when the feeblest of radio beacons has a range of many thousand kilometers?

More important than these somewhat lighthearted comments (which it is perhaps unfair to take too seriously) is Dr. Campbell's discussion of fuel requirements. He admits that his originally assumed exhaust velocity (one mile per second) was too low and raises it to two miles per second—still a rather pessimistic ceiling even for chemical fuels. This at once raises his mass-ratio from 2×10^9 to 4.5×10^4, and this figure could be reduced still further by employing a higher initial acceleration. (Large mass-ratios are extremely sensitive to variations in the number of "g" used.) The figures, however, are still much too high to be practicably realizable, and Dr. Campbell then goes on to discuss a lunar voyage with a refueling stop at a space station five hundred miles from the Earth.

Here, we are grieved to say, Dr. Campbell obtains figures that are quite incorrect, owing to his cavalier treatment of vector quantities. He states that to reach the space station from the Earth's surface the rocket must attain velocities of 4.37 miles per second horizontally and 2.33 miles per second vertically—a total of 6.70 miles per second. But of course, one cannot add directed quantities in this way, and the actual velocity required is much lower. To put it somewhat crudely, the Earth's gravitational field can cause the necessary ninety-degree bend in the ship's path, without the rockets doing any work at all. The correct answer to the problem is found by considering the velocities in an elipse touching the Earth's surface and the orbit of the space station, with one focus at the center of the Earth. Making allowances for transfer velocities, the figure is 4.9 miles per second, which requires less than half the mass-ratio calculated by Dr. Campbell (11.5, against 28.5). The remaining figures are incorrect by the same factor, but as there is little point in considering lunar voyages with a fuel of only two-miles-per-second exhaust velocity, we will not discuss them any further.

Dr. Campbell ends his paper with what is perhaps the unkindest cut of all: "The descriptions available for many of the suggestions are lacking in detail." When one considers the scores of books and hundreds of articles in which every conceivable aspect of interplanetary flight has been discussed (often with Teutonic thoroughness), this remark seems more than a little ingenuous.

A postcript to the paper gives some criticisms of a booklet, "Atoms, Rockets and the Moon," by Dr. Dinsmore Alter, many of which are justified, since Dr. Alter's enthusiasm is almost as ill-informed as Dr. Campbell's lack of it. (The less said about Dr. Alter's spherical spaceships with helicopter screws the better.) But at least he errs in the right direction, even if his aero- and astronautical ideas are somewhat strange.

When we had finished Dr. Campbell's paper, we were haunted by that "I have been there before" feeling, and suddenly we realized why. We dug out our copy of Simon Newcomb's *Sidelights on Astronomy* and read these words written some ten years before the Wright brothers flew, which should be an awful warning to scientists who make prophecies outside their field of special knowledge: "The demonstration that no possible combination of known substances, known forms of machinery and known forms of force can be united in a practicable machine by which men shall fly long distances through the air, seems to the writer as complete as it is possible for the demonstration of any physical fact to be."

Perhaps warned by this debacle, Dr. Campbell is less dogmatic and appears open to conviction. We would accordingly like to point out, as has been demonstrated so many times before, that the whole case for interplanetary flight rests on the possibility of attaining exhaust velocities considerably greater than two miles per second—in other words, on engineering rather than on astronomical considerations. It is therefore very significant that those who have had practical dealings with rockets (e.g., Wernher von Braun at Peenemunde and Frank J. Malina and Fritz Zwicky at the California Institute of Technology, to mention only three) are very much more optimistic than Dr. Campbell. This illustrates the operation of a general law. The extreme enthusiasts and the extreme skeptics almost invariably have only a superficial acquaintance with the subject. Those who have studied it for years realize, far better than most critics, the gravity of the problems that must be overcome; but they do not doubt the ability of the rocket to master them.

The Lackeys

of Wall Street

Prior to the launch of the first artificial space satellite, the Soviet Sputnik I, *in 1957, some of the Soviet literary community debunked Western science fiction as an example of "capitalist corruption." In the following essay, Clarke takes on this Soviet point of view.*

The following exchange of shots at the beginning of the Cold War has an amusing sequel—which I could certainly never have imagined when I was writing in 1949. The "all-out attack on American science fiction" was originally published in *Literaturnaya Gazyeta,* and the translation that appeared in the *British Fantasy Review* for December 1948–January 1949 aroused my ire.

Perhaps it is just as well that I had totally forgotten it by 1982, when I made my one and only trip to the late Soviet Union—because my genial host then was none other than one of my targets, Vasili Zaharchenko! Though I doubt if he had forgotten it, Vasili was too tactful to remind me of our earlier encounter.

And I am truly sorry that—quite unintentionally—I got him into serious trouble two years later, when he published *2010: Odyssey Two* in his magazine *Tekhnika Molodezhi* (Technology for Youth). Though I had warned him that he might have to "sanitize" the novel, he had failed to notice that I had named the crew members of the good ship *Alexei Leonov* after well-known dissidents.

As a result, publication came to a screeching halt after only two installments, and poor Vasili was fired (see the afterword to *2061: Odyssey Three* for details). But all was well eventually, for in 1989 two officials from the Soviet embassy presented me with the November issue of *Tekhnika Molodezhi* containing the complete translation of *2010*, with a commentary on what had happened, which rather quaintly stated: "Today, all these accusations seem to be ridiculous. . . . The decision on the novel was one of the last flights of the stagnation period in our ideology, demonstrating the paradoxical inability of the administrative guide in literature."

Vasili has long since forgiven me for my accidental disruption of his career, and on rereading the following piece after half a century, I feel that many of his criticisms were much better founded than I appreciated at the time. (For a hint, go back to the chapter "The Moon and Mr. Farnsworth.")

The all-out attack on American science fiction by Viktor Bolkhovitnov and Vasili Zaharchenko will, if we know them, have filled fantasy readers with a mixture of indignation, incredulous amazement, and hysterical laughter, in proportions varying according to their political outlooks. It is couched in the elegant language developed by the late Herr Goebbels for the castigation of the decadent democracies; and the writers would appear to have read widely before firing their broadside, their quotations ranging from Russell to Shaver, from Binder to del Rey. From this miscellaneous collection they have attempted to show that "the lackey of Wall Street, in the livery of a science fiction writer, carries out the orders of his bosses: to persuade the reader of the invulnerability of the capitalist system." This will certainly come as a great surprise to readers of *Astounding,* who have long grown accustomed to seeing the capitalist system, and frequently the Solar System, destroyed at least once per issue—and often two or three times for good measure.

That science fantasy has a political bias is quite true; but most of us, in our ignorance, had thought it was a bias to the left, and we have come across dark references to "pinkie science fiction" in certain Midwestern circles not a V-2's throw from the *Chicago Tribune*.

Indeed, in a recent issue of the *American Rocket Society Journal* it was hinted that the science fiction magazines have been toeing the party (i.e., left-wing) line, and for some time we have been expecting the House Un-American Activities Committee to come rampaging up Forty-second Street.

This just shows how our bourgeois prejudices have blinded us to the truth. We never expected, for instance, that Raymond F. Jones's "Rebirth" was, as comrades B. and Z. maintain, "monstrous in its openly fascistic tendency." Nor had we imagined, Trotskyite deviationists that we are, that our "shameless" authors had revealed capitalism's innermost secret, which serious literature only dares to hint at; though we always thought frankness was a good thing in literature—can it be that Mrs. Grundy has taken refuge in Russia?

As one of Wall Street's long-distance lackeys, my views on the matter may, of course, be suspect. But we had been foolish enough to suppose (to take a recent example) that no one of goodwill could possibly object to Theodore Sturgeon's "Thunder and Roses," in which an American survivor of an atomic war deliberately refrains from launching the retribution intended for his country's attackers, since if he does, there will be none left to rebuild human civilization. We had thought that such stories—and there have been several of the kind recently—were above party and nationality; yet we are told that "American science fiction in its unbridled racial propaganda reaches heights which might have made Goebbels envious." It looks, after all, as if Mr. Sturgeon is one of the more cunning of Wall Street's minions; so cunning, in fact, that he had us completely fooled. Or perhaps he is one of capitalism's inherent contradictions?

But let us see if we can get a clearer picture of the party line, before we discuss any further examples. According to the comrades, there are numerous instances of "fascist revelations" in American science fiction, but unfortunately they do not quote a single convincing case. We cannot for the life of us see why a story by Russell containing "an ecstatic description of the adventures of a spy from Mars" should be particularly fascist. Rather, in view of recent Canadian revelations, it seems positively communistic. Then: "To fortify the power of the imperialist war machine, the science fantasts of America unrestrainedly threaten with the atomic bomb." They did—but in a rather different way. We seem to remember that it was usually their own country they blew up first in the bomb-happy post-Hiroshima period. Even poor Adam Link appears a sinister reactionary from the other side of the Iron Curtain; and although we go 90 percent of the way with the critics as far as Mr. Shaver is concerned, we still think his stories are as innocent of politics as of good writing.

The concluding paragraph of their article, which in its substitution of invective for reasoning is indistinguishable from the sort of thing that Julius Streicher was hanged for printing, gives us cause to wonder at the standard of literary criticism in modern Russia, let alone anything else. What is one to make of this paranoic rubbish? Is it worth bothering about at all? We think it is; for it fits so perfectly into the pattern of current Russian behavior. All too clearly it is comrades B. and Z., not the readers of science fiction, who live in "a fearful world . . . in a world of nightmare fantasies." To such "sick minds," even L. Ron Hubbard's fine *The End Is Not Yet,* which had an American big businessman as villain, would be merely one of the subtler wiles of a capitalist dupe.

We have no particular love of American capitalism, which we do not suppose is any more permanent than any other social system; nor do we wish to defend the vast amount of rubbish that appears disguised as science fiction, some of which undoubtedly merits the description of a "screamingly shameless mess." We will grant that in the whole of their tirade the Russians make one valid criticism: that far too many stories of the future "describe worlds constructed according to the American system." But the reason for this, as should be obvious to any sane mind, is nothing more diabolical than laziness or lack of imagination on the part of our science fiction authors.

Every writer is conditioned, consciously or unconsciously, by his environment. Only a genius can imagine and describe a culture completely alien to that in which he lives; and some science fiction authors have made partially succesful attempts to do so—witness Robert A. Heinlein's *Beyond This Horizon* and A. E. van Vogt's "Null-A" stories, with their sociological implications. Unfortunately, few science fiction authors are creative geniuses. But because they describe societies that are reflections of their own, it does not follow that they approve of them. After all, few writers resemble the people they create, or necessarily condone the behavior of their characters.

One of the themes that has run through fantasy since the beginning has been the idea that eventually all races will be united in a world state, in which all will have equal rights. This theme has become more urgent since the advent of atomic power. For every story with a "fascist" tendency (meaningless catchphrase!) one can find dozens that have preached tolerance, the equality of men, and the enrichment of life by the application of science. But perhaps it is only appropriate, now that such charlatans as the Russian geneticist Trofim Denisovich Lysenko are turning Russia against science itself, its poor relation science fiction should come under the same interdict.

Voyages

to the Moon

Literary imagination about interplanetary flight has flourished as long as man
has looked skyward and counted the stars. Clarke, in the following essay,
examines these leanings in the writings of a professor of English
at Columbia University.

Marjorie Hope Nicholson's scholarly book *Voyages to the Moon* was published by Macmillan in 1948, and I am happy to know that she lived long enough to witness the entire first era of manned lunar exploration—and, indeed, the initial reconnaissance of the Solar System by rocket probes.

Miss Nicholson is a professor of English at Columbia University; she has no specialized knowledge of science but is deeply interested in its impact upon literary imagination. This book is a study of the field in which that contact has been most fruitful: but many students of the interplanetary romance will be astonished to discover that Miss Nicholson finishes her investigation—apart from a brief epilogue—at the end of the 18th century. Those who imagine that, except for Verne, no writer of importance appeared in this field before Wells, will be annoyed or amused by Miss Nicholson's contention that after the balloon ascent by the brothers Joseph and Jacques Montgolfier in 1783 the "cosmic voyage" suffered a slow decline and has now become so obsessed with technology that it has lost much of its pristine charm. But however much one may disagree with this thesis, *Voyages to the Moon* will provide a mine of information and entertainment. It is hardly necessary to say that it is well-written, and within its self-imposed limitations it is in every respect superior to the only other book of its type, Bailey's *Pilgrims Through Space and Time*.

The history of the cosmic voyage is inextricably entangled with early dreams of flight, and both received an enormous stimulus from the great astronomical discoveries of the early seventeenth century. The invention of the telescope was responsible for one of the great liberations of the human mind; it made the Moon visible to men's eyes as a world in its own right and revealed for the first time the true scale of the universe.

John Wilkins, in 1638, listed the four possible means of flight as follows: "(1) By spirits, or angels. (2) By the help of fowls. (3) By wings fastened immediately to the body. (4) By a flying chariot." Miss Nicholson uses this classification as a convenient division for her book, and to a modern reader it comes as a considerable surprise to learn that the greatest of the "supernatural voyages" was written by no less a scientist than Kepler. The magical elements in the *Somnium,* however, are concerned only with the voyage itself: when Kepler reaches the Moon, his descriptions are strictly scientific, as far as the knowledge of the time permitted.

According to Miss Nicholson, "Kepler transformed the old Lucianic tradition into the modern scientific Moon voyage. The weight of his scientific preeminence caused his little fictional work to be taken with the utmost seriousness by the learned, and his sense of mystery—part of the mysticism that marked all of his work—appealed greatly to poets and writers of romance."

It is, incidentally, of great interest to see what famous names in science and literature one encounters in Miss Nicholson's survey. Swift, Defoe, Huygens, Bacon (R. and F.!), Boyle, Hooke, Donne, Rousseau, Wren, Voltaire, Poe—all have been interested in flight in or beyond the atmosphere. It was only in comparatively recent times that such scientific speculating became no longer quite respectable.

Stories based on "flight by the help of fowls" and the use of artificial wings became very common as the old superstitions died and some factual basis was needed to replace the convenient "daemons" of earlier stories. With the notable exception of Godwin's *Man in the Moone,* however, these are more concerned with aero- rather than astronautics, and the technically minded reader will be more intrigued by the fourth category of stories, those employing "flying chariots."

It would be impossible to enumerate all the ingenious mechanical devices and plausible-sounding engines that authors have invented to elevate their heroes into the heavens. One may laugh at the vials of dew that Cyrano de Bergerac attached to his body so that he may be drawn toward the Sun; but perhaps the familiar space warp of contemporary science fiction will bear no closer examination.

The climax of this period of sheer literary inventiveness came, Miss Nicholson believes, at the close of the eighteenth century.

> Nothing further seemed left for restless imagination. And, indeed nothing was left to stimulate the kind of imagination with which I have concerned myself. We have reached the end of a chapter. . . . For hundreds, even thousands of years, man had . . . let his fancy play with means of flight both credible and incredible. . . . Those soaring souls . . . had recognized no barriers of time or space, no limitations of plausibilities. . . . But the future is upon us, a future in which no such untrammeled voyages of imagination . . . will continue to be possible. . . . As the balloons of the Montgolfiers . . . symbolize a beginning, so they mark the end of a long period of trial and error. . . . They mark the end, too, of a peculiar form of literature. The cosmic voyage will go on, but after the invention of the balloon it suffers a change into something, I think, less rich and strange. Science has conquered fancy.

Notwithstanding this, Miss Nicholson gives brief accounts of a few more recent stories that seem to her to stand in the classical tradition, or at least stem from it. These include Verne's *From the Earth to the Moon,* Wells's *First Men in the Moon,* and, in our own time, C. S. Lewis's *Out of the Silent Planet,* which Miss Nicholson considers, not without justice, to be "the most beautiful of all cosmic voyages and in some ways the most moving." It is one of the very few modern stories, she considers, that retains some of the charm that technology has banished.

This continually recurring theme of Miss Nicholson's book—the idea that poetry and romance were exorcised by the arrival of exact science—surely will not bear close examination. Miss Nicholson's erudition and width of reading are both phenomenal, and one therefore hesitates to suggest that she has overlooked a large part of modern fantastic literature. But it is rather difficult to judge the extent of her reading in this field, as her chief references to the current scene are to *Flash Gordon* and the comic strips—surely of anthropological rather than literary interest. (One has, incidentally, a delightful mental picture of Miss Nicholson's desk at the beginning of each week being submerged beneath a pile of Sunday supplements collected by her eager pupils.)

It is true that the "technical" story is now prominent, though perhaps not as prominent as Miss Nicholson imagines. Her judgment in such matters can scarcely be taken very seriously, for she makes the quaint remark that Dr. Weston's spaceship in *Out of the Silent Planet* is "as elaborately realistic as you will find in any of the pseudoscientific pulps!" But stories have been written in the last few decades as full of poetry and wonder as any that earlier times can show. There is almost no "technology" in Olaf Stapledon's cosmic novels, but their vision and majesty has never been matched by an earlier writer. For sheer magic and beauty I do not believe the otherworldly tales of Lord Dunsany have been excelled, though it is true that few of them are in the direct "cosmic voyage" tradition. But many of the elements that Miss Nicholson (and this reviewer) most admires are in his tales, as also in the better stories of the American writers H. P. Lovecraft and Clark Ashton Smith. The pity is that the work of modern authors is in danger of being submerged by the flood of trash by which many judge the entire genre. Luckily, good work is now being performed by such publishers as Arkham House in rescuing the best of such stories from oblivion beneath the rising sea of wood pulp.

To lament that "science has conquered fancy" would in any case be to ignore the lessons of the past. Fancy cannot exist without science. The stories that Miss Nicholson cherishes could never have been written without the basic scientific discoveries that inspired them in the first place—as she herself points out. Men had to know that the Moon was a world before they could visit it; and the greater the field of exact knowledge the greater— not the smaller—the possibility of imagination becomes. Where the ancients had only a handful of planets and a single sun, we have entire island universes full of wonders never dreamed of in earlier ages.

It is true that the frontier shifts; science overtakes earlier romances and obliterates them as the advancing tide smoothes down the sand. Five years ago fiction was still being written about the release of atomic energy; now only a few years are left in which to describe the first crossing of space.

There is nothing to regret in this. In many respects science has liberated, not enslaved, the fantasy writer. He need no longer devote the usual (and usually boring) chapter to his means of propulsion; this can be taken for granted and he can get on with the story, concentrating, for example, on the psychological or social implications of the subject. One result of this is that characters have made a shy appearance, here and there, in science fiction; the stories are no longer enacted entirely by lay figures whose emotional reactions to extraterrestrial circumstances have changed remarkably little since Kepler's day.

Such writers as, for example, Ray Bradbury in "And the Moon Be Still as Bright" or "The Million Year Picnic" have made brilliant use of this newfound freedom. Although there is often far less science—pseudo or otherwise—in these stories than in tales written before this century, the best of them have a three-dimensional quality that the earlier stories wholly lack. Both have their own qualities and merits, but to prefer the older type to the exclusion of the new is like insisting that music was ruined when sackbuts, hautboys, and harpsichords gave way to the modern orchestral instruments.

You're on the

Glide Path, I Think

When scientific events overtake the science fiction author, the challenge to the
writer is to move yet another step forward technically. It is hard to
remember the days before radar, but the following essay on
airport glide paths will go a long way in informing
the reader just how the writer does move on.

One day in early 1943, Flying Officer Clarke was summoned to group headquarters and interviewed by Wing Cmdr. Edward Fennessey, who then rashly sent him to join a team of American radar engineers who had just arrived with a top-secret invention. I had to wait until the war was over to describe what happened next; this article from a 1949 issue of *The Aeroplane* was my first, not very serious attempt. Another result was my only non–science fiction fiction, *Glide Path* (1963).

One of these Americans was Luis Alvarez, the inventor of ground-controlled approach, commonly known as GCA, who went on to win the Nobel Prize in physics (as did his equivalent in *Glide Path*) but became even more famous when, with his son Walter, he proposed the now widely accepted theory that the extinction of the dinosaurs was caused—or at least accelerated—by a cosmic impact 65 million years ago. (See my review of Walter's *T. Rex and the Crater of Doom*.)

In 1987, Louie published his autobiography, *Alvarez: Memoirs of a Physicist* (Basic Books), and I was honored when asked to contribute a blurb for the dust jacket:

"Luis seems to have been there at most of the high points of modern physics—and responsible for many of them. His entertaining book covers so much ground that even nonscientists can enjoy it; who else has invented vital radar systems, hunted for magnetic monopoles at the South Pole, shot down UFOs and Kennedy assassination nuts, watched the first two atomic explosions from the air—and proved that there are no hidden chambers or passageways inside Chephren's pyramid?

"And now he is engaged on his most spectacular piece of scientific detection, as he unravels the biggest whodunit of all times—the extinction of the dinosaurs. He and his son Walter are sure they've found the murder weapon in the crime of the eons. . . . But I've no idea what he'll get up to next."

Alas, what Louie did do next, only the following year, was to die—but not before paying me the greatest compliment I have received in my entire life. The copy of *Alvarez* he sent me is inscribed, "For Arthur Clarke, long-time friend and role model."

At the end of 1994, on a brief visit to England, I was kidnapped by the BBC for its notorious *This Is Your Life* program. When I was taken to the studio, I had no idea what was happening, and great was my astonishment to be greeted by Apollo 11's Buzz Aldrin, Cosmonaut Aleksei Leonov—and Sir Edward Fennessey, retired director of Post Office Telecommunications. I was glad to learn that he did not regret his decision of half a century before. Neither did I.

Few people today will understand the reference in the opening paragraph of the following essay, but Gatow was very much in the news at the time; it was the airport through which essential supplies were flown into Berlin, when Stalin attempted to isolate the city in 1948. Without GCA, the Berlin airlift might never have been possible; so Louie's invention may have helped to win the first round of the Cold War.

Now that GCA has become thoroughly established as one of the leading blind-approach systems and has been doing yeoman service all over the world—not least at a little place called Gatow—it is perhaps safe to reminisce about some of the early pioneering days, when the idea of taking pilots down the glide path was regarded with the gravest suspicion.

The first GCA unit was built in 1942–43 at the Radiation Laboratory, Massachusetts Institute of Technology, its inventor being the atomic physicist Dr. Luis Alvarez, who had become tangled up in radar when America entered the war. He was later to be rapidly untangled and directed back into atomic physics when that subject appeared to have some remote connections with warfare.

Most people must by now have seen photographs of the present GCA equipment, even if they have not seen it in actuality, but the prototype was a very different affair. It occupied two vast trucks instead of a single trailer and was much more impressive and very much more of a nuisance to move. It was also considerably more complicated (someone once counted five hundred vacuum tubes before they got tired), and it needed more operators than the present sets.

These were not fundamental objections, as the Mark I was built entirely for experimental purposes, with no thought of serious operational use. In those days it had still to be shown whether radar could track aircraft accurately enough for the information to be used to land them. And also—equally important!—if pilots would do as they were told when they were ordered around the sky by disembodied but apparently omniscient voices.

The Mark I was undergoing tests in the States, apparently without arousing any great excitement, when it was discovered almost accidentally by a visiting British VIB (very important boffin). He at once realized its importance and, by what means we know not, succeeded in "capturing" the whole equipment and loading it aboard a British battleship. He also "kidnapped" Dr. Alvarez and his team, whisking them to the United Kingdom on a priority so high that they crowded Bob Hope and Frances Langford off the flying boat at Shannon.

The equipment was reassembled at Elsham Wolds, then a bomber station, where the first trials were successfully accomplished. Unfortunately, it was not long before some genius decided that the weather at Elsham was altogether too good, and that since GCA was supposed to be a blind-approach system, it ought to go to a station in a state of more or less permanent "clamp." So the unit was moved to Davistowe Moor.

We only saw this airfield in the rainy season, which probably does not last the entire year, but when we arrived on the scene as technical officer U/T, we found the American scientists amplifying their already excellent vocabularies over expiring transformers, and complaining bitterly that their equipment was not built for underwater operation. At night, when the apparatus closed down and cooled off, the all-pervading mist would creep gleefully into every cranny, depositing moisture in high-voltage circuits so that brief but spectacular fireworks displays would ensue in the morning.

As part of the battle against this insidious enemy, electric heaters were installed and switched on at night. One evening, as we were connecting these to the mains in the completely blacked-out hangar, a series of eldritch ululations disclosed the fact that we'd electrified one of the mechanics inside the equipment. We eventually located him by rapid sound-ranging and are still annoyed that we never got a Royal Humane Society medal for hauling him, still in a state of twitch, out of the works. Our tough Canadian flight sergeant, who got bitten more or less regularly by the fifteen-thousand-volt transmitter power supply, couldn't understand what all the fuss was about.

Luckily, the unit was moved to St. Eval, near Newquay, Cornwall, before the whole apparatus became waterlogged, and it was here that the Mark I saw most of its service with the RAF.

Testing an experimental blind-approach system on an operational station had its disadvantages. We had our own little flight of Oxfords and Ansons, which were liable to be making approaches on odd runways—and even downwind on the runway in use—when Flying Control was trying to land a customer!

To make matters worse, since there were no proper hard-standings for the apparatus, the big GCA trucks and their satellite fleet of service trucks, the Naafi canteen supply vans, and visitors' cars had to be sited on one of the out-of-use runways, near the main intersection. All too often a change of wind would demand a hasty retreat by the entire unit—a move that we resisted tooth and nail, since it required adjusting the controls and uncoupling all our cables. There were so many of these—an inch or more thick—linking the vehicles that the site sometimes looked like a rendezvous for amorous squids; but eventually we got so streamlined that we were able to change positions in about twenty minutes.

The Flying Control officers became quite used to seeing what appeared to be an advance party from Bertram Mills's Circus proceeding up the runway, turning onto an intersection, and then proceeding to pitch camp with sublime indifference one hundred yards from the edge of the runway in use. Once we miscalculated and found a squadron of Spitfires taking off behind us. Luckily our twenty-ton trucks did some rapid footwork and skipped onto the grass in time.

Some of the GCA sites were in more reasonable places on outlying pieces of perimeter track, and one was surrounded by a tasteful tableau of crashed Liberators. We were always very careful to explain to visitors that they had managed to get that way without any help from us.

The original American team was still with us for the early part of the time at St. Eval, although Dr. Alvarez had now returned to the States. Incidentally, Alvarez was very far from the popular conception of a high-powered scientist. He had a pilot's license and was one of the best, as well as perhaps the first, of GCA controllers.

According to legend, he would calmly continue talking a plane to landing, even when cathode ray tubes were popping in all directions, frenzied mechanics were crawling under his legs, and smoke was gently curling from the meter panel. Moreover, he was an expert at breaking down "sales resistance"—and there was plenty of it in those days, particularly among exponents of rival systems.

When Dr. Alvarez returned to America, some of us guessed the reason, but we did not know until a long time later that he was one of the atom bomb team on Tinian in August 1945. His deputy, Dr. George Comstock, remained in charge until the rest of the team returned. Our dearest memory of Dr. Comstock is of him, on his last night in England, lying in bed avidly reading something called *The Gamma Ray Murders.*

We were very sorry to see the Americans go, and some of our WAAF operators were quite heartbroken. They were a grand crowd and taught us a great deal. Our discussions were by no means devoted entirely to waveguides, magnetrons, and pulse techniques. We also learned some interesting songs.

With the assistance of the Americans we had trained a team of RAF mechanics, operators, and controllers who were later to form the nucleus of the GCA empire. But we were now very much on our own and could no longer run to the experts when anything went wrong—as it frequently did. It had never been intended that the laboratory-built Mark I should be used continuously, month after month, for training and for innumerable demonstrations, in a foreign country and run by people who hadn't watched it grow up from a blueprint.

Yet it always managed to be serviceable when A. V. M. Blank or A. C. M. Double Blank arrived in his private Proctor.

We sometimes thought that everyone in the RAF above the rank of group captain had visited us at one time or another. They usually went away thoughtful, if not convinced. There were times when the crowd in the control truck was so thick that mere air commodores had to sit outside on the grass waiting their turns. The operators grew quite accustomed to working with a packed mass of humanity breathing down the backs of their necks.

They also grew quite accustomed to the sudden disappearance of all signals as we switched off to forestall some incipient breakdown. If the weather was dirty and we had an aircraft up at the time, that was just too bad; it would have to ask someone else the way home. As we so often pointed out to the OC Flying, it was easy enough to get more aircraft and pilots, but there was only one GCA and we couldn't take risks with it. He rather stubbornly refused to see our point of view.

At St. Eval we made every imaginable mistake, and quite a few others, mastering the technique and developing the patter that has now become universally familiar. Nothing could be taken for granted, and we had to learn by trial and error.

No one, for example, seemed sure of the best glide path; anything between two and five degrees was suggested for different types of aircraft. Changing the glide path involved mechanical rearrangements in a Heath-Robinson apparatus full of gears, clutches, solenoids, and selsyn motors. As the GCA was not sited at touchdown, but well up the runway, the radar operator saw a distorted picture of the aircraft's approach: the glide path, in fact, appeared on screen as a hyperbola instead of a straight line.

This distortion was corrected by most peculiar cams based on a curvilinear spiral coordinate system. These revolved once during every approach, except when they fell off their shafts. Changing a glide path meant changing a cam, but one day the wrong cam was accidentally left in the machine so that we brought a heavy bomber down a fighter glide path. Nevertheless the pilot reported an excellent approach, so we decided not to pamper our clients anymore, and thereafter everyone came down at 3½ degrees, whether they knew it or not.

The biggest operating boob we ever made on the Mark I might have had serious consequences had our aircraft not carried safety pilots who kept an eye open while the driver was obeying our instructions and could break off the approach if anything was obviously wrong. One day the observer in the aircraft was a civilian scientist whose progress from station to station was always marked by the trail of mislaid secret documents he scattered in his wake. (If he reads this, we are of course only kidding.)

It was his first approach on GCA, and any faith he had in the system was somewhat shattered when he found his aircraft descending into the sea some miles off the coast while the controller was saying, "On the runway, one mile to go, coming along very nicely." He stood it as long as he could, then tapped the pilot on the shoulder and suggested that the depressingly damp scenery below bore little resemblance to runway 320.

It turned out later that the inexperienced radar operators had picked up the wrong aircraft and were tracking someone who was making a normal visual approach and was naturally "coming along very nicely," while our own aircraft had been missed altogether. The mistake was, in the long run, a fortunate one, as it focused attention on the problems of identification and resulted in improvements in control technique. But it was some time before we lived it down.

Another slip, which might have had equally sad consequences, received no such publicity, being covered up by a quick screwdriver adjustment between approaches. We have kept very quiet about it ever since, but it may do no harm to mention it now.

In order to check that the radar system was properly lined up, each runway had at its end a metal reflector, or marker, which acted as a kind of radar mirror and gave a fine signal on the screens. One day a marker fell down and we didn't notice its absence as we found a nice echo more or less in the expected place. Unfortunately this echo happened to be caused by a Liberator on an apron one hundred feet from the marker, so that in lining up on it we had slewed our glide path around through several degrees. We discovered that something was wrong when the pilots complained that the approach we gave them passed through the top of a hangar.

At this point it might be as well to reassure any nervous readers by again pointing out that these incidents took place during the training of the first crew and with the first experimental equipment. They were our schoolboy howlers. We learned a great deal from them, and they never did anybody any harm.

The last incident was by no means the only time when we lost a marker; one day we found that workmen were borrowing them to serve as builder's hods, which they strongly resembled. So eventually we sent a mechanic out to the runway marker to wave it up and down while we watched to see what radar signal disappeared from the screen.

One day when the marker was taken down, the echo didn't vanish and we found that the mechanic was sending back quite a juicy signal himself. We can still remember our surprise as we watched him running toward us, visible through the open door of the truck, and also as a small blob creeping across the radar screen. A number of unflattering explanations were offered for this phenomenon, but we never really solved it to our satisfaction. No one else ever gave anything like such a good echo.

We were occasionally troubled by seagulls—though not in the usual way. They gave transient echoes that flickered across the screen from time to time; there was no possibility of confusing them with aircraft response— they were far too feeble—but they puzzled us until we found the explanation.

The resolving power of the Mark I was very great, particularly at short ranges. We had a striking demonstration of this one day when, on the radar screen, we watched our CO cycling absentmindedly down the runway in use, observed the airfield controller emerge rapidly from his kennel, and saw the two signals blend into each other as an animated conversation developed. After a while there was an amoebalike separation of the echoes, the CO proceeded blithely on his way, and the controller went back to his Aldis lamp, no doubt thinking that next time he wouldn't bother if a Walls ice cream tricycle came trundling down the runway.

St. Eval was one of the first airfields to be fitted with FIDO (fog, intensive, dispersal of), and the installation was a colossal one, burning, we are afraid to say how many, thousands of gallons of petrol a minute. Not only was there a double row of burners the whole length of the main runway, but various sheets of flames branched out at right angles as well. When the whole affair was going full blast, it lit up most of Cornwall and caused confusion among all the fire brigades for fifty miles around.

For a long time attempts were made to arrange a combined GCA-FIDO landing, but they were foiled by persistently good weather. At last we got what we wanted—a drizzling fog with practically zero visibility. It was so bad, in fact, that the aircraft could never have taken off without FIDO.

At midnight, all was ready. The scene might have come from Dante's *Inferno*—there were great sheets of fire roaring on either side, clouds of steam rising into the mist, and a heat like that from an open furnace beating into our faces, for we were only a hundred feet from the nearest burners. The aircraft was standing by, waiting to take off with the station commander aboard, and in the GCA trucks the cathode ray traces were scanning normally, building up the radar pictures on the screens. At that precise moment, the turning gear that rotated the plane position aerial decided it had had enough and crunched to a halt, shedding half its teeth in the process.

The search or traffic control system, with its 360 degrees of vision, was thus completely blind; but the aerials of the landing system were still scanning, giving us a picture some 30 degrees wide centered on the runway and pointing downwind. It was decided to risk it, by keeping the aircraft in the narrow 30 degree sector and using the landing system, which was now all we had, both for control and approach.

As soon as the aircraft took off, it of course promptly vanished into our 330-degree blind sector, but we immediately turned it through 180 degrees and it soon reappeared. It was allowed to fly downwind for a few miles—we dared not let it go too far, as the landing system had a range of less than 10 miles—and then whipped around for an approach. The pilot was unable to land on this run; he found himself at the edge of the runway, but the visibility was so bad that he could see only a single line of FIDO burners and didn't know on which side of the runway he was! So the maneuver had to be repeated, and luckily the second approach was successful, despite the attempts of the FIDO-induced gale to push the aircraft off course.

Those were the worst conditions under which the Mark I was ever used. As we took the trucks back to their hangar, visibility was still so bad that the drivers had to be guided by instructions from those on the running board; and we could have done with our own radar to get us around the perimeter track.

That exploit was also one of the last highlights of the Mark I's career. It had already run for six months longer than it had ever been intended to operate, and we are very proud of the fact that in the days before it was finally dismantled it was working as well as it had ever done, thanks to extensive overhauls and partial rebuildings. But the operational Mark II's were now on the way, and the GCA team was moving to a new airfield all (or nearly all) of its very own.

The Mark I made the trip, but was never reassembled and finally perished in a cannibal orgy. A long time later we came across the gutted and derelict vehicles in an MT park, and we had a quiet weep inside them, remembering some of the happiest as well as some of the most exasperating hours of our life. *Requiescat in pace.*

Morphological

Astronomy

It took two tries for Einstein to get it right with his theories of relativity. First
with the specific in 1905 that presented atomic advances, and second with
the general in 1915 that proved the existence of the space-time continuum,
allowing man to launch spacecraft into gravitationally pulled orbits.
Here Clarke shows how Prof. Fritz Zwicky, the celebrated astronomer, was the
kind of scientist who always knew he was right and usually proved it!

The following brief note is a tribute to one of the greatest—and certainly most imaginative—astronomers of the century. As this lecture
amply proves, Fritz Zwicky anticipated many of the spectacular developments and discoveries that were to transform our view of the universe.
Perhaps his most remarkable achievement was the detection of dark matter, still a major mystery.

His ill-conceived awareness of his own genius did not endear
Zwicky to his fellow astronomers, who often took to their heels when
they saw him coming. But I owe him a debt of gratitude, because in

1952, while he was recovering from a heart attack, he conducted me on my one and only visit to Mount Palomar Observatory.

As will be seen in this note, Zwicky was planning to launch space probes as early as 1949—eight years before *Sputnik*. He intended to do this by firing shaped charges aloft in V–2 rockets. However, the only experiment failed because, as he told me darkly, "someone had altered the wiring." I am quite sure that he suspected sabotage.

He may have been right . . . and I would love to know what the Oxford audience thought about his modest proposal for reconstructing the universe.

The Halley Lecture for 1948, delivered by Professor Zwicky at Oxford, "Morphological Astronomy," has now been printed in *The Observatory* of August 1948. It is a long and most stimulating paper, worthy of careful study, and the following are merely some of its highlights.

Morphological thinking is defined as an orderly way of looking at things and considering all the possible relations involved in any specific problem. Zwicky gives examples of its application to telescope design as a simple case, and then discusses the extension of the method to general astronomy. Considering first the observation of celestial phenomena, he discusses possible instrumental developments. The photoelectric telescope can amplify faint images and can also eliminate uniform backgrounds of light, a possibility that does not exist with the optical telescope. Moreover atmospheric disturbances may be eliminated by electronic compensation by image stabilizers, one of which has now been built by the physicist Vladimir Zworykin.

Until recently the location of astronomical instruments has been disadvantageous as they have been limited to the Earth's surface. Instruments have now been taken to heights of two hundred kilometers in V–2 rockets, but this is insufficient for some purposes. "The author, therefore, is working on rockets to reach one thousand kilometers in height," and "the morphology of astronomy thus includes the morphology of rockets." (NB, Professor Zwicky was one of the leaders of American jet and rocket research during the war, and his application of morphological principles to propulsion systems resulted in a number of advances such as the submarine ramjet, the hydrobomb.)

Zwicky then considers recent observational work on the contents of the universe, particularly clusters of nebulae, supernovae, and unusual types of stars. A search is being made at Palomar for very blue faint stars to establish a more representative Russell-Hertzsprung Diagram (which enabled us to help understand the evolutionary development of the stars) and remove the effects of selective observation. The forty-eight-inch Schmidt is to be used in a "determined search" for gravitational lenses produced by compact nebulae and neutron stars. It is thought that the latter may exist as remnants of supernovae; they would have densities of 10^{14} grams per cubic centimeter and "make possible, in principle, a reversal of time."

Recent developments, Zwicky continues, will enable us to experiment with celestial phenomena. Work is in progress (at White Sands, New Mexico) to give small test bodies the velocity of escape by ejecting them from carrier rockets. It is hoped that the collisions of these with the Moon and other planetary bodies can be observed.

Professor Zwicky concludes as follows: "The knowledge gained in astronomy has had wide applications . . . at almost all times. . . . An extrapolation of these applications which lies in the line of morphological thinking, and which we might just as well visualize cold-bloodedly, since it appears inevitable, is the reconstruction of the universe. The reconstruction of the Earth naturally comes first. One of the biggest problems that comes to mind is the nuclear stabilization of the Earth. Since no fundamental principle seems to stand in the way of realizing nuclear fusion on a large scale, the danger exists that the whole Earth might be exploded. . . . In the wake of the realization of large-scale nuclear fusion there will, no doubt, follow plans for making the planetary bodies habitable by changing them intrinsically and by changing their positions relative to the Sun. These thoughts are today perhaps nearer to scientific analysis and mastery than were Jules Verne's dreams in his time."

The Conquest

of Space

*Few are gifted enough to imagine, then visualize, and finally to execute as
paintings full-color impressions of the Solar System. One such person is
Chesley Bonestell, who captured the planets and their satellites in his
artwork, images that have stood the test of time. Clarke writes how
uncanny it was for Mr. Bonestell to have this gift, long
before the new developments in radio astronomy.*

I t seems appropriate that I should end the 1940s—the last Earth-
centered decade of human history—with this review of a book that
perhaps did more than any other to inspire a generation of would-be
space cadets.

It certainly inspired my 1953 story "Jupiter V" (reprinted in *Reach
for Tomorrow*). This was set in the year 2044, for a reason explained by
one of the characters:

"These pictures are nearly one hundred years old. They were
painted by an artist named Chesley Bonestell and appeared in *Life* back in

1944—long before space travel began, of course. Now what's happened is that *Life* has commissioned me to go around the Solar System and see how well I can match these imaginative paintings against the reality. In the centenary issue, they'll be published side by side with photographs of the real thing. Good idea, eh?"

Well, it was a good idea—back in 1953. But it happened in just a quarter century, thanks to the Viking and Voyager space probes. We did not have to wait until 2044.

In 1972—when NASA was planning the grand tour of the outer planets that led to the Voyager missions—I had the privilege of working with Chesley Bonestell on *Beyond Jupiter,* which tried to anticipate what would be discovered. Chesley, who died in 1986 at the age of ninety-eight, still painting furiously, lived long enough to compare many of his visualizations with the reality. I have been even luckier, having now seen close-ups of all the planets except Pluto, and most of their major satellites.

The fears expressed in the last paragraph have, fortunately, turned out to be groundless. The Solar System is even more astonishing—visually—than we ever imagined. Only biologically has it turned out to be disappointing.

But not, I suspect, much longer. Stay tuned.

The Conquest of Space

by Chesley Bonestell and Willy Ley (Viking Press, 1949), 18 fig., 48 plates

This beautifully produced book must be one of the most dramatic "popular" introductions to astronomy ever published. It is interesting to note that the first American edition was sold out in less than a month, thus justifying the publisher's enterprise in putting out so sumptuously printed a book at such a low price.

Chesley Bonestell is an artist who for some years has applied his remarkable talents to astronomical subjects, and his scientifically accurate portrayals of lunar and other landscapes have been very well received in the pages of *Life* and elsewhere. In this book he has cooperated with Willy Ley, the well-known rocket expert, and the result is a book that combines an excellent elementary introduction to astronautics with a trip around the Solar System enlivened by pictures of an almost photographic realism.

The text is in four sections. Chapter 1 opens with a dramatic description of a typical V-2 launch at White Sands, New Mexico, and then goes on to discuss the basic features of rocket propulsion, the problem of escape, and Keplerian orbits. Chapter 2 describes the lunar voyage and conditions that would be encountered on the Moon, while chapter 3 gives a good general picture of the rest of the Solar System. The last chapter, somewhat surprisingly, is devoted entirely to the asteroids and is the best nontechnical account of these "vermin of the skies," as they have been unflatteringly called, that I have encountered.

Of the forty-eight plates, sixteen are in full color and must be seen to be appreciated. (We have known cases of people who mistook them for actual color photos taken on the spot and were mildly surprised that they had read nothing about the matter in the papers!) Particularly striking are the views of the bleak Mercurian landscape, blistering with heat; the beautiful orange light spilling over the Leibnitz Mountains during an eclipse of the Sun by the Earth; the rugged landscape of the lunar pole with the full Earth on the horizon; and the truly magnificent portrayal of the rocket among the mountains, awaiting the return journey. The views of Mars from Deimos, of the surface of the planet looking along one of the canals from the polar cap, and of the great deserts themselves are extremely beautiful. But there can be little doubt that most impressive of all are the pictures of Saturn seen first from Phoebe, its outermost satellite, and then, coming closer moon by moon, finally dominating the sky of Mimas.

The reviewer has only one fear after reading this book. Let us hope that, when we finally reach the planets, they do not prove an anticlimax after Chesley Bonestell's previews.

THE 1950s:

BENEATH THE

SEAS OF CEYLON

Taprobane, February 1971.

Introduction

Looking back at it from almost half a century later, I realize that the 1950–59 decade was probably the most crucial of my life. For in that period, I lost my amateur-writer status when the Book-of-the-Month Club made *The Exploration of Space* its June 1952 selection, triggering my first visit to the United States. Shortly afterward, *Childhood's End* was published, and an annoyingly large number of people still consider it my best novel.

In 1954 my increasing fascination with underwater exploration took me, aboard SS *Himalaya,* to Australia's Great Barrier Reef, with a stop-off for one afternoon in Colombo, Ceylon. While diving off the world's largest coral formation, I heard on the shortwave radio that the United States proposed launching an artificial satellite during the 1957–58 International Geophysical Year.

The Space Age was about to dawn, so I hurried back to New York to research *The Making of a Moon,* and to be the guest of honor at the 1956 World Science Fiction Convention. But the call of the tropical seas was too great, and in January 1956, with my partner, Mike Wilson (who had introduced me to skin diving in 1950), I sailed to Ceylon on SS *Orcades,* to work on a series of books about the Indian Ocean.

In October 1957, I was back in Europe as a member of the British delegation to the International Astronautical Congress at Barcelona. On the morning of the fourth, I was awakened by a phone call from the *Daily Express* in London. The Earth had a second moon; but it was not, as almost everyone had expected, an American one.

The end of the 1950s found me commuting from the Colombo home established by Mike Wilson (and later his wife, Elizabeth) to the United Kingdom and the United States. Though I became more and more reluctant to leave Ceylon, I could not reside there for more than six months in any year without being ruined by the local income tax.

Amazingly, despite all this traveling, during the decade I appear to have written 140 pieces of fiction and 211 of nonfiction, a record never approached by me again. Virtually all of the fiction (sometimes in revised form) is still in print, but most of the nonfiction was ephemeral journalism, long forgotten even by the author. Most that seems worth saving has been published in several earlier volumes of essays, but as these are no longer available, I am reprinting the more important—or, I hope, more interesting—ones here, even at the cost of occasional embarrassment.

The Effect of

Interplanetary Flight

On February 4, 1950, Clarke discussed the effect of interplanetary flight at a London meeting of the British Interplanetary Society. A report on his speech has been provided by Walter H. Gillings, his first professional editor and donor of his first typewriter.

Opening the discussion, A. C. Clarke said that if, when we reached the planets, their environments proved so hostile that we could only establish scientific bases on them, the number of human beings who would venture outside the Earth would never be more than a few thousands; and although spaceflight might well revolutionize our knowledge in other fields, after the initial experiment had worn off, it might prove an anticlimax. If one took a more optimistic view, he supposed that quite large and eventually self-supporting colonies would be established on the Moon and on Mars; but how quickly these groups would become politically important he could not foresee, since a single scientific discovery might transform the situation at any time. For example, if it was found that one could live longer under low gravity, all the over seventies might emigrate en masse to our satellite.

While drawing a parallel between the rise of the United States as a world power and the emergence of extraterrestrial colonies, he emphasized that fundamental differences in developments were involved. We had to face the fact that there was no other world in the Solar System on which men might live without mechanical assistance, and if unrestricted colonization was to be carried out, concessions would have to be made either by us or the planets concerned. He thought it would be easier to change the planets than the human species, and with the powers we should possess in another few centuries we could effect very radical alterations to the atmospheres, temperatures, and even the orbits of the planets, as suggested by Professor Zwicky. He assumed that we would not encounter intelligent life in our explorations; otherwise there would be no question of exploring our peculiar ideas of what constituted an equable climate. But any other really intelligent race in our own Solar System would presumably have reached our world long ago.

Space Travel in

Fact and Fiction

An address to the British Interplanetary Society, April 1, 1950.

There are, it seems to me, two obvious ways of tackling the subject that the title of this paper is so careful not to specify too exactly. The first might be called the "Ph.D. or Bust" method. It would involve the reading of some hundreds of books and thousands of short stories, and a prolonged incarceration under the dome of the British Museum Reading Room. At the end of a few years' labor the patient researcher might, if still sane, be able to produce a comprehensive analysis of the interplanetary story since Lucian of Samos—little knowing what he'd started—first tried his hand at this theme in A.D. 160.

The second approach is the one that I have adopted. It relies simply on the fading memories of a youth that, in retrospect, seems to have been largely misspent in the pursuit and avid consumption of American science fiction magazines, on offensive sweeps through my friends' libraries, and on frequent dips into two quite essential books—J. O. Bailey's *Pilgrims*

Through Space and Time and Marjorie Nicholson's *Voyages to the Moon*—my debts to which I acknowledge herewith.

All I have attempted to do, therefore, is to pick out those ideas and themes in the interplanetary story that have struck my fancy or that seem to me relevant to our present conceptions of astronautics. I have also concerned myself primarily with the technical content of these tales; their literary merits, such as they are, have not been considered here. This means that I will say practically nothing about some of the finest of all interplanetary romances—such as Stapledon's *Last and First Men,* or C. S. Lewis's *Out of the Silent Planet,* which were concerned with social or philosophical rather than technical ideas—but will deal largely with stories at a far lower literary stratum, such as Verne's *From the Earth to the Moon.* Perhaps this self-denial is only appropriate on the part of one who some time ago sacrificed his own amateur status in this field.

The first problem encountered in this survey is that of classification. My interest now being mainly concerned with techniques, I could not use the simple and historical approach and discuss stories of space travel in their historical sequence. Instead I have divided them into two main groups that for convenience may be labeled mechanistic or nonmechanistic. In stories of the first class, some engine or technical device, more or less plausible according to the science of the time, is used to bridge space. The second class contains all those stories in which dreams, supernormal intervention, psychic forces, or the like are invoked. This includes most of the very earliest works, but the division cuts across any historical sequence, since some of the best stories of our own era belong to this category.

It is somewhat curious that the first truly scientific Moon voyage invoked supernatural forces. This was the *Somnium* (1643), written by no less a man than Kepler, to whom astronomy and hence astronautics owe almost as much as to Newton himself. To the modern man, Kepler presents something of a paradox. The discoverer of the laws governing the motion of the planets—and therefore of spaceships—he was both a scientist and a mystic; his background may be judged by the fact that his own mother barely escaped execution for sorcery.

In the *Somnium,* which was not published until after his death, Kepler employed demons to carry his hero to the Moon, but he was careful to make the point that as one leaves the Earth, the air becomes rarefied and breathing can only be carried out by "sponges moistened and applied to the nostrils." Even more significant is Kepler's remark that as the voyage progressed, it would no longer be necessary to use any force for propulsion. Thus three hundred years ago, before the discovery of the law of gravitation, Kepler had foreseen two of the most important features of spaceflight. His description of the Moon, based on the new knowledge revealed by the telescope, was also as accurate as possible, though he assumed the existence of water, air, and life. It is interesting to note that the *Somnium* influenced H. G. Wells, who mentions its ideas in *The First Men in the Moon*.

At the end of Kepler's book, it is revealed that the whole adventure is a dream—an annoying device that has been used all too often in fantastic literature, particularly in stories of this kind. Equally common is the idea that in some trancelike state one's mind or even one's body could travel across space to other worlds, not limited, perhaps, by the miserable speed at which light is forced to crawl along. This device was used in Stapledon's *Star Maker* (1937), C. S. Lewis's *Perelandra* (1944), and in David Lindsay's remarkable but little-known work, *A Voyage to Arcturus* (1920). Descending a few orders of magnitude in the literary scale, it was also employed by Edgar Rice Burroughs to transport John Carter, Prince of Helium, to the bloodstained little planet into whose population he was to make such serious inroads.

Before the age of science, there was good reason to employ such paraphysical means of conveyance because they seemed as plausible as any other in times when an airborne broomstick would have excited far less surprise than a balloon drifting across the sky. On the other hand, when a modern writer uses such methods, it must not be imagined that he is too lazy to think of anything better; he may have very good reasons for his choice. There is, indeed, little alternative if one wants to write a story of cosmic scope, yet assumes that the speed of light can never be exceeded. Some of the most thoughtful of recent authors, such as Jack Williamson in his novel *And Searching Mind* (1948), have suggested that in the long run purely mechanical solutions to the problem of spaceflight will be superseded by paraphysical ones. How far one is prepared to grant this possibility depends on one's assessment of the parapsychologist Dr. J. B. Rhine's latest work (see, for example, *New Frontiers of the Mind*). It will certainly be an irony of fate if the giant spaceships of the next millennia belong to the childhood of the universe—if, after all, Kepler has the last laugh.

In the earliest times, writers who wished their stories to have a certain plausibility, or did not approve of the trafficking with supernatural powers—transactions in which, however carefully one read the contract, there always seemed to be some unsuspected penalty clause—often used natural agencies to convey their heroes to the Moon. (It was, of course, almost always the Moon. We tend to forget that the discovery that the other planets were actually worlds, and not mere points of light on the celestial sphere, is relatively recent. It was not known, for example, to Shakespeare, although it had been guessed by some of the Greeks.)

Natural forces were invoked in the earliest stories of space travel, the misleadingly entitled *Vera Historia* (True History) written by Lucian of Samos in A.D. 160. In this book, the hero's ship, cruising in the dangerous and unexplored region beyond the Pillars of Hercules, was caught in a whirlwind and deposited on the Moon. It is true that no one has much good to say about the weather around the Bay of Biscay, but this must have been a rather rougher passage than usual.

It is an astonishing fact that though Lucian wrote two stories on this theme (his second, *Icaromenippus,* we shall come across later), no one bothered to imitate him for fifteen hundred years. (Though it is, I suppose, no more astonishing than the fact that for even longer men possessed ships yet never sailed them westward. Perhaps Lucian's first story scared them back into the Mediterranean.) At any rate, it was not until after the death of Kepler and the appearance of the *Somnium* that the first English story of a lunar trip appeared: Bishop Godwin's *Man in the Moone* (1638). Godwin's hero, Domingo Gonsales, flew to the Moon on a flimsy raft towed by trained swans. Gonsales had no intention of traveling to the Moon, but accidents will happen even in the best circles, and when he made an emergency takeoff to escape from brigands, he did not realize that his swans were in the habit of hibernating on our satellite. Gonsales's journey lasted twelve days, and he appears to have had no difficulty with respiration on the way; he did, however, notice the disappearance of weight, though this happened when he was still quite close to the Earth. Such a view of the short-range nature of gravity, one might point out, is still quite common among educated laymen today.

The most ingenious use of natural forces was, I think, employed by Cyrano de Bergerac in his *Voyages to the Moon and Sun* (1656). In the first of his several interplanetary expeditions, the motive power was provided by vials of dew around his waist, for Cyrano very logically argued that as the Sun sucked up the dew in the morning, it would carry him up with it. In other voyages, to which we will refer later, Cyrano used more scientific means and, quite accidentally, made some remarkably accurate predictions.

The last story that I shall mention in this group is Verne's *Hector Servadac* (1877), in which a comet grazes the Earth, scoops up Hector and his servant, and takes them on a trip around the Solar System. As they explore the comet, they come across bits of the Earth that it has acquired in the collision, some of them still inhabited. A fragment of the Rock of Gibraltar is discovered, occupied by two Englishmen playing chess and, according to Verne, unaware of their predicament. I doubt this; it seems much more likely that they knew perfectly well that they were on a comet but had come to a crucial point in the game and refused to be distracted by trivialities.

So much for pure fancy. With the development of the scientific method in the seventeenth and eighteenth centuries, and the fuller understanding of what interplanetary travel really implied, authors went to greater and greater efforts to give their stories some basis of plausibility, and as a result the first primitive spaceships began to appear in the literature. They were, naturally, not much like the spaceships of today's fiction; but we had better not be too supercilious, for some of our own conceptions may seem almost as quaint a century or so from now.

The first mechanical attempts at flight—in the atmosphere or above it—were of course made with artificial wings. Since the early writers did not realize that the air extended only for a few miles from the Earth, they assumed that if one could fly at all, then it would only be a matter of a little extra effort to go to the Moon. Lucian of Samos used this idea in his second story, *Icaramenippus,* where his hero removed one wing from a vulture and one from an eagle and, despite the resultant asymmetric thrust, succeeded in reaching not only the Moon but also the Sun.

To Cyrano de Bergerac, however, must go the credit both for first applying the rocket to space travel and, much more astonishing, for inventing the ramjet—a priority which I do not think has hitherto been recognized. In his trip to the Moon (the first attempt, the one using bottles of dew, had been unsuccessful and he had come down in Canada), Cyrano took off from the Earth in a "flying chariot" festooned with firecrackers. No detailed description of the apparatus is available, but from what we now know of mass ratios and exhaust velocities the performance is most remarkable.

Cyrano's last attempt at interplanetary flight is, I think, the most interesting and the most scientific. The flying machine he evolved consisted of a large, light box, quite airtight except for a hole at either end, and built of convex burning glasses to focus the sunlight into its interior. As a result, the heated air in the chamber would expand and escape through one nozzle, continually being replenished through the other. As Cyrano put it:

> I foresaw very well, that the vacuity that would happen in the icosahedron, by reason of the sunbeams, united by the concave glasses, would, to fill up the space, attract a great abundance of air, whereby my box would be carried up; and that proportionable as I mounted, the rushing wind that should force it through the hole, could not rise to the roof, but that furiously penetrating the machine, it must needs force it upon high.

Making allowance for the quaintness of the language, this is surprisingly like some kind of ramjet. However, Cyrano's speculations were no more than brilliant flukes, for he had no real understanding of the forces he was trying to describe, and indeed his idea that "nature abhors a vacuum" made him imagine that it would be the air rushing into the lower orifice that would propel his vehicle upward! But he did at least realize that the thrust would fall off with altitude!

With the discovery of the nonmechanical forces of electricity and magnetism, new possibilities were opened up to writers, but on the whole they seemed to take little advantage of them. Cyrano—who seems to have tried everything once—did make the prophet Elijah ascend to Heaven by taking a lodestone and a "very light machine of iron," sitting in the latter, and throwing the lodestone into the air. The iron chariot was then attracted to the stone, and the prophet repeated the operation until, presumably, St. Peter was able to give him a helping hand.

The most famous of all magnetically driven vehicles is, of course, Swift's flying island of Laputa, 4½ miles in diameter, which was propelled by an enormous lodestone, pointed to give any required direction of flight. Laputa, however, lies outside our terms of reference as it was earthbound and could never fly very far from the mainland beneath it.

The use of magnetism also reminds me of a much later story that I remember reading many years ago (*The Conquest of the Moon*, A. Laurie, 1894). In this an iron mountain was turned into a vast electromagnet for the purpose of pulling down the Moon. I suppose this would count as some sort of interplanetary voyage, though it was certainly a spectacular case of the mountain coming to Mohammed.

The devices mentioned in this section can be classed as "engines" since they do represent deliberate attempts to cross space by mechanical means, however crazy the actual suggestions were in detail. Toward the end of the eighteenth century, writers became more cautious in describing what we should now call spaceships, possibly because the public was becoming sufficiently well educated to see through the proposals they put forward (though looking at some of the things we read in the daily press nowadays this hardly seems a sufficient explanation) and perhaps because the invention of the balloon in 1783 had turned attention toward navigation of the atmosphere rather than the remoter parts of the universe. Whatever the reason, the nineteenth century was well under way before the interplanetary story got into its stride again and steadily proliferated until it now seems that there are very few corners of the cosmos that are not pretty thoroughly explored. In the last century, also, the types of propulsion that are still in common fictional use began to establish themselves and to be worked out in some detail. They fall into three main classes—projectile, antigravity, and rocket—each of which we will now illustrate by some typical examples.

The idea of the space gun does not, as is generally believed, originate with Jules Verne, although he provides us with the most famous—or notorious—specimen of its class. According to Professor Nicholson, the conception first appears in print as early as 1728 in a little-known book by one

Murtagh McDermot called rather originally *A Trip to the Moon*. McDermot traveled to the Moon by rocket, after the style of Cyrano de Bergerac, but came back in true Jules Verne fashion after inducting the selenites to dig a great hole containing seven thousand barrels of gunpowder. Here is McDermot's description of the project:

> We already know, said I, the height of the Moon's atmosphere, and know how gunpowder will raise a ball of any weight to any height. Now I design to place myself in the middle of ten wooden vessels, placed one within another, with the outermost strongly hooped with iron, to prevent its breaking. This I will place over seven thousand barrels of powder, which I know will raise me to the top of the atmosphere. . . . But before I blow myself up, I'll provide myself with a large pair of wings, which I will fasten to my arms in my resting place, by the help of which I will fly downward to the Earth.

The last item provides a distinctly modern touch, with its hint of braking ellipses and hypersonic glides back into the atmosphere.

Jules Verne's *From the Earth to the Moon* appeared in 1865, and its sequel, *Round the Moon,* in 1870. It is difficult to say just how seriously Verne took the idea of his mammoth cannon, because so much of the story is facetiously written. But he went to a great deal of trouble to check his astronomical facts and figures and had the ballistics of the projectile worked out by his brother-in-law, a professor of mathematics. Probably he believed that if such a gun could be built, it might be capable of sending a projectile to the Moon, but it seems unlikely that he seriously imagined that any of the occupants would have survived the shock of takeoff.

The Columbiad, as it was christened, was a nine-hundred-foot vertical barrel sunk in the ground in Florida. It weighed 68,040 tons and was packed with four-hundred-thousand pounds of guncotton (then a new explosive), and the cylindrical shell was made of the recently discovered wonder metal, aluminum. It cost $5,446,675, which in those times was quite a lot of money, though nowadays it wouldn't keep a nuclear physicist in heavy hydrogen.

Ignoring the impossibility of its projection, Verne's projectile must be considered the first really scientifically conceived space vessel. It had hydraulic shock absorbers, an air-conditioning plant, padded walls with windows deeply set in them, and similar arrangements that we now accept as commonplace in any well-ordered spaceship. I need hardly say, however, that the gun itself would not have produced the results predicted by Verne. Willy Ley, in his *Rockets and Space Travel,* disposes of it pretty thoroughly. Not only would the initial acceleration of some 40,000 g have converted the occupants into practically monomolecular films in a few microseconds, but the projectile itself would have been destroyed before leaving the barrel, owing to the air in its path. It is of some interest to note that both Oberth and von Pirquet have attempted to see if there are any conditions under which a space gun could operate (for example, by building it on a very high mountain and evacuating the barrel to reduce air resistance). Even in these circumstances, however, the project seems impossible.

Verne's gun was not by any means the last of its kind, and scarcely less famous was that devised by H. G. Wells for his film *Things to Come* (1936). This caused much annoyance in the British Interplanetary Society, it being generally felt that Wells had let us down badly. The explanation may be that Wells was never very interested in science qua science; he explicitly denied attempting technological prophecy and was always more interested in the impact of science on society. Certainly his space gun was no more impracticable than his antigravity screens, which we will discuss later, yet they aroused no such ire, though the law of the conservation of energy was really quite well understood in 1900. But, of course, there was no BIS in those unregenerate days.

Two much more plausible attempts to use the space gun (in conjunction with rocket propulsion) have appeared in this century. One is in Haldane's essay "The Last Judgment" (from *Possible Worlds,* 1927), but a more thorough treatment was made in the interesting book *Zero to Eighty* (1937), written by the well-known electrical engineer E. F. Northrup under the improbable name Akkas Pseudoman. This book, thinly disguised as fiction and apparently containing some real autobiographical material, was a really serious attempt to show that space travel could be achieved. Certainly it must be the only interplanetary romance with a forty-page mathematical appendix and photographs of the models constructed to test the theories involved!

Northrup, being a practical scientist, realized that human beings could only survive being shot from a gun if the barrel was made immensely long and the acceleration correspondingly reduced, though sustained for a longer period of time. He therefore used an electromagnetic gun (details of frequency, phase, etc., are discussed at some length) stretching for two hundred kilometers along Mount Popocatépetl. Even this did not give the full velocity of escape. And the final impulse was provided by rockets.

We do not often come across space guns in these more sophisticated days, for their fundamental disadvantages are too clearly recognized and are quite unavoidable. Traveling at 5 g acceleration, one must cover a distance of over a thousand kilometers before reaching escape velocity, and any practical launching device could only be a fraction of the required velocity. A track even a hundred kilometers long, for example, would only produce a tenth of escape velocity.

It does not necessarily follow, however, that space guns will never be used, for they may well come into their own for one particular but very important application where they can be employed under ideal conditions. I refer to the projection of fuel from a lunar base to spaceships orbiting either the Moon or the Earth, where the initial velocity is relatively small and there would be no restrictions set by air resistance or acceleration. However, this is a subject I hope to discuss in detail elsewhere.

I am not sure who has the credit, or otherwise, for inventing antigravity, but the earliest reference to this popular method of propulsion seems to be in J. Atterly's *Voyage to the Moon,* published in 1827. Atterly was the pen name of Prof. George Tucker, under whom Edgar Allan Poe was a student at Virginia University, and this work had a considerable influence on Poe's own satirical Moon voyage, "The Unparalleled Adventure of One Hans Pfaall" (1835), not one of that great writer's more successful efforts. Atterly's hero encounters a metal with a tendency to fly away from the Earth (how any of it has managed to stay on this planet neither Atterly nor his numerous successors ever explain), and by coating a vessel with it he is able to make a journey to the Moon.

This idea, of course, foreshadows what developed much more fully in Wells's *First Men in the Moon* (1901), which is still perhaps the greatest of all interplanetary stories despite its inevitable dating. Wells's "cavorite" was, as most of you will recall, a substance impenetrable to gravity just as a sheet of metal is to light. Consequently one had only to build a sphere— or polyhedron—coated with it to fly away from the Earth. Control could be effected by rolling up sections of the cavorite toward the body that one wished to approach. So much simpler than these noisy and alarmingly energetic rockets!

I do not suppose that Wells had ever come across Atterly's book, but I cannot help wondering if he knew of Kurd Lasswitz's *Auf Zwei Planeten* (1897), which has long been very popular in Germany and indeed has just been reprinted in an illustrated edition. As far as I know, Lasswitz's book has not been translated into English, which is a great pity as it is one of the most important of all interplanetary romances. Not only did it include such ideas as antigravity, but explosive propulsion systems (repulsors—the word later used by the German Rocket Society to describe its own early rockets) and, most surprising of all, space stations! All these details were worked out with great care by the author, who was a professor of mathematics at Jena.

As another of the countless users of antigravity—though not for space travel—I cannot forbear to mention no less a scientist than Prof. Simon Newcomb. Professor Newcomb's famous article "proving" that heavier-than-air flight was impossible has often been quoted against him, frequently by us. It is something of a surprise, therefore, to discover that he was the author of a book with the quaint title *His Wisdom, the Defender* (1900), in which he showed how the airplane might be used as a means of abolishing war. (Once again, I fear, the professor proved himself a rather poor prophet!) In this book an antigravitational substance named "etherine" was invented by one Professor Campbell—which, as many of you will know, happens by an odd chance to be the name of a well-known critic of spaceflight.

It is hardly necessary to mention any of the innumerable other stories that have used the apparently plausible device of antigravity in some way or other. And it is hardly necessary to say that it won't work—at least in the way that Wells and company described it. There is, it is true, no fundamental objection to a substance that is repelled by gravity so that it tends to fly away from Earth, and such a substance could, in principle, be used to lift a spaceship. But in that case it would take work to pull it down again—exactly as much work, in fact, as would be required to lift an equivalent mass of normal matter to the same altitude. Thus the only way the travelers could return, or land on another planet, would be to jettison their antigravitational material.

An antigravity screen, as opposed to a substance that gravity repels—is quite a different proposition and can be ruled out of court at once on first principles. A little examination will show that it involves a paradox of the "what happens when an irresistible force meets an immovable object?" category. If such a screen could exist and could be used in the manner so often described, one need only place it under a heavy object, let this rise to a considerable height, remove the screen, and let the object fall—thus obtaining a source of perpetual energy! Looking at it from another angle, Willy Ley has pointed out what a paradoxical situation such a material would produce. Imagine that one had a sheet of it nailed down to the floor. Above it, by definition, there would be no gravity, and therefore the space here would have the same gravitational potential as a point millions of miles from Earth. Thus to step the few inches from outside the sheet onto its surface would require just as much effort as jumping clean off the Earth!

It must be emphasized, however, that there is no fundamental objection to an antigravity device that is driven by some appropriate source of energy and therefore does not produce something for nothing. Presumably this covers those innumerable stories in which the release of atomic power provides propulsion through an unspecified space drive. The chances are that one day it will; but at the moment it shows no signs of behaving in such a convenient manner.

Cyrano de Bergerac, as you will recall, was the first writer to use the rocket for interplanetary travel. Cyrano, of course, had no idea of the rocket's peculiar virtues (or, for that matter, its considerable vices) so he cannot be given much credit for the invention. Nor, I am afraid, can this passage from Defoe's *Consolidator* (1705) be regarded as more than a pure fluke, though it is certainly an uncannily accurate description of a liquid-propellant rocket motor.

> . . . and as the bodies were made of Lunar Earth, which would bear the Fire, the Cavities were filled with an ambient Flame, which fed on a certain Spirit, deposited in proper quantity to last out the Voyage.

I wonder what would happen if one of our rocket engineers specified "Lunar Earth" for a combustion-chamber lining. It might be worth trying.

Although the rocket, or some other form of firework, was often mentioned in the space travel story, it was not until late in the nineteenth century that it began to become prominent. Verne used it in his *Round the Moon* (1870) to alter the orbit of his projectile and understood clearly enough that the rocket was the only means of propulsion that would operate in space; but it never occurred to him to use it for the whole voyage.

Nowadays, of course, it is exceptional to find an interplanetary vessel that is not driven by rockets, and there is no point in listing the modern stories that have used it. As the work of Oberth and the German experimenters became more widely known, so a class of painstakingly accurate stories sprang up—some, indeed, being little more than thinly disguised textbooks. German writers, such as Valier and Gail, were good at this sort of thing, and some of their works appeared in translation in early issues of *Wonder Stories*. I need hardly say that few of these tales were of much literary merit, but they are still very interesting historically. One of the few stories that did have a fairly elaborate and convincing technical background without hurting its entertainment value was Laurence Manning's "The Wreck of the Asteroid" (*Wonder Stories,* 1932–33). Manning was an early member of the American Interplanetary Society, as it was then called. He once introduced the rocket exhaust equation, complete with root signs and awkward exponents, into one of his stories—no doubt to the annoyance of the *Wonder Stories* compositor!

The almost universal acceptance of the rocket has left writers little room for ingenuity and one spaceship is now very much like another. Very few of them will have much resemblance to the ships that, unfortunately, we will have to build for the first voyages into space. Mass ratios and similar inconveniences do not bother the science fiction writer—still less the science fiction artist, who gaily runs rows of portholes the whole length of the hull and depicts thousand-ton rockets racing low over exotic landscapes with no visible means of support.

Certainly the spaceships of recent fiction have very little in common with those designed by the British Interplanetary Society, which rapidly—

though I hasten to add deliberately—fall to bits immediately after takeoff. It should be recorded, however, that the old BIS cellular ship has been mentioned at least once in contemporary fiction—by that talented writer Jack Williamson in "Crucible of Power" (*Astounding Stories,* February 1939).

Going right out on a limb that time will probably saw off behind me, I have suggested that the spaceship of the next century will be so much unlike our contemporary pictures that we wouldn't recognize one if we saw it. Certainly if orbital refueling techniques are developed as we expect them to be, then the spaceships designed for true interplanetary flight would never land on any world, or even enter an atmosphere, and so would have no streamlining or control surfaces. Indeed their natural shape would be spherical, but as the necessity for atomic shielding might rule this out, I have—until I change my mind again—suggested that a dumbbell arrangement has much to recommend it, since the radioactive power plant could then be placed far away from the living quarters. It would be refueled in a free orbit around the Earth, by a winged tanker rocket of more conventional design, which after it had done its job would reenter the atmosphere and make an aerodynamic landing.

In addition to the main categories of spaceships discussed, there are those that I place in a miscellaneous class—like all those vehicles propelled by unspecified rays, tractor beams, force fields, overdrives, underdrives, and just plain drives. Some authors, however, have made serious attempts to evolve new methods of propulsion that at least do no violence to accepted physics, and I would like briefly to mention one or two of these ideas.

Consider a cylinder full of gas. All the molecules of the gas, according to the kinetic theory, are dashing hither and thither at hundreds of meters a second, but because there are so many trillions of them all moving at random, the motions cancel out and there is no resultant tendency for movement. It is not impossible, in theory, that by the laws of chance all the molecules might decide to move in the same direction simultaneously, if one waited long enough. It would have to be quite a wait: according to my very rough calculations, there is about one chance in 10^{22} that all the molecules in a liter of gas would have even a small component of motion in common, and this is almost a "dead cert" against the even more astronomically remote possibility that they would have almost identical directions of movement.

Much of science and technology, however, depends on arranging things—stacking the cards, as it were—so that some operation, not normally probable, becomes in fact the one that actually happens. If therefore by some method of external persuasion one could induce all the molecules in a gas to cooperate and move in the same direction, presumably the container would move, too, with anything that was attached to it. In the process, the gas would give up thermal energy and become very cold, so one would have to supply heat to maintain the movement.

It is difficult to imagine a more attractive way of converting heat into motion, but I fancy that somewhere along the line that old bogey the second law of thermodynamics will step in and show that it can't be done. The system would certainly be ideal for running spaceships among the inner planets, where there is always plenty of heat available from the Sun!

This idea was evolved about twenty years ago by John W. Campbell Jr., now the editor of *Astounding Science Fiction*. Some years later Campbell also produced a number of ingenious spaceships that operated on the principles of wave mechanics and uncertainty. In the uncertainty principle, devised by the German physicist Werner Heisenberg, a particle cannot be said to have a fixed position in space but has a very small, though finite, probability of being anywhere in the universe. All you had to do, therefore, to get an instantaneous mode of transport, was to manipulate Heisenberg's equations until you were more likely to be somewhere else than where you started, and, hey, presto!

Finally a word about ships that don't travel through space so much as make space move past them. It has often been suggested that two points that are a long way apart in our universe may be quite closer in some higher, non-Euclidean or multidimensional space. As an example of this, consider the shape that can be made by taking a piece of paper, giving it a twist of 180 degrees, and then joining the ends—so that you have a loop with a kink in it. You can get from a point on the paper to a point separated from it by the thickness of the material either by going all the way around the loop (if one is restricted to movement on the surface of the material) or by traveling a fraction of an inch through the paper (if one is allowed to move off in another dimension). So the Andromeda Nebula may be a million light-years away in our space—but only across the road if we knew the right direction in which to move. Needless to say, many science fiction writers have found this direction and perhaps one day science may do the same.

For some reason, the space station has attracted very few writers, probably because it is still, to most people, such a novel idea, whose possibilities and implications are not yet fully understood. No doubt we may expect an increasing number of stories on this theme in the near future, and when the first orbital rockets are set up, it may for a while become one of the main preoccupations of contemporary science fiction.

It is generally supposed that the idea of the space station was first put forward by von Pirquet, Noordung, and others in the 1920s. Hence it is extremely surprising to discover a story on the subject as long ago as 1870. Unfortunately, the only information I have about Edward Everett Hale's

The Brick Moon is a short note in Bailey's *Pilgrims Through Space and Time.* According to this, a group of men decided that it would be of great assistance to navigation if the Earth had a second moon, so they decided to construct one. (This also is a surprisingly modern idea. It was put forward quite recently by Dr. Sadler, superintendent of the Nautical Almanac Office, in an address to the Royal Astronomical Society, *Occasional Notes of the R.A.S.,* September 1949.) Until coming across this work of Hale's, I was under the impression that Dr. Sadler had discovered a completely new use for the space station.

The artificial moon was to be projected upward by being released at the required speed from the rim of an enormous rotating wheel, and one would very much like to have the engineering details of this remarkable device!

I suppose that one reason that the space station has been neglected is that it is such a nuisance to have to stop to build one, and most writers are in such a hurry to get on to the planets. But the space station has a good many possibilities that have not yet been fully exploited. Quite recently, in *Astounding Science Fiction,* Hal Vincent (who I believe has training in astrophysics) wrote an interesting story called "Fire-Proof" around the idea that it would be impossible to have a freely burning flame in a space station, since there would be no convection to take away the product of combustion. This fact has recently been demonstrated experimentally by the German physicist Ramsauer by the simple device of filming a candle in a freely falling chamber (see *Ad Astra,* December 1949). This seems to be an interesting case of two people arriving simultaneously at the same rather novel idea.

It would be entertaining to consider the secondary features of the space travel story; to analyze, for example, the types of the social system encountered on other worlds, the difficulties of communication (so often conventionally overcome by telepathy), and above all, the reactions of extraterrestrial beings to their unexpected visitors. It is regrettably true to say that these reactions are usually hostile, or else overbearingly supercilious. The behavior of the terrestrials themselves often leaves much to be desired, for in next to no time they usually get mixed up in local politics.

I would like to end this survey of certain aspects of the interplanetary story by considering a point that is of particular interest to members of this society. What, we may ask, will happen to these tales when space travel actually begins? Will they become extinct?

A test case has already risen in connection with atomic power. Five years ago fiction was still being published about the first release of nuclear energy; though it is no longer possible to write stories about this particular theme, nuclear energy is still a familiar subject in science fiction. Similarly, when space travel is achieved, the frontier will merely shift outward, and I think we can rely on the ingenuity of the authors to keep always a few jumps ahead of history. And how much more material they will have on which to base their tales! It should never be forgotten that without some foundation of reality, science fiction would be impossible, and therefore exact knowledge is the friend, not the enemy, of fancy and imagination. It was only possible to write stories about the Martians when science had discovered that a certain moving point of light was a world. By the time science has proved or disproved the existence of the Martians, it will have provided hundreds of other interesting and less accessible worlds for the authors to get busy with.

So perhaps the interplanetary story will never lose its appeal, even if a time should come at last when all the cosmos has been explored and there are no universes to beckon men outward across infinity. If our descendants in that age are remotely human, and still indulge in art and science and similar nursery games, I think they will not altogether abandon the theme of interplanetary flight—though their approach to it will be very different from ours.

To us, the interplanetary story provides a glimpse of the wonders whose dawn we shall see, but of whose full glory we can only guess. To them, on the other hand, it will be something achieved, a thing completed and done countless aeons ago. They may sometimes look back, perhaps a little wistfully, to the splendid, dangerous ages when the frontiers were being driven outward across space, when no one knew what marvel or what terror the next returning ship might bring—when for good or evil, the barriers set between the peoples of the universe were irrevocably breached. With all these things achieved, all knowledge safely harvested, what more, indeed, will there be for them to do, as the lights of the last stars sink slowly toward evening, but to go back to history and relive again the great adventures of their remote and legendary past?

Yet I think we have the better bargain; for all these things are still ahead of us.

Review:

Destination Moon

*Nineteen years before the first manned Moon landing, science fiction stories
and films were exploring that very possibility. Things will never be the
same again, since one more barrier in man's quest for
interplanetary flight has been breached.*

Though I have a videotape of this 1950 film, I doubt if I will ever
screen it again—it would evoke too many memories, and I would like
to leave untarnished the youthful enthusiasm of my initial review.

Only two years later, in 1952, I was the guest of the *Destination
Moon* author, Robert Heinlein, and his wife, Virginia, in their high-tech
Colorado Springs home. And in 1980 I was able to reciprocate their hos-
pitality by flying them around Sri Lanka, showing them Sigiriya and
Adam's Peak (the main terrestrial locales of *The Fountains of Paradise*).

———

No one seriously interested in astronomy should miss this remarkable, exciting, and often very beautiful film—the first Technicolor expedition into space. After years of comic-strip treatment of interplanetary travel, Hollywood has at last made a serious and scientifically accurate film on the subject, with full cooperation of astronomers and rocket experts. The result is worthy of the enormous pains that have obviously been taken, and it is a tribute to the equally obvious enthusiasm of those responsible.

On a few points of detail the film can be faulted—it is, for example, very unlikely that the first lunar expedition will take place without a series of preliminary approaches and circumnavigations—but the general level of authenticity and realism is astonishingly high. This will not surprise those who have seen the astronomical paintings of Chesley Bonestell, who acted as art adviser and designed the impressive scenes representing the interior of the crater Harpalus, where the landing takes place.

Without going into story details, the theme of the film is the first lunar expedition, financed by a large group of American industrialists, who realize that whatever country owns the Moon will dominate the Earth. This note of interplanetary imperialism, which appears several times, is somewhat to be deplored, and in view of the rather considerable supply problems one cannot help thinking that the military advantages of a lunar base are a little overrated! The spaceship is driven by a nuclear rocket using water as a working fluid—an idea that will probably puzzle the layman a good deal if he stops to think about it. In fact, water is an extremely attractive propellant, if reactor-chamber temperatures of the order of 5,000° K can ever be attained. The exhaust velocity, mass ratio, and other technical details of the spaceship have obviously been worked out with great care by someone who knows what he is doing.

The views of the spaceship under construction are breathtakingly impressive: here is no mere painted backcloth or obvious model—it seems the real thing. The takeoff and the weightless period during free orbit are also very well done, though the physical effects of acceleration are (one hopes!) slightly exaggerated. Out in space, there is an exciting and again technically accurate sequence when one of the crew, going outside to repair a radar aerial, drifts away from the ship and has to be rescued by a colleague riding on an oxygen cylinder that acts as a primitive rocket. My only criticism of these sequences is that there are far too many stars of zero or brighter magnitude! (It is, apparently, extremely difficult to film realistic star fields, at least in color.)

The excitement of the approach over Mare Imbrium and Sinus Iridium, and the actual landing itself over the jagged peaks of Harpalus, is very well conveyed, as is the bleak loneliness of the lunar landscape when the travelers leave the ship in their space suits. These, incidentally, are most convincing—we noted the way in which they inflated and stiffened when the outer door of the air lock opened to the vacuum. This is the sort of detail that very few people are likely to notice and indicates the level of technical accuracy throughout.

It would not be fair to give away the ingenious ending, though one cannot help feeling that it is a little too abrupt—one would have liked to have seen something of the return journey.

On the purely artistic side, the color is some of the best we have seen in films, and the acting is also very good. The dialogue is intelligent and often very amusing. By a daringly original idea that has come off brilliantly, the fundamental ideas of rocket propulsion and astronautics are put across in a most entertaining cartoon sequence.

Other technical points worth noting are the brief but impressive glimpse of the Bush Differential Analyzer going flat out, and the opening sequence showing the launch of a large experimental rocket. This is not a studio mock-up, but the real thing—the first time we have seen a V-2 take off in color.

Destination Moon should not, on any account, be confused with another production called *Rocket Ship X-M,* about which no more need be said than an attempt to reach the Moon results in a landing on Mars. We understand that a whole series of films on interplanetary flight is now in preparation, but we fancy it will be a long time before the standard set by *Destination Moon* is excelled.

Interplanetary

Flight

When taking the time and energy to figure out the often tedious calculus of scientificing extrapolation, readers appreciate, and even welcome, extensive and pleasant narration. Thus it is no surprise to learn how Carl Sagan was influenced by Clarke's early writings.

In January 1947 the British aviation journal *The Aeroplane* printed two articles I had written on the principles of rocket flight. These caught the eye of a senior editor at Temple Press, Jim Reynolds, who was then planning a series of books under the general title "Technical Trends." The result was an advance of fifty pounds for *Interplanetary Flight: An Introduction to Astronautics*.

I wrote this in a couple of months—by no means a record, since *Prelude to Space* took only three weeks. (Such concentration is no longer possible in this Brave New Age of television, fax machines, IDD, E-mail, video games, and other seductive electronic distractions.)

When it was published in May 1950, *Interplanetary Flight* was the first book in English to give readers the principles of space travel in any mathematical detail, though there had been numerous massive volumes on the subject in German, French, and Russian. It was a modest success, on both sides of the Atlantic, and I am proud to have learned what an impact it had on many lives.

Thus the late—and badly missed—Carl Sagan was kind enough to tell my biographer Neil McAleer: "When I was in high school . . . I had not the foggiest notion about how rockets worked or their trajectories were determined. . . . The calculus, it slowly dawned on me, was actually useful for something important, and not just to intimidate algebra students. As I look back on it, *Interplanetary Flight* was a turning point in my scientific development."

And he could never have imagined that, almost a half century after he had read my book, the landing site of the Mars *Pathfinder* would be named the Carl Sagan Station.

The authors of technical works should avoid purple prose, but I have no apologies for *IF*'s concluding paragraphs.

It is fascinating, however premature, to try to imagine the patterns of events when the Solar System is opened up to mankind. In the footsteps of the first explorers will follow the scientists and engineers, shaping strange environments with technologies as yet unborn. Later will come the colonists, laying the foundations of cultures which in time may surpass those of the mother world. The torch of civilization has dropped from falling fingers too often before for us to imagine that it will never be handed on again.

We must not let our pride in our achievements blind us to the lessons of history. Over the first cities of mankind, the desert sands now lie centuries deep. Could the builders of Ur and Babylon—once the wonders of the world—have pictured London or New York? Nor can we imagine the citadels that our descendants may build beneath the blinding Sun on Mercury, or under the stars of the cold Plutonian wastes. And beyond the planets, though ages still ahead of us in time, lies the unknown and infinite promise of the stars.

There will, it is true, be danger in space, as there has always been on the oceans or in the air. Some of these dangers we may guess; others we shall not know until we meet them. Nature is no friend of man's, and the most that he can hope from her is her neutrality. But if he meets destruction, it will be at his own hands and according to a familiar pattern.

The dream of flight was one of the noblest, and one of the most disinterested, of all man's aspirations. Yet it led in the end to that silver Superfortress driving in passionless beauty through August skies toward the city whose name it was to sear into the conscience of the world. Already there has been half-serious talk in the United States concerning the use of the Moon for military bases and launching sites. The crossing of space may thus bring, not a new Renaissance, but the final catastrophe that haunts our generation.

That is the danger, the dark thundercloud that threatens the promise of the dawn. The rocket has already been the instrument of evil and may be so again. But there is no way back into the past; the choice, as Wells once said, is the universe—or nothing. Though men and civilizations may yearn for rest, for the elysian dream of the lotus-eaters, that is a desire that merges imperceptibly into death. The challenge of the great spaces between the worlds is a stupendous one; but if we fail to meet it, the story of our race is drawing to a close. Humanity will have turned its back upon the still untrodden heights and will be descending again the long slope that stretches, across a thousand million years of time, down to the shores of the primeval sea.

The Exploration

of Space

In this essay, Clarke reveals the fact and fiction behind his nonfiction work The Exploration of Space *in 1952 and how he came to know the whats and wherefores about the Book-of-the-Month Club.*

The reception of *Interplanetary Flight* prompted my Temple Press editor, Jim Reynolds, to do a little arm-twisting. "Why not," he suggested, "write a book not just for engineers and scientists, but for the general public?" Translated, that meant leaving out the equations. This was several decades before Stephen Hawking was told that equations would halve his sales.

I was not at all keen on the idea, possibly because I felt that it would interfere with the serious business of writing fiction. But Jim was persuasive, waving an advance of one hundred pounds in front of my eyes, so I capitulated and delivered the seventy-thousand-word manuscript in March 1951.

It was published only six months later and instantly snapped up by Harper's senior editor George Jones for publication in the United States. Shortly thereafter I received what may well have been my first transatlantic phone call, from my New York agent, Scott Meredith. He never quite got over my innocent query "What is the Book-of-the-Month Club?"

And I never quite forgave the distinguished editor Clifton Fadiman for writing, in the flyer that went out to prospective purchasers, "Mr. Clarke does not appear to be a particularly imaginative man." I can only assume that Mr. Fadiman was trying to assure anxious subscribers that the B-o-M Club was not about to start publishing science fiction. (Ironically, in view of my often expressed hatred of such institutionalized cruelty, the alternative selection to *The Exploration of Space* was Barnaby Conrad's *Matador!*)

After its initial 1951 publication in England, *Exploration* was updated several times in increasingly desperate attempts to keep it one jump ahead of history. In 1968, while the Apollo program was under way and the first lunar landing was expected before the end of the decade, I abandoned any further attempts at minor revisions and rewrote it completely as *The Promise of Space*. That, too, is long out of print, and I am quite happy to leave coverage of contemporary and near future astronautics to the small army of specialists who make a living out of it.

But I would like to say farewell to my first best-seller with its concluding words, which will soon be a half century old. I hope that their optimism, however naive it may seem in the light of the latest television news bulletin, will be fully justified in the coming millennium.

We stand now at the turning point between two eras. Behind us is a past to which we can never return, if we wish. Dividing us now from all the ages that have ever been is that moment when the heat of many suns burst from the night sky above the New Mexico desert—the same desert over which, a few years later, was to echo the thunder of the first rockets climbing toward space. The power that was released on that day can take us to the stars, or it can send us to join the great reptiles and Nature's other unsuccessful experiments.

The choice is ours. One would give much to know what verdict a historian of the year 3000—as detached from us as we are from the Crusaders—would pass upon our age, as he looks back at us down the long perspective of time. Let us hope that this will be his judgment:

"The twentieth century was, without question, the most momentous one hundred years in the history of mankind. It opened with the conquest of the air, and before it had run half its course had presented civilization with its supreme challenge—the control of atomic energy. Yet even these events, each of which changed the world, were soon to be eclipsed. To us a thousand years later, the whole story of mankind before the twentieth century seems like the prelude to some great drama, played on the narrow strip of stage before the curtain has risen and revealed the scenery. For countless generations of men, that tiny, crowded stage—the planet Earth— was the whole of creation, and they the only actors. Yet toward the close of that fabulous century, the curtain began slowly, inexorably to rise, and man realized at last that the Earth was only one of many worlds, the Sun only one among many stars. The coming of the rocket brought to an end a million years of isolation. With the landing of the first spaceships on Mars and Venus, the childhood of our race was over and history as we know it began."

Review:

When Worlds

Collide

When Worlds Collide *was a dramatic science fiction film about the end of the world and the flight of a chosen few to a new life on another planet. In this review, the reader learns why this story is not necessarily to be taken seriously.*

In 1947–50 and again in 1953, I was chairman of the British Interplanetary Society and published a good deal of nonfiction in the society's journal, in the form of reviews and comments on the rapidly developing astronautics scene. Here, though it appeared in 1952, is one item that may soon be newsworthy again in the very near future; see closing footnote!

The well-deserved success of *Destination Moon* made the news that the George Pal–Chesley Bonestell partnership was working on further interplanetary films all the more welcome. It is with considerable regret that we must record our disappointment with their second offering, *When Worlds Collide*.

Perhaps one should make clear at once the difference between the two films. *Destination Moon* was, in effect, a straight documentary of the first lunar flight. It was not concerned with any other plot elements, was devoid of romance, and the characters were of secondary importance.

When Worlds Collide, based on the 1932 Balmer and Wylie novel of the same title, is concerned with a far greater theme—nothing less than the end of the world and the migration of a few survivors to another planet. It has a number of well-defined characters and a good deal of love interest. Yet although the film contains moments of great excitement and never fails to hold interest, the final effect is one of anticlimax. This is largely owing to a script of naïveté unsual even for Hollywood.

The story is simple enough—a dying sun enters the Solar System on a course that will eventually cause a collision with the Earth. It has a single planet, and the only hope of mankind is that spaceships can be built in the eight months before the collision, to travel to the solitary planet of the invading star.

The sequences showing the construction of the spaceship are excellently done, and we must give Chesley Bonestell and George Pal full marks for these. The takeoff is from a long ramp which, for some unexplained reason, dips down into a valley and then goes up the side of a mountain. (The only effect of such a dip would be to waste a certain amount of gravitational energy.) The ship itself is winged, so that it can make an atmospheric landing on the new planet—which for some reason has had its name changed from the book's sensible Bronson Beta to Zyra.

Some of the scenes of destruction on the first approach of the invaders are superbly done—genuine disasters being skillfully combined with model work to produce a really terrifying sequence. The takeoff is also very well contrived, and one of the moments that lives in the memory is the last view of the launching site, bathed in a ghastly yellow light from the star that now fills the sky.

By keeping the travelers in this interplanetary Noah's Ark strapped in their seats, the producer neatly avoids grappling with the problem of weightlessness. (Zyra is apparently so close that the whole voyage only lasts an hour or so.) The landing among the mountains of the new world, on a convenient snowfield, is dramatic enough, but what should be the big moment of the film is completely ruined. When the door of the spaceship is flung open (as someone rightly says: "Why test the air? We've got nowhere else to go!"), the snow has mysteriously vanished and we are confronted with a crude backcloth that is more like off-color Disney than vintage Bonestell.

Since the film does not attempt technical accuracy on the scale of

Destination Moon, it is unfair to make detailed criticism of it on this score. One would, however, have imagined that it was obvious even to the most unscientific that the first spaceship could hardly be built from scratch in a fraction of the time necessary to design and construct a perfectly conventional airplane—especially when civilization was collapsing on all sides. One also wonders what the average picturegoer will think of astronomical accuracy when he sees one group of astronomers, equipped with differential analyzers and all modern conveniences, producing a prediction that is flatly denied by another group who presumably have the same facilities. (To anyone who remarks that astronomers are always disagreeing and points to Fred Hoyle in proof, we would testily reply that there is not much room for argument over a simple problem in celestial mechanics.)

On balance, *When Worlds Collide* certainly contains enough interesting material to make it worth a visit. Whether any producer of a lesser genius than D. W. Griffith could have handled this theme properly is a subject on which everyone will have his own opinion.

A 1998 FOOTNOTE

This old warhorse contains many moments that I still enjoy, and soon afterward George Pal took me around the sets where *The War of the Worlds* was filming. I was privileged to watch his special effects team trying to discourage the invading Martians with an atomic bomb; nowadays, of course, no outsider would be permitted to spy on such trade secrets.

The 1932 Balmer-Wylie novel has the same theme as H. G. Wells's classic short story "The Star"—a stray sun enters the Solar System. This could happen, but it would be hard to sustain much suspense over an event that in reality would take hundreds of thousands of years. The human race would have centuries to prepare for evacuation, not the few months given by George Pal's scriptwriters!

Review:

Man on the Moon

In the 1950s, Collier's *magazine played an important role in presenting the idea of spaceflight to the American public. Yet, as Clarke describes here, it seems extraordinary that the men who would actually achieve it, only a decade later, went out of their way to make it appear even more complicated and expensive than necessary.*

Collier's following up its earlier symposium on spaceflight, devoted much of its issues of October 18 and 25, 1952, to a series of beautifully illustrated articles on the first expedition to the Moon. The members of the panel were Willy Ley, Dr. Fred Whipple, and Dr. Wernher von Braun. The series was edited by Cornelius Ryan, and illustrated by Chesley Bonestell, Rolf Klep, and Fred Freeman. Just as the previous symposium appeared later in the book *Across the Space Frontier,* so we understand that this material will be published in volume form around the end of 1953.

The first article, "The Journey," is by von Braun and describes an expedition of fifty scientists in three ships—two passenger ships, and one supply ship, which would be left on the Moon. The vessels would be built and fueled in an orbit 1,075 miles up—an unnecessarily great altitude, one would think, for efficiency. The ships are of the nonstreamlined, open construction type already described in *Across the Space Frontier,* with additional landing legs and many more motors. Nitric acid and hydrazine are used as propellants. The vehicles are 160 feet long, 110 feet wide, and weigh 4,370 tons.

A magnificent double-page spread by Chesley Bonestell depicts the building of the ships in the space-station orbit, and fine cutaway paintings by Rolf Klep show their construction.

A short article, "Inside the Moon Ship," by Willy Ley, is illustrated by Fred Freeman and explains what life will be like in one of the passenger ships. A careful scrutiny of the painting shows Willy Ley having difficulty with a hatch, and Dr. von Braun relaxing in a contour couch!

It is considered that the most suitable place to make a landing is the Sinus Roris, about 650 miles from the north lunar pole. Mare Imbrium fans will take a poor view of this! The main reason given for such a high latitude is that the temperature will not be excessive, though this argument appears of dubious value. The first part of the symposium ends with another Bonestell painting showing the three spaceships descending on their braking rockets.

Part II is mainly devoted to "The Exploration," written by Dr. von Braun and Dr. Fred Whipple. It is concluded that, for safety against meteors, the base should be set up in a deep crevasse. The cargo ship is dismantled and tractor-driven vehicles unloaded for the use of the expedition, which will stay on the Moon for six weeks. Hydrogen peroxide fuel-oil turbines power the tractors.

The work that such an expedition could do is then described: among other things, artificial "moonquakes" would be set off by explosives, so that seismic records could be made. Perhaps the most impressive painting in the whole symposium is another Bonestell double-page spread showing a convoy of tractors moving over a vast lunar plain, illuminated by the greenish earthlight.

Another short article by Willy Ley, again illustrated by Fred Freeman, describes the interior layout of the lunar base. This would be constructed from the cylindrical hold of the cargo ship, which is divided into two sections and then lowered into a crevasse.

The last picture is the view from the abandoned cargo ship as the passenger craft takes off on the return voyage, with the full Earth shining in the sky.

A fuller and more critical review of this material will be made in the *Journal* when it appears in book form, but it will already be gathered that the symposium is most impressive and extremely well presented. Many readers, however, will consider that it is highly misleading to suggest that the first lunar expedition will be on anything like the scale shown in *Collier's*. The enormous amount of equipment used, the special devices, such as lunar tractors, and so on, could only become available after a great deal of experience of lunar conditions had been accumulated. If it was all designed in advance, the inevitable snags that always arise in practice would probably wreck the enterprise.

A truer assessment of the position has been made by von Braun himself in *Across the Space Frontier:* "The first attempt to land on the Moon will be a daring undertaking for a small crew traveling in a single ship."

Flying Saucers

Here Clarke recounts a television discussion, "Flying Saucers," done in
1953, which drew heavily on his experiences garnered on his debut
broadcast on the BBC in 1950 from Alexandra Palace,
just north of his London home.

Esoteric topics aside, my first exposure to the world of television prepared me for an unusual program, "Flying Saucers," I made on December 10, 1953, with Robert Barr. My main target was a fraudulent
book by George Adamski, now deceased, that contained what purported
to be photographs of UFOs. With a little help from Scotland Yard
(which preferred not to take any credit) I was able to refake Adamski's
pictures—and to show that at least one of his saucers must have been inside the tube of his telescope.

Earlier in that same year, I had published my views on this controversial subject in the *Journal of the British Interplanetary Society*. I never
imagined that it would be even more in the news four decades later.

Here are my 1953 comments, followed by the review I published the following year.

Some of our members have previously taken us to task for our determined skepticism on the subject of flying saucers or, to use the less controversial phrase, unidentified flying objects. Let us first be clear on one point—there is no official BIS party line on the subject, and any views expressed in the *Journal* are purely personal. I would like to clear the air, however, by stating my present opinion, and giving some evidence, which will be new to many members.

Before going to the United States in the spring of 1952, I believed that flying saucers probably did not exist, but that if they did, they were spaceships. As a result of meeting witnesses whose integrity and scientific standing could not be doubted, and discussing the matter with many people who had given it serious thought, I have now reversed my opinion. I have little doubt that UFOs do exist—and equally little doubt that they are not spaceships! The evidence against the latter hypothesis is, in my opinion, quite overwhelming for reasons that are given below. If this explanation is ruled out, however, it is hard to see what they can be. Only recently did I come across the classic sighting by Walter Maunder at Greenwich Observatory in 1882 that undoubtedly accounts for some of the most spectacular cases and proves the existence of a phenomenon that, a priori, one would not have believed possible.

I think it must be said at once that there is no single explanation for all UFOs. Even the reports that are left after a thorough weeding out of hoaxes, misinterpretations, etc., vary too much among themselves. And it is worth mentioning that misinterpretation is much easier than generally believed even among trained observers. I had a fine example of this while flying in Ohio in broad daylight at about ten thousand feet altitude over broken clouds. Suddenly I became aware that a brilliant light was following the plane, some miles away and apparently a few thousand feet below, appearing and disappearing through gaps in the clouds. It agreed perfectly with some of the saucer reports I have read, and I am quite convinced that many observers would have been deceived. The illusion was so perfect that it was several minutes before I discovered the explanation, and had the clouds closed over shortly after the sighting, I might not have known to this day what I actually saw.

My own flying saucer turned out to be fairly close and a short distance below, but at ground level and ten or twenty miles away. It was sunlight reflected from a roof or window—yet another example of the total impossibility of judging the size and distance of an unfamiliar object.

But let us return to the "genuine" UFOs for which no explanation is available. Two facts, in my opinion, prove almost conclusively that perhaps the largest and most representative class are not spaceships. In the first place, they have been observed to travel at accelerations that no material body could stand. One may postulate a technology that has anti-acceleration fields, but even so there seems no purpose in such violent maneuvers—which are, however, characteristic of purely immaterial phenomena such as optical effects and electric discharges.

The second fact is even more significant. Despite the enormous speeds reported, no sounds are ever heard from any UFOs. As anyone who has ever heard a sonic bang will admit, even a small body moving at speed through the atmosphere can produce a considerable amount of noise. Some of the UFOs reported would have caused nothing less than concussions that would have blasted hundreds of square miles.

This, in my opinion, leaves no doubt that they are not material bodies. Of if they are solid, they must be above the atmosphere—an explanation that is ridiculous as it would make the accelerations even more fantastic and would mean that some of the objects were many miles in diameter.

As it is now well known, attempts have been made (notably by Dr. Menzel) to explain UFOs as a form of mirage—the reflections of bright lights at ground level from layers of air. Such mirages are not uncommon in desert areas and undoubtedly account for some of the phenomena. (Perhaps the famous Lubbock lights, for example.) But they cannot explain them all.

Nevertheless, the erratic and purposeless behavior of the phenomena, their nonmaterial character, and above all the fact that they have been observed in exactly the same form for at least 150 years suggests overwhelmingly that we are dealing with some hitherto unexplained natural occurrence. The analogy with fireballs has often occurred to me. Here is something that undoubtedly exists, which parallels on a small scale the behavior of UFOs and has never been explained. Of course, the fact that UFOs have been seen for so long has been taken as evidence that Earth has been under observation for all this time—a hypothesis that cannot be disproved! But it is far simpler and more logical to accept the natural explanation. There would never have been any scientific progress if people had always accepted the most complicated and improbable theories to account for events.

With this idea in mind, I had for some time been making a desultory search of the literature when I came across the Royal Astronomical Society's publication *The Observatory* of May 1916. For this anniversary issue, the well-known astronomer Walter Maunder (who had been secretary of the RAS for 1892–97) had been asked to describe the most remarkable phenomenon he had ever witnessed in his career as an observer. His mind went back to November 17, 1882, to "an experience that stands out from its unlikeness to any other."

There had been a violent magnetic storm during the day, and as soon as dusk fell, Maunder—who was then at Greenwich—took up his position on the roof of the observatory, where he could have an uninterrupted view of the sky. He rightly believed, though it was not proved conclusively until later, that magnetic storms were associated with auroral displays and hoped to see something interesting. He was not disappointed.

As soon as the sunset hues faded away, a rosy auroral glow spread itself over the NW and began to exhibit the usual rays. However, at first there were no features of special interest. Let Maunder tell the rest of the story:

"Then, when the display seemed to be quieting down, a great circular disk of greenish light suddenly appeared low down in the ENE, as though it had just risen, and moved across the sky, as smoothly and steadily as the Sun, Moon, stars, and planets move, but nearly a thousand times as quickly. The circularity of its shape when first seen was merely the effect of fore-shortening, for as it moved, it lengthened out, and when it crossed the meridian and passed just above the Moon, its form was almost that of a very elongated ellipse, and various observers spoke of it as 'cigar-shaped,' 'like a torpedo,' or a 'spindle' or 'shuttle.' Had the incident occurred a third of a century later, beyond doubt everyone would have selected the same simile—it would have been 'just like a zeppelin.' [Remember that Maunder was writing in 1916, when zeppelins were very much in the news—even more so than spaceships are today!] After crossing the meridian its length seemed to contract, and it disappeared somewhat to the south of the west point. Its entire passage took less than two minutes to complete.

"I watched for several hours longer, but no repetition of the phenomenon occurred. A pale greenish glow fringing the upper edge of the great London smoke cloud in the north was observed but showed little, if any, structure or movement.

"The 'torpedo,' on the other hand, was many times brighter than this northern glow . . . and it had a clearly defined outline, but a plain and uniform surface. The greatest length which it presented was about thirty degrees; its breadth was from two degrees to three degrees. But in color the light of the 'torpedo' was evidently the same as that of the auroral glow from the north, and this showed me in the spectroscope the familiar auroral line in the 'citron green.' . . .

"This 'torpedo-shaped' beam of light was unlike any other celestial object I have ever seen. The quality of its light, and its occurrence while a great magnetic storm was in progress, seem to establish its auroral origin. . . .

"It appeared to be a definite body, and the inference which some observers drew from this was that it was a meteor. . . . But nothing could well be more unlike the rush of a great meteor or fireball with its intense radiance and fiery train than the steady—though fairly swift—advance of the 'torpedo.' There was no hint of the compression of the atmosphere before it, no hint that the matter composing its front part was in any way more strongly heated than the rest of its substance—if substance, indeed, it possessed. The gleam of a searchlight, focused on a cloud and swept steadily along it, is a more accurate simile for the impression which the appearance produced upon my own mind."

This remarkable phenomenon was seen by hundreds of people, and an article in the *Philosophical Magazine* for May 1883 summarized twenty-six reliable observations, from which it was concluded that the object traveled at ten miles a second and was at an altitude of 133 miles. This would make it at least fifty miles in length.

The great value of this report, which should be better known than it is, is that it proves conclusively that a purely natural—yet still completely unexplained—phenomenon can produce such extraordinary effects. If this torpedo was seen today, there would probably be no way of convincing half the population that they had *not* seen a spaceship. But Maunder was living at a time before saucer hysteria had made the evaluation of UFOs as difficult as it is today, and he was also able, by the spectroscope, to establish the auroral nature of the object beyond doubt. (Though we wonder if even the most credulous of saucer addicts would swallow a spaceship fifty miles long!) It appears highly probable, therefore, that many of the luminous objects recently reported in the sky may have some such origin. However solid, symmetrical, and artificial UFOs may appear, that is no proof that Nature is not up to one of her little tricks.

This particular explanation would not account for a number of daylight sightings, which one may or may not take seriously. But it should teach us to beware of jumping to conclusions about even the most surprising of reports.

The only thing to do, therefore, is to maintain an open mind until the evidence is overwhelming, one way or the other. I consider it still quite *possible*—though unlikely—that UFOs may turn out to be of intelligent, extraterrestrial origin. But in that event, as was pointed out in a British paper recently, they come from a world where scientific progress is very much slower than it is here—since the designs have not altered in almost two hundred years! Or they may, of course, have reached perfection.

But if they *are* artificial and come from other planets, it is fairly certain that they are not spaceships—their nonmaterial nature proves this. They will be something very much more sophisticated.

Review:

Flying Saucers

Have Landed

The following review by Clarke of the book by Leslie and Adamski was published in the Journal of the British Interplanetary Society, *with the admonition that most readers would wonder why the society was bothering. The answer was to save other members from wasting their money.*

The book is in two parts. Book One is by Desmond Leslie and is an incredible hodgepodge of saucer reports going back to ancient oriental writings and obviously the result of much reading. We hesitate to say *scholarship* in a book that cannot even get its short bibliography in alphabetical order and that refers in one place to Arthur Clark and Willy Lee, though we did not let this prejudice us unduly. It must be admitted that many of the passages that Leslie has dug up from Indian religious writings are quite fascinating. If they prove one thing, it is that science fiction has a more venerable antiquity than even its most devoted advocates imagined. No doubt some diligent reader in two or three thousand years will employ Mr.

Leslie's technology to prove, from ancient files of *Amazing Stories,* that the early twentieth century had spaceships, heat rays, antigravity, and robots.

But let us leave these ancient writings and Mr. Leslie's comments upon them, most of which will only interest those whose brains have already been addled by occultism. We are even prepared to grant the author's premise, which is that the persistence of such accounts down the ages proves that flying saucers have always been with us.

Coming to our more modern times, there are lists of sightings, including many from contemporary newspapers, which at first make an impressive case. Some of these are undoubtedly genuine—unexplained—saucers, but such indiscriminate listings are totally worthless without careful evaluation. Even from the scanty information that Leslie gives, it is possible to eliminate many at once, and many others prove yet again the complete inability of untrained observers to describe what they are seeing. Again and again altitudes and speeds are given, despite the fact that these can seldom be estimated even approximately by a single observer. Such phrases as, for example, "faster than a jet" are utterly meaningless, since a jet or its vapor trail (the undoubted explanation of many cases) can move across the sky at almost any apparent speed.

A report is given of a saucer fleet over southwest France in October 1952 (the date is important) which rained down "bright, whitish filaments like glass wool." Many eyewitnesses gathered whole tufts, but unfortunately they disappeared before arriving at the laboratory.

From this report, Leslie conjures up a whole superstructure of speculation involving ectoplasm and celestial circuses. Anyone with an elementary acquaintance with natural history could have told him, on the other hand, that this beautiful phenomenon happens every autumn over many parts of Europe; although this reviewer has seen it only once, he will never forget the sight of whole fields draped with a continuous carpet of glittering, almost invisible threads.

It did not occur to him that it was ectoplasm, which was a pity. How much more fun if he had never heard of the gossamer spider! It would obviously be impossible to explain away all the sightings and phenomena reported by Leslie, and even if one did account for 99 percent of them, he would still cling desperately to the remaining hundredth. There is a type of mind that will believe anything if it is sufficiently fantastic, and it is a waste of time arguing with it. No one has ever received much thanks for exposing credulity.

The second part of this book consists of a report by George Adamski, whose small observatory-cum-wayside-café can be seen nestling at the foot of Mount Palomar, to the considerable annoyance of the people on the summit. Adamski's hobby is photographing flying saucers, and he is undoubtedly the most successful exponent of this interesting art.

There are several close-ups of saucer spaceships, leaving no doubt that they are artifacts. The uncanny resemblance to electric light fittings with table tennis balls fixed underneath them has already been pointed out elsewhere. At first sight, indeed, one may almost conclude that Adamski's spaceship photos are so unconvincing that they aren't faked. To us, the perspective appears all wrong, and though this is a qualitative impression perhaps not susceptible to rigorous proof, the pictures seem to be of a small object photographed from very close up and not a large object seen through a telescope. (Many people, including we suspect Adamski, do not realize that a large object seen through a telescope bringing it to within say twenty feet looks quite different from the object itself twenty feet away.)

We have a much more serious comment, however, to make on one photograph, which purports to show a fleet of saucers taking off from the Moon. Alas, something has gone wrong here. We would like Adamski to account for the fact that one of the saucers appears to be inside his telescope.

This would not be apparent to anyone who was unaquainted with lunar geography, but an inspection of the background shows that the line of the saucers is not clear of the Moon's edge, as appears at first sight, but extends off the field of view of the lens altogether. It is odd to say the least that Adamski's discriminating telescope is able to see a saucer and to ignore the Moon shifting around it. . . .

There is one plate that is undoubtedly unfaked and is an insult to the public. It is a night scene in New York and shows a street lamp with a "saucer" hanging above it.

If Messrs. Adamski and Leslie had bothered to show this to any competent photographer, they could have learned at once that this was nothing more than the internal reflections in the camera lens, the almost inevitable result of having a bright light in the field of view. Even the characteristic curves due to spherical aberration are clearly visible, tangential, as they should be, to the line pointing to the source of light. The same camera, at the same place, will always produce the same saucer. If there is another edition of this deplorable book, at least the publishers should have the honesty to label the photograph for what it is.

But Adamski is not content to merely photograph saucers. With his friends, he succeeds in meeting the representative of saucerian civilization, when he lands his vehicle in the—presumably—totally uninhabited state of California. The account of this meeting should succeed in blasting any fragments of belief the more credulous readers may have in Adamski's reliability. It is as ludicrous as the idea that for years saucers have been flitting around Palomar—of all places!—invisible to everyone except Adamski and his friends.

This encounter with the Venusian pilot raises a very intriguing question. In his comprehensive account of saucer phenomena, Leslie makes no mention of the crashed spaceships described in Frank Scully's book *Behind the Flying Saucers*. (Though he does mention Scully in passing.) Can it be that there is something that Leslie does not believe in? Or can it be because Scully's three-foot-high Venusians do not agree too well with Adamski's five-foot variety? And what will he do when the seven-foot model turns up—as is only a matter of time? I hope he will be sufficiently sporting not to ignore the rival versions as they accumulate.

Why are we so annoyed with Adamski and Leslie? For a long time we were completely skeptical of saucers, and it was quite a shock to us when we discovered that they did exist and that for many of them no reasonable explanation has yet been forthcoming. Moreover, despite some serious objections, the explanation that they are in fact the products of extraterrestrial intelligences is stimulating, has small but definite probability, and should certainly not be dismissed with scorn—least of all by members of the British Interplanetary Society. But this conclusion is of such over-whelming importance that it cannot be accepted without a degree of proof that it would be unreasonable to demand in a case of lesser significance.

Books like *Flying Saucers Have Landed* do a real disservice by obscuring the truth and scaring away serious researchers from a field that may be of great importance. If flying saucers do turn out to be spaceships, Leslie and Adamski will have done quite a lot to prevent people of intellectual integrity from accepting the fact.

Undersea

Holiday

*In a departure from his usual articles on space travel, in 1954 Clarke was
commissioned to write an undersea article for* Holiday *magazine. The
following piece depicts an undersea holiday resort of the future—
one that would no doubt require reservations way in advance.*

Most prestigiously (and profitably) I became associated with *Holiday*
magazine; a farsighted editor realized that interplanetary tourism was
inevitable, and I obliged with some useful hints for travelers in "So
You're Going to Mars." These, too, I fear, will now be rather out-of-
date.

Holiday also commissioned an article about my other main interest,
scuba diving. When it appeared in the August 1954 issue, it was entitled
"Undersea Holiday," but I had originally called it "Underwater Safaris." A
decade later, I helped to establish a company of that name in Ceylon . . .
and the circle was nicely closed when it took the Apollo 12 astronauts diving off the East Coast in 1970, on their return from the Moon.

The revised version of the August 1954 *Holiday* article, which was entitled "The Submarine Playground" and used as chapter 12 of *The Challenge of the Sea,* follows.

Man does not live by bread alone, and that will still be true even when most of his nourishment comes from the sea. In this book we have talked about all the practical and scientific uses of the oceans. Now the time has come to complete the picture, and to recognize that the sea will be one of the great playgrounds of tomorrow—a place for relaxation, amusement, and adventure.

To some extent, of course, it has always been so, but until a few years ago all our enjoyment of the sea was confined to its surface. Swimming, surfing, pleasure boating—these are activities that involve only the first few feet of water. And they will continue into the future as far as our imagination can go. When there is no more surface transportation on our planet and all our traffic moves through the skies, there will still be myriad white sails billowing above the waves. Ten thousand years from now—unless men have changed out of all recognition—our descendants will still be building boats whose only purpose is to give their masters that mingling of power and peace that comes when one is driving silently before the wind, over a sparkling sea.

It is our great good fortune to be born in the age that opened the door into the sea. Yet we have never really escaped from it. The salts in our bloodstream still reflect its chemical composition. Our minds still respond to the call, heard clearly across a billion years of time, from the empty oceans of the dawn in which all life began. And now at last we have discovered how to reenter, for as long or short a period as we wish, the element in which we were born.

We might have done so several centuries ago. It's an ironic thought that the mask, breathing tube, and flippers of the modern skin diver were all depicted in the notebooks of the fantastic genius Leonardo da Vinci. Yet until our generation, for want of a piece of glass and a few square inches of rubber, men have swum blind amid beauty, have been strangers peering through a distorting mirror into a world that they could never fully enter. For the human eye is unable to focus when it is in direct contact with water, and a diver without a mask is almost blind. But when he wears a face mask, his eyes are in air and can therefore focus normally. He can see everything around him, just as clearly as if he is looking into an aquarium.

No sport in history has ever grown so rapidly as underwater exploring. At first much of the emphasis was on spearfishing, but now many are content to enter the submarine world as peaceful and passive spectators, wishing only to observe its wonders and keep on good terms with its inhabitants.

Everyone who goes underwater becomes an amateur scientist, for so much is new and unexplained in that blue world where the sea plants wave, like trees forever tossing before a silent gale.

Many old myths vanished as men saw with their own eyes where before they had merely guessed. We now know that most of the sea's dangers have been grossly overrated, for men have swum and hunted safely in waters that were once supposed to hold sudden death. Today there is no reason why anyone in normal health should not, after a little instruction, enter the world of wonder and beauty that lies so near at hand, on the other side of the waves.

Only ten years ago it would have seemed a wild fantasy to talk of underwater tourism. But today, guides are conducting sightseers over reefs and wrecks in the Caribbean, the Mediterranean, the Florida Keys, and along the coast of Ceylon. With this beginning, the next stage is obvious. Let us look just a few years ahead, to the time when the techniques and equipment that already exist have led to their logical conclusion. We will pay a visit to the underwater resort of the 1960s.

The hotel seems to grow out of the reef, concrete blending with coral both above and below the waterline. But you have to see it from the air, when you look down into the clear blue depths, before you realize that the greater part of the building is beneath the surface. Ten feet down, running completely around the hotel so that both lagoon and ocean can be surveyed, are the wide observation windows. They are crowded with spectators by day and night, for after sunset, fish in countless thousands are attracted by the glare of the undersea lights.

Nothing is more restful than sitting here in the cool submarine twilight of the observation room, sipping a drink and watching the strange shapes that come and go on the far side of the glass. But for the greatest thrill that the Reef Hotel can provide, you must become a fish yourself.

There's no danger. The equipment is practically foolproof, and there's never been a serious accident with it. The guides have gone out with tourists of all ages, from six to ninety. Some of the sightseers have come back with mild headaches—but they've all wanted to go out again.

The normal-size group for an underwater trip is six. Any guide would have difficulty in handling a larger party. The excursions take place around noon, when the sun is at its height and the submarine landscape is most strongly illuminated. There have been a few night expeditions, using powerful searchlights, but these are for the experts only.

Before you dive, you'll have to undergo a routine check by the hotel physician. He'll be mainly interested in your nose and ears, for sinus trouble can result in severe pain as pressure increases, and wax in the ears can also be dangerous. Very few people, however get eliminated from the trip.

The diving chamber—Neptune's Lobby—is a large, bare room that has to be entered through an air lock. It's thirty feet below the surface, and a wide flight of steps leads from it down into the water. You may have to swallow hard and blow briskly while holding your nose, before you get used to the pressure. The air here has to be compressed to twice its normal density to keep the water from filling the chamber.

Your guide is a muscular, barrel-chested youngster who is quite prepared to submerge for several minutes without any breathing gear at all. Slowly and carefully he briefs the party, making sure that everyone understands each piece of apparatus as it's fitted on.

First the flippers. You'll have learned how to use them in the shallow water of the lagoon and will already know the enormous feeling of confidence it gives to be able to torpedo effortlessly through the water.

Next, the face mask that makes you look and feel like a spaceman. It covers your eyes, nose, and mouth, and there's a tiny microphone built into it so that you can talk to your companions. You can hear them through a speaker strapped behind your ear, pressing against the bone.

Finally comes the harness carrying the two small fiberglass cylinders containing air under enormous pressure. These are coupled to the regulator, which automatically supplies you with the exact amount of air you require at any depth.

The tiny underwater radio, with a range of about one hundred yards, is also part of the harness. It's not actually a radio, of course, although everyone calls it that. Radio waves will not travel through water for any distance, so this equipment works on very high frequency sound waves. The guide will use his transmitter to give you instructions and point out objects of interest, but he asks you not to talk back unless he calls you. The sea is no place for idle chatter.

The last adjustment is made. The air begins to hiss through the regulator. At a word from your guide, you slowly and awkwardly waddle down the steps into the still, blue water.

Your weight ebbs away to zero as you submerge. To many people, breaking through the surface of the water while still continuing to breathe requires a definite effort of will. They try to hold their breath until their lungs are bursting, unable to believe that the air supply is still uninterrupted. But after a very few minutes they have forgotten all these fears and are reveling in thier newfound freedom.

There is a wide concrete shelf at the foot of the stairs overlooking the exit to the diving chamber. You will spend some minutes here, only a few inches from the reassuring safety of air, while the guide makes his final check and tests the radios. One by one you reply, confirming that your sets are working and that you feel fine. There are some last-minute adjustments to weight belts—a stout gentleman has difficulty in staying down and needs a couple of extra pounds of lead. Then you are ready to go.

The guide launches himself forward with a slow but powerful kick, and the rest of you follow with varying degrees of skill. You are swimming under the wall that traps the air in the diving chamber. Now you are out in the sunlight—in open water, thirty feet down. Above you, like a silken roof, lies the frontier between sea and air. It is very still, for there is hardly any wind. In one place the image of the sun winks and dances in slow explosions of light with the gentle movement of the water.

Effortlessly, with a steady beating of your flippers, you follow the guide. So clear is the water that it's easy to imagine you are really flying, surrounded not by fish but by birds. The illusion is heightened by the sea plants beneath you, which look exactly like small trees or shrubs.

You are now well clear of the hotel. Looking back, you can see the underwater windows from which your friends are doubtless watching you—although you cannot see into the darkened observation chambers. Fish are beginning to swarm around, sometimes swimming right up to your mask to peer inquisitively into your eyes. They have absolutely no fear, for hunting is rigidly forbidden near the hotel. It would scare away the creatures who are the resort's main attraction.

It is true that the guide carries a gun that fires finned, jet-propelled spears with explosive heads. But this is purely for defense in an emergency—he's never had to use it yet. When large fish become too inquisitive, he can always drive them away with his electric tickler—a slender rod carrying two probes with a few hundred volts between them.

"We should have company in a few seconds," the guide announces. "They're usually here the moment we leave the diving chamber—in fact, sometimes they come into it and escort us out. Oh, here they are!"

Two gray, streamlined shapes come shooting toward you at an incredible speed. Your heart skips a beat before you recognize Joe and Jill, the hotel's tame porpoises. They circle the expedition several times, then dart swiftly to the surface. Air-breathing animals, they can drown in a few minutes. You realize, rather smugly, that with your Aqua-Lung you can stay underwater ten times as long as they can. On the other hand, you can't swim at forty miles an hour.

There is something about porpoises that restores one's faith in humanity. They actually like people, and while they are around as protection, there's no need to worry about sharks. Not that anyone does, for the much vaunted tiger of the seas is a complete coward who will usually run for his life at the first sign of determined opposition. Usually there are enough exceptions to provide that variety that is the spice of life.

With the porpoises as escort, you glide slowly over the coral, resisting the temptation to pluck the beautifully colored stone flowers that are not flowers at all, but the homes of myriad marine animals.

"Don't touch the live coral," your guide warns. "It can sting badly—

particularly that red variety. No one knows what purpose these colors serve, by the way. They still exist a couple of hundred feet down, where the fish can never see them, as the red light has all been filtered out by the water."

Beneath you now is a carefully landscaped coral garden, with ranks of multicolored sea anemones laid out in a pattern that is just regular enough to be pleasing to the eye, but not so regimented that it is obviously artificial. It is a fine example of a new art—underwater gardening. The garden is silent, but never still. The tentacles of the anemones are continually waving with a slow, hypnotic rhythm. The fish move, sometimes languidly, sometimes in sudden darts, back and forth among the branches of the coral. Although the colors here are as brilliant as any on the surface of the Earth, if one brought them up into the alien air, they would swiftly fade. They can be seen only by those willing to go down among them.

At one end of the coral garden is a concrete bench, shaded by a trellis-work that has been overgrown by marine plants. You sit here in a solemn line, wavering slightly in the gentle current, while your guide gives you another briefing. The Sun's rays, streaming through the water overhead, are broken into bands of light that move unpredictably in all directions, covering the painted coral with shifting zebra patterns of light and shade.

"Even with flippers," says your guide, "it's tiring to travel any considerable distance underwater—as you have doubtless noticed. Yet there's some fine scenery around here that we'd like to show you, so we've arranged transport."

He starts to glide away, suddenly notices something, and dives swiftly toward a wall of coral. Thrusting his hand into a crevice, he pulls out a wiggling mass of tentacles and holds it in front of the nervous spectators.

"They're shy beasts," he says, neatly foiling the octopus's desperate attempts to escape. "Look at all the color changes he is going through— they express emotions, and I guess he is pretty frightened now. Anyone else like to play with him?"

There's a profound silence. Nobody seems anxious to volunteer. So the guide releases his captive, which promptly squirts its way back into its hole, leaving behind a cloud of ink as a token of disapproval.

Your guide disappears behind a large, obviously artificial mound, and a moment later a throbbing drone reaches you through the water. A long, slim torpedo slowly rises into sight. The guide is riding it like a jockey and adjusting the controls on its tiny instrument panel.

"This is one of our hydrojets," he explains. "You can clip yourselves to these tow ropes and then we'll be ready to go."

Amidship, two horizontal rods project from the vehicle, trailing towing lines so that three passengers can attach themselves on either side. As you clip the buckle to your harness, you feel rather like a husky in a dog team—except for the fact that the traces will be towing you.

When everyone has been securely attached, the pilot eases the throttle forward. The vibration of the water jet can be clearly felt, but it's not powerful enough to cause discomfort. With the two porpoises still as escorts, you slowly rise from the seabed and begin your effortless exploration.

Your radius of vision is more than one hundred feet, although distances are notoriously hard to judge underwater. At the limit of visibility, objects fade into a blue-green haze that first blurs fine detail then obliterates everything. But up to fifty feet, you can see as clearly as in the open air.

Now a cliff is approaching—a vertical wall climbing out of the gloom below and reaching almost to the sunlit surface. From top to bottom, as far as the eye can see, it is an unbroken sheet of multicolored life. The pilot cuts his motor, and you drift along the face of the cliff while he points out some of the creatures who have made it their home.

Here the distinction between plant and animal has been lost. Here are plants that move in search of prey, and animals that spend their lives rooted to a single spot. And there are extraordinary partnerships—crabs with anemones growing on their claws, large fish with tiny scavengers swimming unmolested in their mouths.

Quite suddenly darkness falls. For a moment you think that a cloud has passed across the Sun. Then, with a shock that almost freezes your heart, you see that an enormous, shadowy shape is floating above you. Before alarm can grow to panic, the guide's voice sounds in your speaker.

"There's nothing to worry about—that's only a whale shark. They are absolutely harmless—they live on minute sea plants and can't even bite. Some of them grow up to sixty feet in length. Notice that characteristic mottled skin. They are true fish, not mammals, like whales—see those huge gills opening and closing? If I harpooned this character, he wouldn't even fight back—he'd just swim slowly away. Let's go up and have a better look at him."

You seem to be drifting past a submarine. Surely no living creature could be this big! The shark takes not the slightest notice of the hydrojet and its passengers. The great beast cruises slowly along with bovine indifference, gills opening and closing like vast venetian blinds. A few tiny pilot fish ride its bow wave, and the monster's body is so encrusted with barnacles that it resembles the hull of a ship. You're now too interested to be scared. You can really believe that the largest fish in the sea is also the most harmless.

The great shadow drifts away, and the Sun emerges from eclipse. It's time to turn for home. Already, the expedition has been gone for more than an hour. In a wide arc, banking like an airplane, the hydrojet swings around and the shoals of fish scatter before it. The living cliffs of coral seem to be toppling—you have lost all sense of up or down as you swing at the end of your towline. Then you notice the direction of the Sun and realize that you are falling along an invisible slope into deeper water.

"I'm going down to one hundred feet," says the pilot. "Just keep breathing steadily, and swallow hard every few seconds. If anyone's ears start to hurt, we'll come up at once."

The light is changing around you, becoming blue and, curiously, more intense—although that is really an illusion. The reds and oranges that give warmth to the upper world are being leached away by the thickening layers of water overhead. Now you are in the cold twilight on the very frontier of the sunless abyss. It is cold, too, in more senses than one. You have passed out of the warm surface waters and are heading down toward the level where it is never more than a few degrees above freezing.

The pressure over your body is now more than one hundred tons—yet you feel no discomfort and can still breathe perfectly normally, thanks to the regulator on your back. With similar equipment, breathing helium-oxygen mixtures, men have descended more than one thousand feet. Your ambitions, however, fall far short of that. At just below one hundred feet, the hydrojet flattens out and cruises on a level keel for a few minutes.

You can still see the wrinkled surface of the water far overhead, with the mock suns dancing in it. But there are different fish and different colors around you now. Life in the ocean changes with depth, obeying the general rule that each layer of the sea feeds the one below it.

You feel a long way from the world of sun and air, yet the greatest depths of the ocean lie farther below you than the peak of Mount Everest towers above. Down there, under a pressure of a thousand tons to the square foot, life still exists. And as the hydrojet begins its slow climb back to the hotel, a sudden wild fancy strikes you.

A century ago, this expedition would have seemed a fantastic dream. But once men have started on a road, they will follow it to its end. So sometime in the 2100s, a guide may be standing in front of a group of tourists, protected by equipment you cannot imagine, and saying, 'Well, folks, here we are at the bottom of the Mariana Trench. For your information, there are approximately seven miles of water above us. Now, if you'll just stand back while I deal with this giant squid, we can proceed to the next point of interest."

You are still chuckling over this fantasy—yet is it a fantasy? When the coral gardens come into sight again, another party of tourists is sitting on the bench, waiting to take over the hydrojet. Above them two large rays are circling with slow beats of what you can only call their wings, for they look exactly like giant, spotted birds flapping through the sky.

The hydrojet settles down beside its concealed garage, you uncouple your harness, and once more relying on your own muscles, follow your guide back to the hotel. As you pass the observation windows, you wave gaily to any invisible spectators who may be watching you from inside—

and you hope someone's taken your photograph so that you can send it home to your friends.

The hotel's already thought of that, naturally. The prints are waiting for you as soon as you emerge.

And so, in holiday mood, we will take leave of the sea. In this book we have touched upon many of the practical uses of the oceans in the years ahead, and we have glanced at some of the mysteries still awaiting us in the unexplored two-thirds of our planet. Vast industries as yet undreamed of, scientific discoveries that will shake the world—these and many other things will come from the sea to make a happier and richer future for mankind.

But beyond this, the sea has something else to offer. For centuries, it has inspired the greatest deeds of heroism and the greatest works of art. From Homer's *Odyssey,* the Norse sagas, the tales of Melville, Stevenson, Conrad, and later writers like Herman Wouk or C. S. Forester—how much of the world's literature we owe to the sea! Yet the poets and novelists of the past saw only one of its faces. What lay beneath the waves was as unknown to them as the far side of the Moon.

Perhaps as our knowledge grows, the sea will lose some of its mystery as magic—but I do not think so. As far ahead as imagination can roam, there will be unexplored depths, lonely islands, endless leagues of ocean upon which a lost ship could wander for weeks without sighting land. When the continents have been tamed from pole to pole, when all the deserts have been irrigated, the forests cleared, the polar ice cap melted— much of the sea will still remain an untouched wilderness.

Let us hope that it will always be so. In the sea, as nowhere else, a man can find solitude and detachment. There are times when each of us needs this, just as there are times when we need action and adventure— which the sea can also give in abundance.

The sea calms the most restless spirit, perhaps because of its own perpetual but never-repeating movement. Men who will relax nowhere else will sit for hours on a beach, or upon the deck of a ship, watching the waves weave their endless patterns. The cares and turmoils of everyday life seem unimportant when we contemplate the sea.

Like all other things, the sea will not endure forever. But by our standards it is eternal. As we look across its moving surface, remembering that it has scarcely changed since the first man saw light of day, our minds wash clear of the petty ambitions and jealousies and meannesses that form so large a part of everyday existence. From the waters that first gave us life we may draw not only food for our bodies and raw materials for our factories, but also refreshment for our spirit.

The sea is our greatest heritage. We are only now beginning to realize its value. Let us use it more wisely than we have used the land.

The Exploration

of the Moon

It took many scientific papers and extrapolations to point the way to a successful
Moon landing, as Clarke points out in the following essay. Reaching the Moon
was achieved in 1969 after President Kennedy had pledged in 1961 and then
boasted of a manned landing on the Moon and a safe
return to Earth by the end of that decade.

One of the most active members of the British Interplanetary Society, before and after World War II, was the multitalented engineer-artist Ralph Smith. He, together with his friend Harry Ross, an electronics expert, wrote some pioneering papers on space-station and orbital techniques during the 1950s.

Ralph's designs and illustrations contributed greatly to the success of *The Exploration of Space,* and in 1954 we aimed at a more specific target. In *The Exploration of the Moon,* I was the minor partner, writing a few paragraphs of text to go opposite Ralph's thirty-seven black-and-white and eight color plates.

We assumed that refueling and rendezvous, both in Earth and

Moon orbit, would be employed to divide the mission into easy stages. This would have been the sensible way to go—and the way we will go, when we return to the Moon.

Though it is quite possible that the first manned expeditions to the other planets may await the development of nuclear power sources, robot vehicles could be sent to Mars or Venus with existing chemical fuels. But if round-trips are to be considered, the total fuel requirements become very high, and the complete journeys last two years or more. As far as can be foreseen today, atomically powered rockets will only be able to work in the vacuum of space, and under low gravities, so the Moon will play a vital role whatever means of propulsion our ships use ultimately. But it must never be forgotten that the Moon is not the goal of astronautics; it is merely the first objective. The space captains of the future will look upon it much as the commodore of the *Queen Elizabeth* regards the Isle of Wight or Staten Island.

From the nature of things, astronautics can have no final goal; there will always be new frontiers beyond the farthest range of man's explorations. But here at the heart of the twentieth century, we are probably midway between Columbus and the day when the last of the Sun's planets is reached by our survey ships. Indeed, this may be a highly conservative statement, for a really efficient nuclear drive would bring even the remotest planets within a few weeks of Earth. Nor must it be assumed that the rocket principle will always remain the sole means of spaceship propulsion. We are now prying into the treasure chest—some would say Pandora's box—of the atomic nucleus. No one can say what powers and forces we may have at our disposal when we have unlocked its secrets.

But it is one thing to show how spaceflight may be achieved; it is quite another to show why. When they see the enormous efforts that will have to be expended to install a few men on the Moon, a great many people will ask, "Why should we bother about space travel when there is so much to be done on Earth? Can the human race afford it, anyway?"

The last question can be answered very briefly. The total cost of the first lunar expedition, including nonrecurring development and research, would be about a thousandth of the wealth squandered in World War II. Yes, humanity can afford space travel, if it really wants it.

Certainly there is a great deal still to do on Earth. However, even on the purely material level, spaceflight will increase the resources—the real wealth—of mankind. It must never be forgotten that scientific research is the most profitable of all investments, though no one can tell in advance when the dividends will start to arrive.

But material considerations are not all-important. To take only the most recent of countless proofs of this, consider the ascent of Everest. No one pretended that the climbing of Earth's highest mountain would be of any practical value or would contribute to the sum of human knowledge. Yet how it lifted everyone's heart—and how much greater the impact of the first landing on the Moon will be! The spirit of curiosity and wonder is the driving force behind all of Man's achievements. If it ever fails, the story of our race will be coming to an end.

There will always be plenty of people who will stay at home and do the jobs their fathers did; without them, civilization would not survive. Yet there must also be those who are never content with things as they are, who will not rest while any new horizons remain uncrossed. It is not safe to keep them at home; in their frustration, they will cause mischief. But let them go, and they will become the great explorers and discoverers, opening up new worlds of mind and matter for those who come after them.

They are the ones who, in the ages now opening before us, will lead the human race out of the nursery in which it has played for long enough. For no healthy society can stand still; a civilization that has no problems, no challenge to try its strength, must eventually stagnate. It may be pleasant for the individual to dream of stability and the end of striving, but when a race seeks such things, its doom is already upon it.

Beyond the atmosphere that challenge awaits. In the closing years of this century, men will go out to meet it and in so doing will change the history of more planets than their own.

It is time to come back to Earth and the present, and to remember that a long road still stretches ahead before any of the events depicted in this book will come true. And let us also be clear on one point—the further we try to see into the future, the more our predictions will depart from the reality when it comes about. We can be fairly confident of the early stages of the conquest of space, since they are being planned in today's laboratories and drawing offices. But we can also be sure that time after time unexpected discoveries will transform our views and either increase our powers or present us with unforeseen obstacles. The spaceships that will ultimately fly to the planets may resemble those shown in this book no more than the fanciful flying machines of the 1800s resemble the smooth jets of today.

We can be sure that those who come after us will think of much better ways of doing these things and will wonder at our conservatism and our quaint, old-fashioned ideas. And they in their turn will be laughed at by those who come after them, when the Moon is only a suburb of the Earth, and the real frontier is far away among the planets. . . .

That does not matter; all we have tried to do here is to show that, even with the knowledge we now possess, there are no insuperable obstacles on the road to the planets. We could—with great expense and difficulty, it is true—reach the Moon even with today's technologies.

But the scientific and technological resources of mankind are increasing out of all recognition every generation—almost every decade. Atomic power was first released on Earth in 1942 on the laboratory scale; by 1945 the death knell of Hiroshima had echoed around the world. By 1952 the hydrogen bomb had increased the energy of nuclear explosions a thousandfold. That stupendous multiplication of power is still only a beginning, but what it aready means can best be realized by the thought that the energy released by a single H-bomb could carry ten *Queen Mary*'s to the Moon and back again to the Earth. We can now unlock from a few pounds of metal enough power to circumnavigate the Solar System. We cannot yet control it, but we will; and when that day comes, the heavens will be opened up.

Eclipse

Eclipses come and go, but not really that often. So when rare opportunities arise to get good looks, they should be jumped at. Following are some interesting highlights in the pursuit of this sport.

I count myself extremely fortunate in having seen three total solar eclipses under perfect conditions—first from the air, then at sea, then on land.

The last occasion (India, 1980) provided a dramatic opening for the television series *Arthur C. Clarke's Mysterious Universe*. And in 1973 I sailed from Miami, together with astronauts Wally Schirra and Rusty Schweickart and *Starlog*'s founder-editor, Kerry O'Quinn, to view a Caribbean eclipse from the deck of SS *Cunard Adventurer*. It was a tricky business, finding a cloudless site in the narrow band of totality, and I remarked at the time that we would have a chance of filming the first keelhauling in two hundred years, if the ship's navigator let us down. Luckily, he didn't.

Nor did the American Airlines pilot—though he had a somewhat easier job, being able to climb above the weather.

My grandstand view of the 1954 eclipse was by courtesy of the Hayden Planetarium and American Airlines. At considerable expense, American Airlines had taken its only DC-4 out of service and modified it—by ripping out the seats near the escape hatches and turning one of the toilets into a darkroom—to serve as an eclipse special. There were about forty of us aboard, mostly astronomers and photographers from the planetarium, and a good collection of press representatives. One of these had an enormous camera with a forty-eight inch focus lens, mounted like a cannon, and on the average there must have been two cameras per person on the plane. I was using a Bell & Howell 16-mm camera with a turret head, and two Leicas (5-cm Summar and 9-cm Elmar), all loaded with Kodachrome. I had intended to employ my third Leica, but it was suffering from indigestion, having swallowed a dose of seawater while I was taking close-ups of barracuda twenty feet below the Gulf of Mexico. You will see that my photographic experience on my last American visit was somewhat varied. . . .

We took off around midnight from New York and landed at Ottawa, in the small hours, to refuel. Our plan was to orbit just outside the band of totality during the earlier phases of the eclipse, and then to fly across it at right angles during totality. In this way, all the windows on the starboard side of the plane would be facing the eclipse, and with the Sun at an elevation of about fifteen degrees, we should have a perfect view.

Our rendezvous was over the extreme southern tip of Hudson Bay, and when we left Ottawa, we were somewhat depressed by the weather forecast. A front had parked itself exactly along the track of the eclipse! However, we pressed on hopefully and managed to get clear of the clouds at around 10,000 feet. There were occasional cloud peaks towering to 15,000 feet or so, but luckily we had left these behind and so we were able to have a completely unbroken view of the whole eclipse.

We leveled out at 13,500 feet and opened all the emergency hatches on the starboard side, after having first wrapped ourselves up like Eskimos. To my surprise, it did not feel at all cold, though it was rather noisy and drafty. The press photographer with his six-foot-long cannon had been lashed to the nearest stanchion to prevent being blown starboard, but this precaution was really unnecessary.

My main plan had been to take movie shots of the Moon's shadow moving across the Earth, but I had forgotten that what actually happens in an eclipse is that it just gets darker and darker and there is no clear-cut line of shadow. So this plan failed completely, and I concentrated on the still cameras. The shots of the partial phase were all taken with exposures of

1/100 second at f/8, through a Kodak 4.0 neutral filter. Unfortunately, the recommended exposure of ½ second at f/2.8 for totality ruled out the chance of getting steady views of the climax of the eclipse—though as it turned out, I could have managed with a much shorter exposure.

I was also particularly interested in the changing light and color values on the cloudscape below us. As the eclipse progressed, the clouds seemed to become more and more solid, and the hollows filled with fascinating shadows so that they looked like black lakes.

I missed the onset of totality, as I was observing on the port side of the plane, trying to see if I could detect the passage of the Moon's shadow. I could see nothing of the sort and had quite a job groping my way across the suddenly darkened interior of the plane to my window.

This was the first total eclipse I had seen, and I was surprised at the brilliance of the ring and the absence of the corona streamers that are such a feature of most eclipse photographs. The sky was not black, but a very deep blue, and I saw no stars; nor, as far as I know, did anyone else on the plane. There was a considerable glow on the horizon, presumably from the clouds outside the band of totality.

I shot off a few frames, more in hope than in expectation, and then settled down to enjoy the spectacle—which seemed to be over all too soon. When the Sun flashed out again, we all felt like excited schoolboys because everything had worked out so perfectly. In particular we felt very grateful to our navigator, who had carried out a tricky job without sight of the ground, in an area where there were few radio aids, and where the magnetic compass was not too reliable.

Astronautical

Fallacies

True or false: The Earth is round? It took Copernicus, and then Columbus, sailing west, to make this belief come true. And what about Galileo, who got arrested for insisting that the Earth orbited the Sun and not the other way around. Clarke writes here that nothing should be taken for granted, since many people are still deluded about which way is up.

During its brief history as a science, astronautics has managed to accumulate a remarkable collection of fallacies and delusions. Some of these have been trivial and would never have deceived anyone who had studied the subject seriously; others have been a little more subtle and have crept into the literature; and some, we cannot help feeling, may not be revealed until we get into space.

Fallacies are always amusing and are often extremely instructive, for it is by exposing them that one can obtain a better understanding of the facts that they conceal. So let us look at some of the prize specimens that have crept into the domain of astronautics.

The classic example, of course, is the "rocket won't work in a vacuum where there is nothing for it to push against" argument. This is still heard sometimes and occasionally crops up in the correspondence columns of the *Poona Evening News* or the *West Durban Gazette*. There was a time, however, when lecturers on spaceflight could count on it appearing with monotonous regularity, even when they believed that they had given a lucid explanation of the rocket's functioning.

Incidentally, one can only regard with awe the Olympian conceit of some of the critics who brought this argument forward. Apparently they were under the impression that, despite all the thought that a generation of scientists had given to the subject of spaceflight, they were the first to wonder how a rocket really worked when it left the atmosphere.

There were two variations on this theme by people who had avoided the main fallacy. Some could understand (or thought they understood) how the rocket was propelled, but were unable to see how it could be steered—still less how it could get home again. It seems amazingly hard to get across the idea that the rocket can go back the way it came.

The second fallacy was rather more interesting. Those who quite agreed that a rocket would work in a vacuum sometimes wondered if an additional push could be obtained at takeoff by letting the blast impinge on a wall. In practice, the change in thrust would be negligible—but the damage produced by the reflected blast would be nothing of the sort.

An argument that used to be employed against spaceflight by professional pessimists about twenty years ago ran something like this: "It takes 20 million foot-pounds of energy to lift one pound out of the Earth's gravitational field. The most powerful propellant contains 5 million foot-pounds of energy per pound. Therefore no propellant can even lift itself out of the Earth's gravitational field. Therefore spaceflight is impossible. QED."

It is surprising how a few figures, quoted with a show of authority, can prove a case. (There is no need for the figures to be right as long as there is no one around who can contradict them.) In the example above, the figures were correct—at least approximately—but the interpretation was hopelessly wrong.

The propellants a rocket carries obviously do not have to "lift themselves out of the Earth's gravitational field." If they can be burnt and induced to give their energy to the rocket while it is still close to the Earth's surface, they are scarcely lifted at all. In the extreme case of a projectile fired from a perfectly efficient gun, just four pounds of the propellant quoted would send a one-pound missile clear away from the Earth.

The most extraordinary suggestion we have ever seen concerning rockets, however, was made in perfect seriousness in an old astronautical journal for whom Sir Isaac Newton had obviously lived in vain. The proposal was to catch the exhaust gases, by means of a funnel behind the rocket motor, and use them over again. This scheme reminds one irresistibly of those cartoons of becalmed yachtsmen blowing furiously at their limp sails with a pair of bellows, and forgetting the incurable equality of action and reaction.

It is not surprising that some of the most interesting space fallacies have involved gravity. We will not discuss here the whole range of fictional antigravity devices, many of which have their own built-in contradictions, but will concern ourselves with some proposals that at first sight seem quite sound and reasonable.

One of the first involves the question of jumping on a planet of low gravity, such as the Moon. On Earth, it is sometimes stated, a skilled jumper can clear six feet. On the Moon, where gravity is a sixth of Earth's, he could therefore jump thirty-six feet.

This is an error, owing to the fact that the greatest height a man can really jump on Earth is a mere two feet! For a jumper, at the moment of takeoff, starts with his center of gravity almost four feet from the ground, and his muscular effort merely lifts it another two. Assuming that he could put forth the same effort on the Moon, therefore he could lift his CG twelve feet—and could hence clear a bar about sixteen feet from the ground.

Many years ago we read a science fiction serial in which one of the characters jumped off Phobos, the inner moon of Mars, and was in danger of falling onto the planet below. This involves a fallacy to which we will return later, and it can be said at once that Phobos (diameter about ten miles) is too large a body to permit human beings to escape by muscle power alone. However, there is no doubt that a man could jump off some of the smaller asteroids. The limiting diameter, for an asteroid made of ordinary rock, is about four miles. Deimos, the other Mars moon, may be near this limit, and the thought raises interesting possibilities for future athletic contests. Interplanetary high jumping, however, could be almost as tedious to watch as cricket, since it would be many hours before the slowly rising contestants had sorted themselves out into those who had achieved escape velocity and those who were falling back again. . . .

If anyone did succed in jumping off Phobos or Deimos—even if he jumped directly toward Mars—there would be no possibility whatsoever of his falling onto the planet. In the case of Phobos, for instance, he would still possess the satellite's orbital velocity, which is almost five thousand miles an hour. All that his jump would have done would be to compound this speed with the few miles an hour that his own muscles could provide. His velocity vector would therefore be virtually unchanged, and he would still be a satellite of Mars, moving in an orbit very slightly different from that of Phobos. At the most, he would recede a few miles from Phobos—and if he waited three hours and fifty minutes (half a revolution), the two orbits would intersect again and he would return to the surface of the little moon!

A number of writers have fallen into another gravitational trap by proposing that space travelers should use asteroids or comets to give them free rides. Some asteroids, they point out, have passed within a few hundred thousand miles of the Earth and then gone on to cut across the orbits of the other planets. Why not hop aboard such a body as it makes its closest approach to Earth, and then jump off at a convenient moment when passing Mars? In this way your spaceship would only have to cover a fraction of the total distance. The asteroid would do all the real work.

The fallacy arises, of course, in thinking of an asteroid in terms of a bus or escalator. Any asteroid whose path took it close to Earth would be moving at a very high speed relative to us, so that a spaceship that tried to reach and actually land on it would need to use a great deal of fuel. And once it had matched speed with the asteroid it, would follow the asteroid's orbit whether the asteroid was there or not. There are no circumstances, in fact, where making such a rendezvous would have any effect except that of increasing fuel consumption and adding to the hazards of the voyage. Even if there was any advantage in such a scheme, one might have to wait several hundred years before there was a chance for a return trip. No, interplanetary hitchhiking will not work. . . .

The commonest of all gravitational fallacies is that enshrined in the words "How high must you go before you get beyond the pull of gravity?" In the public mind, the idea of weightlessness is inseparably tied up with escaping from the gravitational field of the Earth. It is extremely hard to explain in understandable language (a) that one can never "get beyond the pull of gravity" and (b) that one can be completely weightless while still in a gravitational field.

Perhaps the most complete misunderstanding of conditions in space that we have ever encountered was shown by a comic-strip artist who depicted two spaceships being coupled together by suction pads. Presumably he had grown a little tired of seeing magnets and had to do something original.

It seems fairly certain that the fate that would befall the human body when exposed to the vacuum of space has been greatly exaggerated, though that is not a fallacy since it can easily be disproved by logic. Certainly there is no question of the body's being severely damaged—or even exploding, as some writers have imagined. Men have been subjected, in explosive decompression tests, to greater pressure changes than would be experienced in a spaceship whose cabin walls were ruptured. It seems likely that the pressure in a spaceship cabin would be about a third of an atmosphere, perhaps less. Normal air contains only a fifth part of oxygen; the remaining, nitrogen, is so much ballast. By breathing pure oxygen at 150 mm pressure instead of air at the normal 760 mm, the crew would still be getting just as much oxygen into their lungs as they would on Earth—and the structure of the pressure cabin could be very much lighter because of the reduced stresses.

Actually it is not quite as simple as that, and some helium or other inert gas might be included in the artificial atmosphere of a spaceship. But in any case, the total pressure would not be much more than 200 mm— say one-third of an atmosphere.

Men have withstood almost instantaneous pressure drops of this magnitude: if you want to know what it feels like, go down ten feet in a swimming pool and come up as quickly as you can. It won't do you the slightest harm.

Designers of the five-mile-long interstellar spaceships launched by the more ambitious science fiction writers would certainly have to allow for this effect, since the atmospheric pressure difference between the prow and the stern of such a vessel would be as great as that between the base and the summit of Everest. However, one fancies that this would be among the least of their worries. . . .

We will end by reluctantly demolishing an illusion that everyone must be very sorry to lose—and that is the popular belief that on the Moon, even in daytime, the sky would be blazing with stars and that the Sun's corona would be seen stretching out from it like a glorious mantle of milky radiance. The lunar landscape by daylight will be spectactular enough—but it will be as starless as the day sky of Earth.

When a large amount of light enters it, as is normally the case during the daytime, the human eye automatically cuts down its sensitivity. At night after a period of some minutes, it becomes a thousand times as sensitive as during the day. It loses that sensitivity at once when it is flooded with light again—as any motorist who has been blinded by an approaching car will testify.

During daylight on the Moon, the eye would be constantly picking up the glare from the surrounding landscape. It would never have a chance of switching over to its "high sensitivity" range, and the stars would thus remain invisible. Only if the eye was shielded from all other light sources would the stars slowly appear in the black sky.

You can put this to the test quite easily by standing well back from the window in a brilliantly lit room one night and seeing how many stars you can observe in the sky outside. Then remember that the light reflected from the walls around you is less than a hundredth of the glare that the lunar rocks would throw back.

This situation poses a pretty problem to the artist attempting to illustrate lunar scenes. Shall he put in the stars or not? After all, they are there and can be seen if one looks for them in the right way. Besides, everybody expects to see them in the picture. . . .

The Sun's corona would be invisible to the naked eye for the simple reason that it would be impossible to look anywhere near the Sun without dark glasses. If they were sufficiently opaque to make the Sun endurable, they would cut out the million-times-fainter corona altogether. It could, however, be seen without difficulty if one made an artificial eclipse by holding up a circular disc that exactly covered the Sun.

The errors and misunderstandings discussed in this article have ranged from the trivial to the subtle, and some of them have involved important principles. The lesson that can be learned from all this is that before we can conquer space, we must not only have a clear picture of all the factors involved, but we must also empty our minds of preconceived ideas that may color our conclusions. We don't want any future astronauts looking at each other with blanched faces and saying, "Someone should have thought of that. . . ."

The Star

of Bethlehem

*This is a subject that triggers a flurry of articles in newspapers and magazines
every Christmas. Clarke views this contribution to the December 1954 issue
of* Holiday *with affection, since it was directly responsible for the
short story many consider his best, "The Star."*

"Where is he that is born King of the Jews? For we have seen his star in
the east, and are come to worship him."

Go out of doors any morning this December and look up at the eastern
sky an hour before dawn. You will see there one of the most beautiful sights
in all the heavens—a blazing, blue-white beacon, many times brighter than
Sirius, the most brilliant of the stars. Apart from the Moon itself, it will be
the brightest object you will ever see in the night sky. It will still be visible
even when the Sun rises; indeed, you will be able to find it at midday, if
you know exactly where to look.

It is the planet Venus, our sister world, reflecting across the gulf of
space the sunlight glancing from her unbroken cloud shield. Every nineteen

months she appears in the morning sky, rising shortly before the Sun, and all who see this brilliant herald of the Christmas dawn will inevitably be reminded of the star that led the Magi to Bethlehem.

What was that star, assuming that it had some natural explanation? Could it, in fact, have been Venus? At least one book has been written to prove this theory, but it will not stand up to serious examination. To all the peoples of the Eastern world, Venus was one of the most familiar objects in the sky. Even today, she serves as a kind of alarm clock to the Arab nomads. When she rises, it is time to start moving, to make as much progress as possible before the Sun begins to blast the desert with its heat. For thousands of years, shining more brilliantly than we ever see her in our cloudy northern skies, she has watched the camps struck and the caravans begin to move.

Even to the ordinary, uneducated Jews of Herod's kingdom, there could have been nothing in the least remarkable about Venus. And the Magi were no ordinary men; they were certainly experts on astronomy and must have known the movements of the planets better than do ninety-nine people out of a hundred today. To explain the Star of Bethlehem we must look elsewhere.

The Bible gives us very few clues; all that we can do is to consider some possibilities, which at this distance in time can neither be proved or disproved. One of these possibilities—the most spectacular and awe-inspiring of all—has been discovered only in the last few years, but let us first look at some of the earlier theories.

In addition to Venus, there are four other planets visible to the naked eye—Mercury, Mars, Jupiter, and Saturn. During their movements across the sky, two planets may sometimes appear to pass very close to each other—though in reality, of course, they are actually millions of miles apart.

Such appearances are called conjunctions; on occasion they may be so close that the planets cannot be separated by the naked eye. This happened for Mars and Venus on October 4, 1953, when for a short while the two planets appeared to be fused together to give a single star. Such a spectacle is rare enough to be very striking, and the great astronomer Johannes Kepler devoted much time to proving that the Star of Bethlehem was a special conjunction of Jupiter and Saturn. The planets passed very close together (once again, remember, this was purely from the Earth's point of view—in reality they were half a billion miles apart!) in May of 7 B.C. This is quite near the time of Christ's birth, which probably took place in the spring of 7 or 6 B.C. (This still surprises most people, but as Herod is known to have died early in 4 B.C., Christ must have been born before 5 B.C. We should add six years to the calendar for A.D. to mean what it says.)

Kepler's proposal, however, is as unconvincing as the Venus theory. Better calculations than those he was able to make in the seventeenth century have shown that this particular conjunction was not a very close one, and the planets were always far enough apart to be easily separated by the naked eye. Moreover there was a closer conjunction in 66 B.C., which on Kepler's theory should have brought a delegation of wise men to Bethlehem sixty years too soon!

In any case, the Magi could be expected to be as familiar to such events as with all other planetary movements, and the biblical account also indicates that the Star of Bethlehem was visible over a period of weeks (it must have taken the Magi a considerable time to reach Judea, have their interview with Herod, and then go on to Bethlehem). The conjunction of the two planets lasts only a few days, since they soon separate in the sky and go once more upon their individual ways.

We can get over the difficulty if we assume that the Magi were astrologers (*magi* and *magician* have a common root) and had somehow deduced the birth of Jesus from a particular configuration of the planets, which to them, if to no one else, had a unique significance. It is an interesting fact that the Jupiter-Saturn conjunction of 7 B.C. occurred in the constellation Pisces, the fish. Now though the ancient Jews were too sensible to believe in astrology, the constellation Pisces was supposed to be connected with them. Anything peculiar happening in Pisces would, naturally, direct the attention of oriental astrologers toward Jerusalem.

This theory is simple and plausible, but a little disappointing. One would like to think that the Star of Bethlehem was something more dramatic, and not anything to do with the familiar planets whose behavior had been perfectly well known for thousands of years before the birth of Christ. Of course, if one accepts as literally true the statement that "the star, which they saw in the east, went before them, till it came and stood over where the young child was," no natural explanation is possible. Any heavenly body—star, planet, comet, or whatever—must share in the normal movement of the sky, rising in the east and setting some hours later in the west. Only the polestar, because it lies on the invisible axis of the turning Earth, appears unmoving in the sky and acts as a fixed and constant guide.

But the phrase "went before them," like so much else in the Bible, can be interpreted in many ways. It may be that the star—whatever it might have been—was so close to the Sun that it could be seen only for a short period near dawn and so would never have been visible except in the eastern sky. Like Venus when she is a morning star, it might have risen shortly before the Sun, then been lost in the glare of the new day before it could climb very far up the sky. The wise men would thus have seen it ahead of them at the beginning of each day, and then lost it in the dawn before it had veered around to the south. Many other readings are equally possible.

Very well then—can we discover some astronomical phenomenon, sufficiently startling to surprise men completely familiar with the movement of the stars and planets, which fits the biblical text?

Let's see if a comet would answer the specification. There have been no really spectacular comets yet this century—though there were several in the 1800s—and most people do not know what they look like or how they behave. They even confuse them with meteors, which any observer is bound to see if he goes out on a clear night and watches the sky for half an hour.

No two classes of object could be more different. A meteor is a speck of matter, usually smaller than a grain of sand, which burns itself up by friction as it tears through the outer layers of Earth's atmosphere. But a comet may be millions of times larger than the entire Earth and may dominate the night sky for weeks on end. A really great comet may look like a searchlight shining across the stars, and it is not surprising that such a portentous object always caused alarm when it appeared in the heavens. As Calpurnia said to Caesar:

> When beggars die, there are no comets seen;
> The heavens themselves blaze forth the death of princes.

Most comets have a bright, starlike core or nucleus, which is completely dwarfed by their enormous tail—a luminous appendage that may be in the shape of a narrow beam or a broad, diffuse fan. At first sight it would seem very unlikely that anyone would call such an object a star, but as a matter of fact in old records comets are sometimes referred to, not inaptly, as hairy stars.

Comets are unpredictable; the great ones appear without warning, come racing in through the planets, bank sharply around the Sun, and then head out toward the stars—not to be seen again for hundreds or even millions of years. Only a few large comets—such as Halley's—have relatively short periods and have been observed on many occasions. Halley's comet, which takes seventy-five years to go around its orbit, has managed to put in an appearance at several historic events. It was visible just before the sack of Jerusalem in A.D. 66, and before the Norman invasion of England in 1066. Of course, in ancient times (or modern ones, for that matter) it was never very difficult to find a suitable disaster to attribute to any given comet. It is not surprising, therefore, that their reputation as portents of evil lasted so long.

It is perfectly possible that a comet appeared just before the birth of

Christ. Attempts have been made, without success, to see if any of the known comets was visible around that date. (Halley's, as noted, was just a few years too early on its appearance before the fall of Jerusalem.) But the number of comets whose paths and periods we do know is very small compared with the colossal number that undoubtedly exist. If a comet did shine over Bethlehem, it may not be seen again from the Earth for a hundred thousand years.

We can picture it in that oriental dawn—a band of light streaming up from the eastern horizon, perhaps stretching vertically toward the zenith. The tail of the comet always points away from the Sun; the comet would appear, therefore, like a great arrow, aimed at the east. As the Sun rose, it would fade into invisibility; but the next morning, it would be in almost the same place, still directing the travelers to their goal. It might be visible for weeks before it disappeared once more into the depths of space.

The picture is a dramatic and attractive one. It may even be the correct explanation; one day, perhaps, we shall know.

But there is yet another theory, and this is the one that most astronomers would probably accept oday. It makes the other explanations look trivial and commonplace indeed, for it leads us to contemplate one of the most astonishing—and terrifying—events yet discovered in the whole realm of nature.

We will forget now about planets and comets and other denizens of our own tight little Solar System. Let us go out across real space, right out to the stars—those other suns, many far greater than our own, which sheer distance has dwarfed to dimensionless points of light.

Most of the stars shine with unwavering brilliance. Century after century. Sirius appears now exactly as it did to Moses, as it did to Neanderthal man, as it did to the dinosaurs—if they ever bothered to look at the night sky. Its brilliance has changed little during the entire history of our Earth and will be the same a billion years from now.

But there are some stars—the so-called novae or new stars—which through internal causes suddenly become celestial atomic bombs. Such a star may explode so violently that it leaps a hundred-thousand-fold in brilliance within a few hours. One night it may be invisible to the naked eye; on the next, it may dominate the sky. If our Sun became such a nova, Earth would melt to slag and puff into vapor in a matter of minutes, and only the outermost of the planets would survive.

Novae are not uncommon; many are observed every year, though few are near enough to be visible except through telescopes. They are the routine, everday disasters of the universe.

Two or three times in every thousand years, however, there occurs something that makes a mere nova about as inconspicuous as a firefly at noon. When a star becomes a supernova, its brilliance may increase not by a hundred thousand but by a billion in a few hours. The last time such an event was witnessed by the human eyes was in 1604; there was another in 1572 (so brilliant that it was visible in broad daylight); and the Chinese astronomers recorded one in 1054. It is quite possible that the Bethlehem star was such a supernova, and if so, one can draw some very surprising conclusions.

We'll assume that Supernova Bethlehem was about as bright as the supernova of 1572, which is often called Tycho's star after the great astronomer who observed it at the time. Since this star could be seen by day, it must have been as brilliant as Venus. As we also know that a supernova is, in reality, at least a hundred million times more brilliant than our own Sun, a very simple calculation tells us how far away it must have been for its apparent brightness to equal that of Venus.

It turns out that Supernova Bethlehem, if that was the case, was more than three thousand light-years or, if you prefer, 18,000,000,000,000,000 miles away. That means that its light had been traveling for at least three thousand years before it reached Earth and Bethlehem, so that the awesome cataclysm of which it was the symbol took place five thousand years ago when the Great Pyramid was still fresh from the builders.

Let us, in imagination, cross the gulf of space and time and go back to the moment of the catastrophe. We might find ourselves watching an ordinary star—a sun, perhaps no different from our own. There may have been planets circling it; we do not know how common planets are in the scheme of the universe, and how many suns have such small companions. But there is reason to think that they may not be rare, and many novae must be the funeral pyres of worlds, and perhaps races, greater than ours.

There is no warning at all—only a steadily rising intensity of the sun's light. Within minutes the change is noticeable; within an hour, the nearer worlds are burning. The star is expanding like a balloon, blasting off shells of gas at a million miles an hour as it blows its outer layers into space. Within a day, it is shining with such brilliance that it gives off more light than all the other suns in the universe combined. If it had planets, they are now no more than flecks of flame in the still expanding shells of fire. The conflagration will burn for weeks before the dying star collapses into quiescence.

But let us consider what happens to the light of the nova, which moves a thousand times more swiftly than the blast wave of the explosion. It will spread out into space, and after four or five years it will reach the next star. If there are planets circling that star, they will suddenly be illuminated by a second sun. It will give them no appreciable heat, but will be bright enough to banish night completely, for it will be a thousand times more luminous than our full Moon. All that light will come from a single blazing point, since even from its nearest neighbor Supernova Bethlehem would appear too small to show a disk.

Century after century, the shell of light will continue to expand around its source. It will flash past countless suns and flare briefly in the skies of their planets. Indeed, on the most conservative estimate, this great new star must have shown over thousands of worlds before its light reached Earth—and to all those worlds it appeared far, far brighter than it did to the men it led to Judea.

For as the shell of light expanded, it faded also. Remember, by the time it reached Bethlehem it was spread over the surface of a sphere six thousand light-years across. A thousand years earlier, when Homer was singing the song of Troy, the nova would have appeared twice as brilliant to any watchers farther upstream, as it were, to the time and place of the explosion.

That is a strange thought; there is a stranger one to come. For the light of Supernova Bethlehem is still flooding out through space; it has left Earth far behind in the twenty centuries that have elapsed since men saw it for the first and last time. Now that light is spread over a sphere ten thousand light-years across and thus must be correspondingly fainter. It is simple to calculate how bright the supernova must be to any beings who may be seeing it as a new star in their skies. To them it will still be far more brilliant than any other star in the entire heavens, for its brightness will have fallen only by 50 percent on its extra two thousand years of travel.

At this very moment, therefore, the Star of Bethlehem may still be shining in the skies of countless worlds, circling far suns. Any watchers on those worlds will see its sudden appearance and its slow fading, just as the Magi did two thousand years ago when the expanding shell of light swept past the Earth. And for thousands of years to come, as its radiance ebbs out toward the frontier of the universe, Supernova Bethlehem will still have power to startle all who see it, wherever—and whatever—they may be.

Astronomy, as nothing else can do, teaches men humility. We know now that our Sun is merely one undistinguished member of a vast family of stars and no longer think of ourselves as being at the center of creation. Yet it is strange to think that before its light fades away below the limits of vision, we may have shared the Star of Bethlehem with the beings of perhaps a million worlds—and that to many of them, nearer to the source of the explosion, it must have been a far more wonderful sight than ever it was to any human eyes.

What did they make of it—and did it bring them good tidings, or ill?

Editor's Note: Whatever origin is finally accepted for the Star of Bethlehem, mankind will have a long time to wait for the definitive proof. Meantime, years after this essay was written, astronomers in 1987 had their first opportunity since 1604 to witness the birth of a supernova, appropriately designated Supernova 1987A. Their observations were relatively close, a mere 160,000 light-years away.

Capricorn

to Cancer

Taking a roundabout route from Australia to Sri Lanka, Clarke and his partner,
Mike Wilson, delivered their stunning photographs of Australia's Great
Barrier Reef to their Manhattan publisher.

On a dismal rainy day, our Australian friends gathered to see us off at Sydney airport. They looked disapprovingly at the *Constellation,* which was to carry us across the Pacific. "She looks a bit skinny in the rear," said one helpful character, though not actually using the word *rear.* "Why are they wrapping Scotch tape around the tail?" shouted another. But despite this encouragement, we got aboard and soon saw for the last time the great steel bow of the Harbour Bridge, under whose shadow we had lived for the past few weeks of 1956.

After treating romantic Fiji in a rather casual fashion, we decided to make more of Honolulu, spending three days there, mostly lazing on Waikiki Beach. There we met Val Valentine, whose sports store was a rendezvous for all serious divers. Val drove us all over the island of Oahu, took us on a conducted tour of his favorite hunting spots, and did his best to make our brief stay enjoyable.

Val's charming wife, Lorraine, is a decorative addition to the rapidly growing international society of women divers (notice I didn't say "drivers"). In our travels we have met some marriages where both partners were equally at home underwater, others where the wife could occasionally be persuaded to take a dip but would rather not, thank you—and some where the wife regarded the sea as an active rival. One of these days we may run into a marriage where the wife spends all her time underwater and the husband is scared of getting wet, and at any moment we expect to hear of excessive skin diving as being grounds for divorce.

Also at Val's we met the well-known diver Jim Oetzel, who has been able to combine his hobby with his business in a very efficient manner. By working as a flight attendant with Pan American, Jim has managed to spend a good deal of his spare time on most of the interesting islands in the Pacific.

The last lap of our journey to the States provided an experience that I shall treasure all the more because it is unique to our generation, being something that the past never imagined and the future will have left behind. Determined to arrive refreshed and full of energy in San Francisco, we had booked sleeping berths and drifted dreamlessly in our comfortable cocoons while the dawn broke across the ocean four miles below. I watched entranced as, below, the morning traffic poured along the maze of freeways; I felt like Zeus reclining on a cloud and felt sorry for the travelers of the jet age, who would never stay aloft enough to spend a night sleeping among the stars.

Once in New York and having acquired all our gear, mostly at Richard's Sporting Goods, the problem then arose of shipping it safely to Ceylon. There were scores of agents who would gladly undertake our commission, but we wanted one who had proved that he could handle delicate scientific equipment. Luckily our friends Frank Forrester and Joe Chamberlain, the energetic directors of the American Museum's Hayden Planetarium, answered that question. Earlier in the year they had organized a trip to Ceylon to observe the longest solar eclipse for more than a thousand years, and this had involved shipping out a fortune in astronomical equipment. We gladly handed over our problem to the museum's agent.

One of the greatest mysteries of New York is the way in which one can spend hours, days, and eventually weeks there without actually doing anything. Looking back on the seven weeks spent there, I can account for few of them; only a few unrelated cameos come to mind:

Leaving my briefcase with all the priceless Barrier Reef color slides in a hotel lobby and not discovering the loss until we were on the other side of Manhattan; dining at Toot Shor's as a guest of the "Woman Pays" Club; attending the world premiere of Sam Goldwyn's *Guys and Dolls,* and rounding off the evening at the original Lindy's; watching Antonio and his dancers being introduced to the *Omnibus* cameras in the CBS Studios by Alistair Cooke; seeing Diane Cilento in Christopher Fry's translation of *Tiger at the Gates* and remembering the witty vote of thanks her father, Sir Raphael, had made when we gave the farewell screening of our Australian films.

After returning to England—our home, but now only temporarily— we settled the important problem of transport in Ceylon by buying a Land Rover, the British equivalent of the jeep. We soon discovered that it was the largest vehicle one could comfortably drive in metropolitan traffic or insinuate through the tortuous alleys of Mayfair.

There then remained only one matter to settle. We had our still cameras and our "dry" movie camera, but what about underwater movies? After looking at the available models, we decided that much the best one at a price we could afford was the electrically driven Beaulieu in the case designed by Dimitri Rebikoff, the famous French underwater engineer and photographer. This gave Mike an excuse to fly to Paris, and no sooner had he got there than Dimitri inveigled him down to Cannes to see his workshop and to watch the Club Alpin *sous-marin* at work.

The main problem of underwater color photography—once the technical difficulties of making a watertight case have been overcome—is not the faintness of light at depths, but the fact that it is essentially monochrome. The fast modern color films have enough speed to take satisfactory photographs at least one hundred feet down in natural light, but such photographs will contain nothing but blues and greens and might almost as well have been taken on black-and-white film.

The reason for this is that even the purest water acts as a blue filter, so that all the red and orange rays are lost as soon as one descends a few feet below the surface. Many a diver, sixty feet down, has been disconcerted to discover that he bled green blood when he cut himself.

To reveal the true beauty of the submarine world, therefore, it is necessary to take a source of light underwater to restore the missing reds. In the case of still photography, this can be done by using flashbulbs—and it seems surprising to land-based photographers that ordinary flashbulbs can be fired and changed underwater without any protection from the medium around them. For movie photography, the problem is more difficult, as a continuous source of light is needed. Rebikoff has solved this with his photographic torpedo, a streamlined housing with powerful headlights at the front, a high-capacity battery inside, and the movie camera at the rear, so arranged that the lens looks toward the area illuminated by the lamp. Switching on the electrically driven camera automatically turns on the light.

No one can have failed to notice the fluorescent or Day-Glo paints that have appeared on advertising boards and rear bumpers. These paints use the property that some chemicals possess of absorbing light at one wave-

length and reemitting it at a longer one. In a sense, they are color transformers. By painting our Aqua-Lung tanks with red fluorescent paint, we
were able to ensure that they still looked bright red even at depths where
most of the natural red had vanished. This was because the blue and green
light—of which there was plenty—was being "converted" by the paint of
the tank. The effect was quite striking, and often very beautiful, providing
color and warmth where it would otherwise have been missing.

Our greatest fear while in London was that the English winter would
catch us before we could escape to Ceylon. Our tickets were booked on
the ocean liner *Orcades* for the first week in January, which seemed to be
cutting it close. But we were in luck; it was still mild and pleasant when
we saw the sun setting behind the cranes at Tilbury dock and knew that
blue seas and palm-fringed beaches were only two weeks away.

Keeping House

in Colombo

Ceylon is the right size and in the right place, moreover, to satisfy the whim of even the most dedicated underwater photographer. Clarke and Mike Wilson had made friends there who showed them around, imparting their enthusiasm and sharing the most promising underwater discoveries. In addition, interest had been whetted in the old pearling fisheries of Ceylon, and Clarke & Company were eager to discover the skills of the Ceylonese divers, who used no equipment whatsoever.

In the long run, I determined that what really drew us to Ceylon was a hunch that both Mike Wilson and I shared, but which we found hard to analyze. Our first brief visit to Ceylon, before we went to Australia, was enough to convince us that we had fallen in love with the place and that nothing would satisfy us until we had returned for a prolonged stay. Having realized this, we had a year in which to concoct valid excuses for going to Ceylon, and long before we had left Australia that particular task was accomplished to our complete satisfaction.

There were times, however, when we had qualms. We had gambled all the money we had earned and a lot that we hadn't on this trip. Suppose the monsoon decided to come a few months early and conditions underwater were hopeless? Suppose customs didn't allow our equipment and Land Rover into the country. . . . Suppose our color film had been stored near the ship's boilers and was cooked into . . .

Despite these nagging doubts, we had an enjoyable enough journey across the Mediterranean, spending much of the time building up that protective suntan without which no one can dive or swim in tropical waters. We looked at the Rock of Gibraltar and calculated what an H-bomb would do to it; we looked at Pompeii and saw what Vesuvius had done to it. The ruins were impressive, the vistas of roofless walls reminding me of the wartime photographs of gutted cities. We never imagined that in a few months we would be looking at the far greater ruins of a civilization of which we had never heard.

Moving slowly down the Suez Canal, Mike and I observed landmarks he remembered from his days in the British army. From time to time, he would make nostalgic comments like "That's where we blew up the police station" or "You should have seen the fight we had in that café," all of which enabled me to understand more clearly why the British are so well loved in the Middle East.

The great oil refinery at Aden, its storage tanks and distillation towers gleaming like silver against the utterly barren background of craggy mountains, always seems to me like an illustration from a science fiction story. The feeling that I was on a strange planet was heightened when I went ashore and drove into the town, which is situated inside the crater of a huge volcano. If the sky were black instead of blue, the illusion that one was on the Moon would be almost perfect. I could not imagine what it must be like in summer, when the vertical Sun shines straight down into the crater cup and the reflected heat is concentrated onto the town. But my sister, who had been there over a year as an RAF officer, said that the place wasn't at all bad—which in the Clarke vocabulary is high praise indeed.

The final lap of our voyage passed swiftly enough, giving me a chance to catch up on arrears of reading. There is nothing like a long sea journey for dealing with *War and Peace* and similar overwritten classics. The library of the *Orcades* was excellent, fortunately.

Colombo is one of the great crossroads of the world, for every liner on the England–Australia run touches here, and shipping lines radiate to all the countries of the Far East. The number of travelers who disembark here is small compared with the number of tourists who spend a few hours in the city on their way to somewhere else—as we ourselves had done on our first visit.

We were met by Rodney Jonklaas, one of the world's leading un-

derwater hunters, whose advice before and after we came to Ceylon was
invaluable. By one of those coincidences that have convinced me that the
world is nothing like as big a place as it is supposed to be, Mike and Rodney
discovered that they had met before. On his way to Australia two years
previously, Mike had spent a couple of days in Colombo and had not missed
the opportunity of doing a little skin diving. He was swimming around a
reef off Mount Lavinia, the popular bathing beach of Colombo, when he
met two other divers and waved them a friendly greeting. Rodney, who
keeps a log of every dive he makes, was able to confirm that he was in the
area at the time and distinctly remembered meeting this lone swimmer and
wondering who he was.

At the time of our arrival, Rodney Jonklaas was working in the Department of Fisheries, having previously been deputy superintendent of the
Colombo Zoo—and, in his own words, chief exhibit in the anthropoid
section. He had managed to arrange his vacations and his official duties so
that he could conduct us to the most promising underwater sites off Ceylon.

His first problem was to assist us in getting all our gear ashore. Luckily,
the amount of our diving and photographic equipment so overwhelmed
the customs officials that they threw up their hands and in the end merely
made us pay a couple of dollars as license fee for our transistor pocket radio.
They also relieved Mike of his Luger, though he got it back after a certain
amount of form filling. Small side arms are not popular with the authorities,
who are justifiably scared that they might enhance what is the highest murder rate in the world. However, there is no reason why this should worry
the visitor; the Ceylonese are very clannish in these matters and keep all
their quarrels inside the family.

In the midfifties Ceylon was very short of good hotel accommoda-
tions, and in any case we would not have been popular with the other
residents when our air compressor started to pump up the Aqua-Lungs. To
our great relief, the indefatigable Rodney had found us a small but pleasant
apartment and a first-class cook-servant, all at an inclusive cost of about $90
a month.

As soon as we had moved into the apartment, the garage floor immediately disappeared beneath flippers, underwater-camera housings,
weight belts, pressure gauges, snorkel breathing tubes, and the countless
other items without which no one is really well-dressed beneath the sea.
There was no hope of using the garage for its proper purpose: the Land
Rover had to take its luck outside.

In addition, cameras, developing tanks, processing kits, spare lenses,
thermometers, flash units, movie editors, slide projectors, and stocks of film
invaded bedroom, kitchen, and bathroom. Since Mike was one of the un-
tidiest people I know (I, of course, am one of the tidiest), life in the apartment was to be a constant battle against chaos. Equipment was constantly

vanishing; with each day's mysterious disappearance, or equally mysterious reappearance, we had another proof of the second law of thermodynamics. It needed no mathematical treatise on entropy to tell us that the disorder of the universe tends to increase.

Our ally in this ceaseless war was our houseboy, Carolis—though *boy* was a slightly misleading term as by the time he entered our employment he was forty and had nine children. An alert, dark Sinhalese who barely topped the five-foot mark, he arrived one morning with a letter of introduction from Rodney and a document that might well be adopted in the few remaining countries where it is still possible to obtain domestic help. This "Servant's Pocket Register" was nothing less than a logbook of all Carolis's previous positions, with dates, salaries, and the comments of his employers. A government-issued document, not unlike a passport in size and layout, the register appears to have been invented by the British for self-protection in 1870, and every servant in the island was once supposed to have one.

We had no hesitation in engaging Carolis when we read his previous employers' testimonials, beginning with the one that concluded regretfully: "He leaves me to try and get better pay. Present salary 20 rupees" ($40 a month). I am glad to say that he did get better pay with us.

In the East, of course, a servant is a necessity, not a luxury. Carolis probably saved us more money when he went shopping than we paid him in wages. The local storekeepers would have made quite a killing had we two British innocents dealt with them directly.

The Reefcombers'

Derby

Colombo, not a particularly beautiful city but a spacious one, soon became home to Clarke and Mike Wilson. They rented an apartment four miles from the business center and near the beach in Bambalapitiya. In no time they were properly organized and ready to move around.

Without our Land Rover, we would have been in a sorry plight, for the public transport system of Colombo had to be seen and heard to be believed. Apart from the rickshaws, which we never patronized because we felt it wrong for men to be employed as beasts of burden, plus we would look silly bobbing along the street perched up in the air, the choice lay between buses and cabs. The cabs are easy enough to get ahold of, about half the price of those in London or New York, which means that they are all right for occasional use but ruinous if one has to employ them constantly. As for the buses, they were unspeakable. Elephants are supposed to have a mysterious graveyard where they all go to die, and the same is true of London Transport's unwanted derelicts. That graveyard is Colombo.

To make matters worse, the buses are run by so many fiercely competing firms that they present an incredible chaos of inefficiency, dirtiness, and unpunctuality. The bus companies are so busy fighting one another that the unfortunate customer is often ignored in the battle. When you stand at a bus stop, there is no guarantee that the driver will condescend to pick you up; he may be racing another company, or in a hurry to see his girlfriend—or he may just dislike the look of your face.

There are also trams in Colombo, which are worse than the buses. I never got aboard one or even understood how it was physically possible for another passenger to insinuate himself into the seething mass of humanity. Luckily, improvement came as the trams were replaced by trolley buses.

The main commercial area of Colombo, known as the Fort, occupies the quarter square mile immediately adjacent to the harbor and is well laid out with wide streets and modern buildings. Almost any day of the week it is crowded with sightseers from the liners that carry two great streams of humanity in either direction across the equator. There will be wealthy sheep barons from western Australia going for a visit to England—and eyeing with a shock of recognition their almost indistinguishable counterparts from the cattle ranches of Texas, touring the mysterious East with Mom and Junior; there will be sturdy Italian farmers on their way to seek better livings in the rich soil of Queensland than they can find in their own country; there will be young rocket engineers and electronic engineers, barely out of college, heading for the desert secrecies of Woomera; there will be administrators, businessmen, retired civil servants, Unesco officials—an entire cross section of mobile humanity.

Most of these transients will see no more of Ceylon than the purely European area of Colombo, but some will take trips farther afield and may penetrate as far inland as Kandy, the romantic hill capital with its Temple of the Tooth and the 2,240-year-old sacred bo tree grown from a cutting of the very tree under which Buddha received his revelation. Everywhere they go they will be fascinated by the inextricable mixture of East and West, perhaps best exemplified by the rickshaws and bullock carts that share the streets with the Fiats and Volkswagens.

Though the climate of Colombo has only two phases—hot and dry, and hot and wet—we soon grew accustomed to it and very seldom felt uncomfortable. That was partly because we were able to dress sensibly; if like most Western visitors we had stuck to shirts and long trousers, we should often have been very happy indeed. We had soon grown so acclimatized that we would start to shiver and reach for our warm clothes when the thermometer dropped below eighty-five.

One consequence of the climate is that no houses in Colombo have any form of heating except in the kitchen. Having cold showers instead of baths was no great hardship when the water was no more than ten degrees below body temperature, and after the battle to keep warm in higher latitudes I think the single feature of the Ceylonese home that I most appreciated was its freedom from fires.

After a year in Australia, during which time we had barely touched the eastern seaboard and come to grips with rather less than 1 percent of the Great Barrier Reef, Ceylon seemed incredibly tiny and compact. Indeed, we soon decided that as a country it was just about the right size. Since it is less than three hundred miles from the extreme north to the extreme south, and half this in width, no spot in the country is more than a day's travel from any other, and most journeys require only a few hours' motoring over excellent roads.

It was partly because of this scale factor that, whereas in Australia it had taken us four months to get from Sydney to the Great Barrier Reef, in Ceylon we were operating two days after coming ashore. Or to be strictly accurate, Mike was; I remained on land with the camera to record the results.

On the first Sunday after our arrival in Ceylon, the local underwater hunters had their annual spearfishing "derby," and Mike went out with their contestants to observe their techniques. Spearfishing, as practiced by the Ceylon Reefcombers, is certainly a rugged sport and seems to break many of the safety rules regarded as sacrosanct in the rest of the world.

The favorite spear gun among the Reefcombers is the simple but highly effective Cressi "Cernia," which is powered by the compression of a long spring. When a Reefcomber gets a new gun, he promptly throws away the spear that goes with it—which he contemptuously regards as a toothpick—and gets one made locally of one-half-inch steel. To this he attaches twelve feet of steel cable having a breaking strain of five hundred pounds—but, very wisely, he does not fasten this cable to the gun, but to sixty feet of strong nylon line. The junction of the two lines is then wired to the tip of the gun in such a way that it will stand a pull of perhaps forty pounds.

In this manner, if a Reefcomber hits a moderate-sized fish, he can handle it with the twelve-foot steel cable alone; the long nylon line remains coiled up out of the way. But if he spears something big, which might tow him down into the depths, the fastening at the end of the gun breaks and the whole seventy feet of line are available. He can then remain near the surface while the fish tires itself out in fruitless dashes around the seabed.

Using Aqua-Lungs for hunting is regarded by the Reefcombers, as by most spearmen, as being in very bad form, and indeed is forbidden by club rules. There is no sport in sitting on the bottom waiting for the fish to come up to you and then shooting them at your leisure, though it is only fair to point out that it will not always work this way. Hampered with his heavy equipment, an Aqua-Lunger has no chance of catching many fish which an unencumbered skin diver can easily land. So using a lung for hunting is not only bad sport—it is often bad sense.

Just as a pointer, all the books on underwater hunting emphasize that you should never swim with dead fish tied to your waist, and to carry bleeding fish in this fashion in shark-infested waters is regarded as nothing short of suicidal. I record this fact for the information of hunters elsewhere; for at least ten years, Rodney Jonklaas has been spearfishing, often alone, and miles out to sea, in the middle of a gory circle of victims. Several times he has had fish taken from his waist, but the sharks have always stopped before they come to him. Perhaps they regard him as a public benefactor; he has known them to hang around, waiting for him to make a kill, and then rush in before he could secure his catch.

Being one of those people who are much too fond of fish to stick harpoons into them, I at first regarded this sort of activity with some distaste. However, I soon learned that—whatever might happen in other countries—the hunters in Ceylon were not indiscriminate piscicides. Everything they caught was eagerly snapped up by the dealers, and quite a number of the Reefcombers relied on the sale of their catches to enable them to pursue their hobby. Despite the fact that Ceylon is surrounded by fish and has thousands of fishermen, fish are expensive and in short supply, and any additional source is welcomed.

Rest Houses,

Catamarans, and

Sharks

Clarke's and Mike Wilson's first major expedition to the Ceylonese sea was on a hot and brilliant Saturday afternoon when they took the coast road running south from Colombo, winding their way through mingled files of bullock carts and Volkswagens.

Two hours after leaving Colombo we arrived at the little village of Ambalangoda and more or less took over the local Rest House. It was our first contact with one of those national institutions, which outside the half dozen main towns provide the only accommodation for travelers. They serve the same purpose as motels in the United States—but there the resemblance ends.

Some of the Rest Houses, government controlled and owned, have been operating continuously for two centuries, often with no more than minor repairs to the original structure. The Dutch were splendid builders, and as far as design and construction are concerned, much of their handiwork still looks perfectly modern. The only important improvements that our age has added are plumbing and electricity.

The Ambalangoda Rest House had the latter, but the former was still being installed while we were there. When we wanted a shower, we had to go to the well in the courtyard and pull up the bucket, then get a friend to put the icy water over our heads. The house was beautifully situated on a headland overlooking the sea and had a picturesque, rock-enclosed bathing pool below it. It was a large, single-story building divided into a dozen bedrooms, dining room, kitchen, and accommodation for the staff. The bedrooms were high and well ventilated, though the loosely curtained windows provided minimal privacy.

Although the Rest Houses are under state control, each establishment is run by a keeper (usually with the help of his large family), whose personality stamps itself upon the place and who is responsible for the catering. The price of the accommodation is fixed; that of the meals isn't, and this gives the keeper interesting opportunities for private enterprise. In our case, this worked both ways, since we often sold our catch to the houses we frequented—and sometimes were sorry we had done so when the fish course came on at the next meal.

When we started to unload our photographic and diving equipment, it would have taken dynamite to prize them away from us. We had to keep a constant watch on the many small children who were itching to play with the shiny knobs of our cameras and the intriguing taps on our air cylinders. By unceasing vigilance, we were able to cut our losses over a period of five or six visits down to one pair of flippers, a diver's knife, and an exposure meter. We particularly regretted the loss of the meter, which had been bought only the week before.

Since the reef we proposed to visit was half a mile out at sea, our first problem was to hire boats. Without Rodney, this would have been an expensive undertaking. After torrents of Sinhalese had flowed in both directions, we secured two boats for ten rupees ($2) each. This did not seem like much of a bargain when we had a good look at the boats.

The fishing vessels that are the standard craft of the southern part of the island are primitive catamarans, designed with no concessions to comfort whatsoever. The main hull was so narrow that it was impossible to get inside it; the crew had to squat on top of the two vertical planks, about a foot apart, which form the sides. The outrigger float consisted of a stream-lined log lashed at the end of a couple of curved poles, which flexed and twisted disturbingly in rough water.

Getting away from the beach was not an easy operation. Only fifty feet out, a wall of coral, barely submerged, ran parallel to the shore. In one spot a passage had been blasted through this natural barrier; to reach the open sea, it was necessary to run this gauntlet. The four rowers who provided the inadequate power for each boat had to paddle frantically in the face of the advancing waves, but eventually we were through the gap and heading in the general direction of Africa.

Our goal was the Akurala Reef, its location marked by an occasional efflorescence of breakers half a mile out at sea. It took us forty minutes to make the journey—forty minutes of trying to find a comfortable resting place on planks that seemed to grow more sharp-edged every minute. But we had really nothing to grumble about on this lovely, calm morning; the unfortunate fishermen who relied on these craft for their livelihood were sometimes out in them in rough weather for the entire night.

We wasted no time getting into the water, leaving the Aqua-Lungs in the boats while we did our reconnaissance. Before anyone could join him, Rodney had already speared his first fish—a fifteen-pound caranx or horse mackerel. Though ours was primarily a photographic, not a hunting, mission, there was a good reason for doing this; within seconds, the vibrations of the captured fish had brought sharks racing toward us from all directions. And so, by a fantastic stroke of luck, the very first movie footage I was to shoot in Ceylon was a lovely sequence of a shark swimming around Mike—something we might have waited a lifetime to get.

It was not long before we became almost indifferent to the sharks sharing the water with us. If they were close enough to make good photographic subjects, we would dive on them in an attempt to make them fill the camera viewfinder before we clicked the shutter. This direct approach, however, seldom worked; the shark would not stay to see if our intentions were peaceable, but would vanish with an effortless flick of its tail.

Once the sharks had gone, we saw few large fish, with the exception of a splendid two-hundred-pound parrot fish that Rodney said was rather rare. After two hours' continuous swimming and diving, I'd decided that I had had enough for one day and resumed my precarious perch on the catamaran. By this time many of the villagers on the shore had been unable to restrain their inquisitiveness any longer and had put out to sea to find out exactly what we were doing.

Onshore at last, we could hardly wait to see what our cameras had caught. As if to discourage us, the skies opened soon after we had left Ambalangoda and were headed back to Colombo. We had neglected to bring the Land Rover's canvas hood, so that by the time we reached the city we were completely sodden. We had spent a large part of the day immersing ourselves in salt water, but did not enjoy the experience of having fresh water poured upon us in large quantities.

The Akurala Reef gave us results—our first satisfying shark close-ups—that we had never been able to obtain during a whole year in Australia. But as totally exhausted as we were, we were not content until we had begun to plan our next venture in Ceylon!

The First

Wreck

Any expedition, however carefully planned in advance, can only succeed with
the cooperation of the local authorities. Many places sought after around
Ceylon could only be visited by government vessels; other goals, such
as the pearl beds, were strictly out-of-bounds unless one had
official permission to go there.

One of our first moves on reaching Colombo was to call upon the friendly and energetic director of the Government Tourist Bureau, Mr. D. C. L. Amarasinghe—and this is as good a place as any to mention an important point concerning Ceylonese names. Ceylon suffers from an unusual shortage; there simply aren't enough names to go around. Since most of the population appears to be called Fernando, Perera, or de Silva, the habit has grown of referring to people by their initials and dropping their last name if it is too common to serve as any identification. What happens when the initials duplicate themselves, I do not know; presumably the situation is resolved as in the American examples of "Engine" Charlie Wilson and

"Electric" Charlie Wilson, which may in turn have derived from the Welsh usage of "Mrs. Jones the Post Office."

DCL was not only helpful to us in his capacity as head of the Tourist Bureau, he turned out to be another camera enthusiast and promptly introduced us to one of Ceylon's leading photographers, J. O. Ebert, whose studies have been exhibited and published in photographic journals all over the world. It was Jo to whom we were continually running when strange things happened in the processing tanks, and he always had the answer.

But Jo was not merely invaluable to us as a photographic consultant, badly though we needed one. He was also the shipping master for Colombo, holding court in a little subterranean empire beneath the white customs building through which passed most of the ocean-borne arrivals to Ceylon. Even when large liners were hooting impatiently in the harbor, Jo was never too busy to attend to our woes and to give us advice or information.

And he had some information that we could have obtained nowhere else—a complete register of all the ships that had gone down in Ceylon waters during the last hundred years. Mike made a special study of that register, in an attempt to find interesting and accessible wrecks.

We had not come to Ceylon to look for wrecks; we just happened to run into them, and thereafter they grew on us. We started with small freighters and eventually worked our way up to the fifty-thousand-ton class.

We found our first wreck on our second visit to Akurala Reef, the weekend after our meeting with the sharks. Once again, we descended upon the Ambalangoda Rest House and took the place over.

The slow journey out to the reef was as tedious as ever, but I had now learned how to fit myself into the nooks and crannies of the boat. I even discovered the art of lying flat on my back, balanced on the narrow edges of the two vertical planks forming the sides of the catamaran. A medieval philosopher was once asked by one of his pupils whether a good man could be happy on the rack. His reply, "Yes, if he was a very good man and it was a very bad rack," kept coming to my mind as I teetered on my twin knife-edges.

The water around the reef was very clear, and we wasted no time throwing ourselves into the comfortable embraces of the sea. As usual, Rodney had shot his first fish within a few minutes, and we waited hopefully for the sharks to arrive. But there was no sign of them; we swam around for half an hour and finally decided that we were out of luck. Why they were so shy this time, when they had been all around us the week before, is just another of the many mysteries of shark behavior.

From well below the surface, the barnacled rocks that formed the highest part of the reef looked like a low hill whose summit was

intermittently hidden by storm-tossed clouds. Those clouds were the breakers—the white water that even though the sea was calm was smashing
with considerable force over the nearly exposed rocks. It was exciting to
swim into it, and to become lost in a dazzling white mist as the surge sucked
one helplessly through the channels between the great boulders. With reasonable care, this amusement was perfectly safe; if you relaxed and let the
water carry you, you automatically avoided the rocks and their frieze of
razor-edged barnacles.

I spent some time diving under the surge around the reef. And looking
up at the brilliant white fog of bubbles that formed and reformed above
me, sometimes hiding the reef, sometimes unveiling it as a mist may roll
away from hills in the Highlands. It is fascinating to watch breakers from
below, for the foaming water gives an unforgettable impression of the sea's
power and restlessness.

Below us was an endless vista of gray, cracked rocks, sparsely peopled
with fish. And then, quite abruptly, we were swimming over the ruins of
man's handiwork. Set upright in the bed was the single vertical blade of a
propeller that had been buried up to the hub; the remaining blades appeared
to have been welded into the rock. This solitary metal tombstone, we later
discovered, marked the resting place of the *Earl of Shaftesbury,* which had
run onto the Akurala Reef in 1893.

A short distance away was the end of a boiler, lying on the seabed
like the lid of a giant saucepan. Near that was a huge crankshaft—huge,
that is, by any but marine standards. And that was all that was left of the
lost ship.

We swam around these pathetic relics for ten minutes, disturbing a
large moray eel, which slithered through the rocks like a hideous black
snake. The sight of the broken propeller blade, containing as it did many
hundreds of dollars' worth of high-grade bronze, roused our cupidity, but
we could think of no way in which we could salvage this small fortune.
Perhaps one day we will be able to do something about the numerous
propeller blades we have located. It represents quite as real a treasure as any
sunken bullion and could be turned into cash with fewer complications.

It was while swimming around this wreckage that I noticed, at unpredictable intervals, a strange groaning noise that seemed to fill the waters
all about me. When I surfaced beside Rodney, I asked him if he could
account for this sound. Like most mysteries, it had a simple enough explanation. I was hearing the huge boulders beneath me protesting audibly as
the swell of the ocean shifted them on their foundations.

A few minutes later, I realized that the Akurala Reef, which seemed
so calm and peaceful now, had claimed more victims than the lost *Earl of
Shaftesbury.* We had swum out to deeper water, and the seabed below us
was now invisible. But lying on it, apparently rolled over on its side, was

the barnacle-encrusted hull of a fair-sized ship. The highest part of the wreck was about thirty feet below the surface, so I was able to skin-dive down to it without much difficulty.

The 3,300-ton *Conch,* which was one of the Shell Company's first oil tankers, was dragged onto the rocks by a powerful current on the night of June 3, 1903, and quickly broke in two. On a later reconnaissance trip to Kalkuda, on the east coast, we were fortunate enough to come across what must be one of the few surviving eyewitnesses of the disaster. Hubert Paterson, now a well-known planter, had then been a young Lloyd's assessor at the port of Galle, about ten miles south of the Akurala Reef.

Early on the morning of June 4, news came that a ship had run aground farther up the coast. Paterson and his colleagues at once hurried to Akurala village and bargained with the fishermen for their boats, just as we were to do more than a half century later. They had to pay no less than fifty rupees a boat—and in 1903 the rupee, like most other currencies, was worth at least ten times as much as it was in the 1950s, so the fishermen drove a hard bargain. One cannot really blame them, for they were being asked to risk their lives and thus the welfare of their families.

Five boats set out through the narrow gap in the wall of coral fringing the shore, and the rowers fought their way against seas that were so high that both land and wreck were often invisible. After battling for half an hour, they reached the enormous patch of oil now spreading around the ship; it seemed, as far as they could tell, to have no calming effect on the water, and waves were breaking entirely over the ship.

After some dangerous and delicate maneuvering, the catamarans were brought up to the wreck and the crew was taken off—even the ship's cat being rescued! The sailors were in a pitiable condition; because of the oil all around them, they had been unable to use any lights and had passed the night in complete darkness, with the knowledge that their ship was rapidly breaking up.

This was the vessel that, fifty-three years later, was being visited by men again. It had only recently been rediscovered by Rodney and in the next few weeks was to become one of our favorite wrecks, as well as the scene of some of our most successful, exasperating, and sometimes hair-raising adventures.

On that first visit, Rodney and I merely made a few dives along the gently curving sides of the overturned hull. As we were not using Aqua-Lungs, we could not enter the wreck, and it was very difficult to decide how large a vessel she had been. Not until some weeks later did Mike, rummaging through the records in the office, discover her name and tonnage.

I was beginning to feel tired and did not object when Rodney started to swim back to the catamaran. Once or twice, just to remind me that I had been in the water long enough, I felt a twinge of cramp. This is perhaps the most serious danger confronting the lone diver on an extended mission, especially if he overexerts himself. In Ceylonese waters the surface temperature hovers in the eighties and one can swim all day without feeling the cold. But even twenty feet down, it can sometimes feel quite chilly, and attacks of cramp through cold and exhaustion have to be guarded against. These twin enemies of the diver can creep up on him unawares, especially when the fascination of his surroundings has made him forget the passage of time.

This had certainly happened to me. When I got back to the catamaran, I had been swimming continuously for three hours, often at depths of over thirty feet. Even a Ceylonese fishing boat felt comfortable as I lowered my protesting muscles into it and dozed lazily away while the rowers took us back to land after a total time of five hours at sea.

Ceylon is a country of many races, and the simple word *Ceylonese* is almost as uninformative as the equally omnibus word *American*. The Sinhalese are in the great majority; they are predominantly Buddhist, though there are many families that have been Christians for generations, and now that the British are gone, the Sinhalese are the leading power in the land.

The next largest group is the Tamils, who are found in the northern part of the island and have close associations with India. They tend to be bigger and darker than the Sinhalese; we were often told that they were harder workers, though we never had a chance to put this to the test. The Tamil religion is Hinduism.

Ceylon has been progressively occupied by the Portuguese, the Dutch, and the British. Only the British left under their own power, and each of the Western interlopers made a permanent mark on the country—the Portuguese their religion, the Dutch their architecture and law, the British their commerce and language.

In Ceylon today, there are also Indians, traders from the Arab countries, British, Chinese—almost any race one cares to mention. English is spoken everywhere, and almost everyone, from the cities to the remotest settlements, understands it.

A Clear Run

to the South Pole

Exactly one hundred miles from Colombo, and almost on the extreme southern
tip of Ceylon, is the little town of Matara. In the seventeenth century, the
Dutch made it one of their main bases and built a small fort, which
stands almost unchanged to this day.

Rodney Jonklaas had chosen Matara as one of the Clarke main bases,
though our intentions—except to the fish—were more pacific than those
of the Dutch invaders. On second thought, that is hardly fair to the energetic
and practical Hollanders. The Dutch were after trade, not empire or con-
version, and were highly adept at avoiding bloodshed. When they realized
that the time had come to bow gracefully from the scene, they did so
without any death-or-glory nonsense, as the unscathed state of their fort
testifies.

The Matara Rest House was attractively situated overlooking the sea,
and one of our first acts on arriving was to lay out all our equipment—
Aqua-Lungs, compressor, cameras, rubber dinghy, Land Rover—on the

lawn so that we could photograph it before it became too battered and bent to make a good picture. The final result looked as if a sports store and a camera shop had combined forces for the spring sales, and the local inhabitants were impressed.

I was up with the others when we set off to the exact southern tip of Ceylon, Dondra Head, just five miles farther along the coast. A fine, white-painted lighthouse stands proudly here in well-kept grounds, overlooking the rocky bay from which Rodney proposed we should start our swim.

We had parked our cars in the shadow of the lighthouse and were adjusting our equipment when suddenly there was a brief, muffled explosion. Rodney and Rupert looked at each other and simultaneously said the same word: "Dynamiters!" Without wasting any more time we grabbed our cameras and hurried to the water's edge.

The beautiful little bay was the scene of great activity. Half a dozen catamarans, full of small boys with an occasional adult to direct operations, were milling around in the center. More small Ceylonese were swimming in the water, diving to the bottom and returning with handfuls of fish, which they passed over to their companions in the boats. They were collecting their illegal harvest from the ocean.

There was quite a large audience watching the busy scene, and no one seemed to resent our arrival. The Sinhalese are so fond of having their photographs taken that they appear not to mind if the shutter closes when they are doing something illegal. The people on the shore moved politely aside to let us photograph their friends and relatives at work; even the man who was obviously in charge of the whole proceedings and had probably supplied the dynamite made no attempt to keep out of our way.

Mike was in the water almost at once, swimming out to the nearest canoe. I thought that this was a little rash; though no explosives would be dropped on top of him—for there were still plenty of divers around collecting their fish—it seemed likely that he might get a knock on the head with an oar if he became too inquisitive. However, he soon made friends by helping to pick up the catch. This was probably compounding a felony, but since the fish had already been killed, there was no point in leaving them to rot.

I followed a bit later with one of the still cameras. The seabed beneath the boats was sprinkled with silvery corpses, but the underwater visibility was so poor that it was impossible to get a good photograph. We swam around for some time until most of the catch had been picked up, then decided that there was nothing further to see here and it might be a good idea to get out of the way before more dynamite arrived. So we swam around the headland and out to sea, where the no-longer-landlocked water at once became rough and we were soon tossed about in the waves.

Since the water at Dondra was not as clear as we had hoped, and it did not appear to be a healthy place for skin divers, we decided to head north. But we spent one more night at Matara, and once again I walked the battlements because it was too early to go to sleep. I did not stay up there for long, however, for to my great surprise a famous and familiar sound soon reached me on the evening breeze. Who on Earth, at this time of night, would be playing Beethoven's Choral Symphony at the lonely southern end of Ceylon?

The last thing that I expected to find in a remote Rest House, which is exactly what its name implies and where very few people stay for more than one or two nights on their way to somewhere else, was a hi-fi enthusiast with a small library of classical records. But there was one at Matara, and I spent some time lurking outside his door while Beethoven's last will and testament filled the night air around me.

The next morning we moved up the coast to the Rest House at Weligama Bay, one of the most famous beauty spots in Ceylon. The great curve of the shoreline, two or three miles in length, embraces a group of lovely islands, one of which—Taprobane—can be reached by wading from the mainland. The word *Taprobane* is the old European name for Ceylon; it had been applied to this island by a romantic exile, the Count de Mauny, when he settled on it early in this century.

We were to visit this modern Taprobane later; for the moment we were busy securing our base at the Rest House and getting our equipment once more in readiness so that we could operate anywhere we chose up or down the coast.

It was probably true to say that no Rest House was ever the same after we had stayed there; Weligama certainly wasn't, as Mike nonchalantly demolished the front steps when we finally drove away. Land Rover and rubber dinghy occuped the lawn; cameras and Aqua-Lungs filled a couple of bedrooms; film-developing tanks and beer bottles full of processing chemicals overflowed into the bathroom. It was not altogether surprising that, despite all our efforts at tidying up, we always seemed to lose something at each stop we made, even it was only a pair of tennis shoes.

We had just finished organizing ourselves and disorganizing the Rest House when a large American car drove into the courtyard. Two men festooned with Leicas and Canons, and followed by a caddy with more cameras, started to take photographs of the fishing boats ranged along the beach. By a brilliant feat of deduction, we decided that was a team from *Life* magazine, which we knew to be in the country, so we went over and introduced ourselves.

Jumping off from a new section of coast is always an adventure; it is a gamble whose outcome can never be foreseen, for even if one is familiar with a given area, the underwater conditions are continually changing and so, to a large extent, is the underwater population. The spot Rodney had chosen for our next dive was the tiny village of Gintota, five miles north of the port of Galle. It was another quite typical fishing village, with palm trees leaning along the beach and catamarans drawn up on the shore. We managed to hire two of these at the standard ten-rupee rate, but we decided to take the rubber dinghy with us as a second and more convenient base for the divers to work from.

Rodney and I each took a catamaran, but Mike and Rupert Giles decided to go out under their own steam. Into the dinghy, despite our warnings, they piled three Aqua-Lungs and a couple of cameras and started to paddle out to sea. They had not traveled more than thirty feet when a large and hostile wave picked up the dinghy and emptied its loose contents into the boiling surf. Fortunately, everything of value had been tied on, and only assorted flippers and face masks went overboard. There was a brief delay while crowds of volunteer searchers scrabbled in the sand for our lost property, amazingly enough, every item was recovered, both from the sea and from the searchers.

Nothing daunted, Mike and Rupert tried again. Paddling furiously, they switchbacked through the surf and this time managed to reach the smooth water without mishap. Our oddly assorted little fleet of two catamarans and one rubber dinghy set bravely out to sea.

We had traveled only four hundred yards when we came to the resting place of the motor vessel *Elsia,* a Danish cargo-passenger ship that had caught fire and been abandoned in 1939. She lay in about forty feet of water, but unfortunately visibility was so poor that we could see only about fifteen feet. The superstructure of the burnt-out ship was just visible from the surface as a dim shadow beneath the waves, and although any photography was out of the question we dived down to see what we could find.

It takes considerable determination to plunge down into a green mist that may conceal anything, and in which you may even lose your way back to the surface because you can no longer distinguish between up and down. When the underwater horizon is only a few yards away, you become lonely and lost; you cannot see your companions, or the reassuring glint of the waves overhead. Anything may come swimming toward you out of the gloom, and even familiar objects are distorted and made mysterious by the haze.

I took a deep breath and plunged down toward the enigmatic blur lying beneath our boat. Before I had run short of air, I found myself inside a ship. The *Elsia* appeared to have split open, revealing all of her internal machinery. A big electrical generator, overgrown with weeds and barnacles, lay surrounded by large air cylinders, most of which had been completely eaten through by corrosion. Pipes and cables snaked everywhere, sometimes looking alarmingly like giant tentacles.

Because visibility was so limited, it was impossible to get any idea of the true size of the ship. She seemed enormous, for wherever we went we kept on discovering more of her. After working steadily across the deck in a series of brief exploratory dives, I finally went over the rail and all the way down to the seabed. The entire side of the ship had been torn off and lay on the sand forty feet down according to my depth gauge. I could just reach it before having to head back to the surface for air.

One of the dangers of diving around such a wreck, in conditions of poor visibility, is that it is altogether too easy to swim under an obstacle without being aware of it until you start to hurry back to the surface. There were also, almost certainly, some very large fish in the *Elsia;* the giant grouper was most unlikely to miss such a desirable residence. After half an hour of exploring, therefore, we decided to try our luck farther out to sea.

But when we tried to leave the *Elsia,* however, we found that she did not want us to go. The heavy stone used as an anchor had caught in the wreckage and could not be dislodged. Feeling heroic, I grabbed a knife and went down to see what could be done.

My earlier diving had left me a little tired, and it seemed a long time before I reached the trapped anchor. It was clearly visible, wedged between two plates, and I tried to pull it clear. But I was thirty feet down, at the end of my operational range, and there is nothing more totally exhausting than hard work underwater. When I finally gave up the struggle, I barely had enough air to make it back to the surface. A miserable lump of stone was hardly worth this effort, so Mike went down and cut the line. It was then that I discovered I'd made my descent without the five pounds of lead I normally used to neutralize my buoyancy, so it was not surprising that the bottom seemed a long way away. I could have got down easily with the lead, but would I have got up again after fighting with the anchor?

Later, at Gintota, we obtained boats for ten rupees. This time we had a sail and bounded out to the reef in fine style. We bypassed the ill-fated *Elsia* and in half an hour were a couple of miles out to sea. Now, for the first time, I could take some interest in the spot I had visited, without noticing it, the day before.

The Rala Gala Reef is also known as Wave Rock—a perfect description of the place. Even on a calm day, a great surge of water perhaps ten feet high may suddenly rear up out of the sea, giving the only indication—perhaps too late—of the danger lying beneath the surface. The fishermen anchored their boats at a respectful distance from the reef and left us to find our way out there under our own power.

The water was a revelation. It was the first time I had ever encountered visibility of over one hundred feet in the open sea, and there appeared to be absolutely nothing between me and the rocks fifty feet below. We swam slowly toward the reef, until Wave Rock loomed up ahead like a small undersea mountain.

Rodney, plunging like a hawk down the mountainside, had soon shot his first pompano, and Rupert quickly followed suit. Mike and I were carrying the cameras, in the hope that sharks would be attracted by the struggles of the captured fish, and very soon our expectations were fulfilled. The slim, blunt-nosed shapes began to appear from nowhere; sometimes we would glimpse them out of the corners of our eyes and they would be gone before we could photograph them, but sometimes we would spot them first as they hugged the rocks below us and would dive-bomb them with our cameras. They would usually put on speed and get out of range in a couple of seconds.

Most of these sharks were small and shy; we had been diving around the reef for about an hour when we met one that was neither. I was swimming with Mike and Rodney over a rather dull and uninteresting wilderness of rocks and had been spending most of my time about twenty feet down, gliding down to this depth with as little effort as I could and letting myself drift back to the surface whenever I felt like it. The whole secret of underwater sight-seeing—and indeed of diving in general—is summed up in what is known in physics as the principle of least action. Never exert yourself unnecessarily; the universe doesn't.

I was relaxing about twenty feet down when I saw a large shark coming straight toward me, almost on a collision course. Such a confident and assured approach was most unusual, and I quickly whipped the camera up to my eye, praying that my air would hold out until the shark reached me. He was a beautiful beast, about nine feet long, with the inevitable sucker fish latched onto his back just behind his big, white-tipped dorsal fin. I could see his little beady eye and the vertical slits of his gills; he seemed to be moving utterly without effort as he grew larger and larger in my field of vision.

At the last minute he swerved to my left; I clicked the shutter and then realized that I was completely out of air. There was nothing to do but to race back to the surface as quickly as I could. When I emerged, spluttering, Mike popped up beside me. "Why didn't you wait until I could get into the picture as well?" he grumbled. I had to explain that I'd no idea he was anywhere near me; nothing but the shark had existed in my field of consciousness.

A place like Rala Gala was bound to have a wreck, but it was a couple of hours before we found it. Nothing was left of the *King Lud* but a huge propeller with ten-foot blades, and an old-fashioned anchor surrounded by a few twisted lumps of iron. The sea had done its job so thoroughly that one could swim right over the debris without noticing it; even the jungle cannot match the ocean in the speed with which it disguises the works of man.

When we returned to Gintota after five hours at sea, we had a good haul of fish as well as of photographs. A brisk auction sale quickly developed, and in a few minutes we had raised seventeen rupees, leaving us only three rupees out of pocket for the entire trip. I am afraid there is something about life in the East that makes one prone to haggling and bargaining even if it is not really necessary; certainly we were rather proud of the fact that this excursion had been almost self-supporting.

The Isle of

Taprobane

Clarke met his energetic partner, Mike Wilson, in 1951, when Mike was a teenage intruder in the pub frequented by the London science fiction fraternity. Every Thursday about fifty fans used to meet to discuss the books and stories they had read, and those they would write whenever they had the time. Clarke later immortalized those days in Tales from the White Hart.

A smoke-filled saloon bar off wet and foggy Fleet Street was a strange place to learn about skin diving on coral reefs. Mike, who had done this during a spell in the British Merchant Navy, infected me with this enthusiasm, and I was soon learning to use flippers and face mask in a London swimming pool. After a few chilly dips in the English Channel (including an Aqua-Lung dive to eighty feet in midwinter), we decided that this was a hobby for the tropics. I traveled west to Florida, Mike east to Australia, where we eventually rendezvoused on the Great Barrier Reef—the thousand-mile-long belt of islands and shoals that runs up the Queensland coast from the tropic of Capricorn almost to New Guinea.

I have described our first small expedition, which took place in 1955, in *The Coast of Coral;* it is significant here because both Mike and I passed through Ceylon on our way to Australia, and both decided independently that we would like to come back and spend more time on the island. We were able to do this in 1956, and the first result was *The Reefs of Taprobane.*

Anyone who thinks that professional diving is a glamorous occupation should be disabused by what happened to Mike and Rodney Jonklaas. The mainstay of their business was cleaning the water inlet and sewage outlet grilles of ships in Colombo Harbor. These essential orifices get rapidly clogged, especially when a ship is anchored for some weeks in tropical waters, and divers can reach them much more easily using Aqua-Lungs than wearing the clumsy helmet and rubber suit. As Rodney put it, he and Mike were the highest-paid lavatory coolies in Asia.

Even less attractive was the task, which arose all too often, of retrieving corpses in various stages of disrepair. One day there was an SOS from a large engineering firm, whose new reinforced-concrete bridge had collapsed into the swollen, muddy waters of a local river—carrying a workman down into the depths. A hostile crowd of more than a thousand villagers had gathered, demanding that the body be found and properly buried before work on the bridge proceeded; so Mike, with a U.S. Marine Corps friend as backup diver, descended into the tangle of twisted reinforcing bars and broken concrete. Fighting a strong current and working by touch in absolutely zero visibility, he managed to retrieve the victim. For thus risking his life, he charged $60—all of which he gave to the unfortunate man's widow. There was, clearly, no future in this sort of thing.

The most ambitious and spectacular job that Mike and Rodney tackled during their professional partnership was at Castlereagh Dam, a great hydroelectric project four thousand feet up in the central mountains of Ceylon. This involved going eighty feet down into the cold, dead waters of an artificial lake, making a vertical U-turn, and then climbing up a slot only eighteen inches wide. The massive steel doors which closed the tunnel leading to the turbines, a quarter of a mile farther down the mountain, were supposed to move up and down in this slot, but they were blocked by protruding metal bars which Mike and Rodney had to cut away.

A helmet diver, in his bulky suit, could not possibly have reached the site; nor could an Aqua-Lung diver carrying a tank on his back. The only solution was, in effect, to split the Aqua-Lung into two parts, keeping the tanks on the surface and connecting them to the diver by a pipeline. After four weeks of hard and dangerous work with this "hookah" system, Mike and Rodney had cleared the slots and earned enough to buy a Volkswagen van; but they had also decided that there must be much easier ways of making a living.

Rodney found his answer in the fish business; I don't mean fish to eat, but fish to watch. The reefs around Ceylon are crowded with beautiful marine butterflies in an unbelievable variety of forms and colors; Rodney knows exactly where to look for any given specimen, and equally important, the right technique for catching it. His knowledge of the waters round the island is quite uncanny; let me give an example from personal experience.

One day the three of us were taking an experimental fiberglass boat (a design of Mike's with which he hoped to revolutionize the local fishing industry) up a deserted stretch of the east coast, miles and miles from anywhere. The sea became rather rough, so we decided to put into a small creek—a narrow river mouth surrounded by jungle. It was easy to believe that we were the first human beings ever to visit this isolated spot. Presently, as we nibbled sandwiches on the shore, Rodney pointed to a group of rocks out to sea and remarked casually, "There's a family of scorpion fish living under those. I think I'll try to catch them." And he did.

We had chosen, I suppose, the worst possible time to settle down in Ceylon. In 1956, the Western-oriented, conservative government of Sir John Kotelawala (which had lost contact with the people, as all conservative governments eventually do) was annihilated at the polls and replaced by an unstable left-wing coalition headed by Mr. S. W. R. D. Bandaranaike. "Banda," as everyone called him, was typical of the Oxford, London School of Economics intellectuals who have played so important a role in transforming the British Empire into the British Commonwealth—not always to the comfort of the British.

While I cannot account for it, no place other than Ceylon is now wholly real to me. Though London, Washington, New York, Los Angeles, are exciting, amusing, invigorating, and hold all the things that interest my mind, they are no longer quite convincing. Their images are blurred around the edges; like a mirage, they will not stand up to detailed inspection. When I am on the Strand, or Forty-second Street, or at NASA Headquarters, or the Beverly Hills Hotel, my surroundings are liable to give a sudden tremor, and I see through the insubstantial fabric to the reality beneath.

And always it is the same; the slender palm tees leaning over the white sand, the warm sun sparkling on the waves as they break on the inshore reef, the outrigger fishing boats drawn up high on the beach. This alone is real; the rest is but a dream from which I shall presently awake.

The Great Reef

*Clarke's adventure with Mike Wilson and Rodney Jonklaas took place in a spot
that few Ceylonese have seen and fewer still have visited. It is a wave-swept
line of submerged rocks, running parallel to the south coast of Ceylon at
a distance of about six miles, and bearing the curious name Great
Basses. This comes from the Portuguese* baxios *(a reef or shoal),
so Great Basses simply means "Great Reef." Here Clarke
describes this mysterious reef—and some of its
more dangerous inhabitants.*

A glance at the map shows that the Great Basses—and its sister reef, the
Little Basses, a few miles to the east—might have been specially designed
to snare ships rounding the southern coast of Ceylon. This double trap lies
directly across one of the main trade routes of the oriental world; every ship
of any size passing the longitude of India has to contend with it. (The
narrow passage to the north of Ceylon, full of shifting sandbanks and almost
blocked by the submerged land ridge known as Adam's Bridge, is not

navigable except by very small craft at certain seasons.) For at least three thousand years the twin reefs must have taken their toll of shipping; it is conceivable that there has been a greater concentration of wrecks here than anywhere else in the world outside the Mediterranean or the Aegean.

Even in daylight, during perfectly calm weather, ships have come to grief on the Basses. And at night, the two reefs must have presented an appalling hazard to ships, especially when they were being driven landward by the winds of the southwest monsoon.

Not until the middle of the nineteenth century could anything be done about this danger. By then, however, progress in engineering had made it practical to construct lighthouses on such remote, wave-swept rocks, and the government of Ceylon, alarmed by the continual losses in this area, gave orders for work to proceed.

Mike Wilson and Rodney Jonklaas made their first trip to the Great Basses Reef in 1958, taking out all their equipment from the mainland each day in the small fiberglass dinghy we had christened *J. Y. Cousteau*, after our patron saint. This twenty-mile trip across dangerous waters in a light, open boat leaded with diving gear was a hair-raising (not to mention hare-brained) performance which I am very glad to have missed. When we went out again the following year, matters were much better organized; thanks to the courtesy of the Imperial Lighthouse Service, we lived in the lighthouse and used that as our base.

Since I have devoted *Indian Ocean Adventure* to this expedition, I will mention only the highlights of the week we spent exploring the world of the Great Basses Reef. We had chosen our time—April—with care, for it is only possible to dive on Great Basses for a short period every year. For almost ten months out of the twelve, the weather is so rough that it is difficult even to approach the lighthouse, and quite impossible to land. The rock on which the tower is built is not much larger than a tennis court and is nowhere more than three feet above the waterline. Even in calm weather, waves are liable to break over it at any moment.

Fortunately, the good season can be predicted with a fair degree of accuracy. This is because the Indian Ocean has a mysteriously regular system of winds—the famous monsoons—which come and go almost according to the calendar. Between October and March the wind blows from the northeast; then it slackens, and there are about two months of calm weather. But between April and May the wind switches to the other direction; the southwest monsoon sets in, with heavy rain and violent storms. The only time that operations on the reef are possible, therefore, is through March and April; and we could not always count on this, for the monsoon is not absolutely reliable. But in 1959 we were lucky; we had almost perfect weather.

Both the Great and Little Basses lighthouses are serviced by a relief boat based at a little fishing village with the lovely name of Kirinda. At Kirinda, the Imperial Lighthouse Service maintains a stoutly built power-boat, the *Pharos,* with a crew of skilled seamen who know how to get her out to the reef, unload a couple of tons of stores, and bring her safely back.

Because the *Pharos* sailed before dawn, when the sea was at its calmest, we had to be on Kirinda Beach at the hideous hour of 4 A.M., and our Aqua-Lungs, underwater cameras, compressed-air tanks, drums of gas, canned food, weight belts, air compressor, and other gear were loaded in the dark, being carried out on the heads of husky Kirindans wading through the surf. It was a clear, starry night, and above us the constellations of the equator hung at angles I had never seen before, being too addicted to sleep to be a good astronomer.

Out on the horizon, ten miles away, the bright ruby flash of the lighthouse flickered every forty-five seconds. As we drew away from land, I could not help wondering when we would get back to civilization. The relief boat was due to pick us up again in five days, but there had been times when it was delayed for weeks by bad weather. If this happened, we would certainly not starve, but we might get rather tired of eating fish.

We were several miles out to sea when dawn broke, not as quickly as it usually does in the tropics, but almost slowly through low banks of clouds. A few minutes before the Sun finally appeared, its beams fanned out across the sky in great luminous spokes, like those of a slowly turning wheel. It was a spectacle I had seen on such a scale only once before, when I was sailing back to the mainland of Australia from the Great Barrier Reef. Then it had been a symbol of farewell, but now it heralded a new adventure. And I thought how perfectly Homer had described the sight three thousand years ago, when he wrote of the "rosy-fingered dawn."

An hour later, the coast of Ceylon was only a low, misty band far behind, dimpled bluely here and there by the inland mountains. It no longer seemed to have any connection with us; all that mattered now was the white column of stone rearing starkly from the waters ahead. Around the base of the lighthouse the waves were breaking continuously at the end of their long march from Antarctica, the nearest land to the south. Every few seconds the exposed reef would be completely hidden by foam, so that only the lower platform of the lighthouse was visible. It was difficult to imagine how we could land—or, having landed, get off again.

For the next ten minutes we watched with anxious interest while the crew maneuvered to get us and our equipment onto those wetly gleaming rocks. The *Pharos* had been towing a large surfboat, in which we traveled with our stores, and while she stood by at a respectful distance, this was rowed through the bucking waves until it came to within fifty feet of the reef. Then two anchors were let out, spaced well apart to keep the boat from being dragged onto the rocks.

The men on the lighthouse had been waiting, and as soon as our boat had come close enough, they threw us a rope. The other end of this passed over the pulley of a sturdy crane bolted to the stone platform at the foot of the lighthouse.

Now the surfboat was secured at three points; by pulling on one or more of the ropes, the crew could hold it at a fixed distance from the reef. There was nothing they could do, however, to prevent it from rising and falling in the swell. At one moment we would be several feet above the level of the lighthouse platform; a second later, in the trough of a wave, we would be several feet below.

So that we could travel along the aerial ropeway now linking us to the rock, wooden bars had been lashed beneath it to act as seats. Perched on these, we were hauled up to the lighthouse, swinging back and forth with our feet just above the waves. And up that same rope, during the next two hours, went all our equipment, carefully packed in watertight containers.

Then followed the hardest work I have ever done (or will ever do) in my life. Merely living in the lighthouse was a strange and often exhausting experience. The temperature was always in the nineties, and we had to keep climbing up and down a spiral stairway a hundred feet high as we moved from one room to another. After a while I began to develop a fellow feeling for such creatures as snails and the chambered nautilus. Everything was curved; it was impossible to walk more than a few paces in a straight line, and I had to learn to sleep in the arc of a circle, for the bunks were neatly tucked into the yard-thick walls.

We had arrived at the lighthouse in midmorning, and as soon as we had unpacked our equipment and had a meal, we prepared to dive. And this involved problems, for the Great Basses is certainly no place for beginners. Indeed, we had been told that it was utterly impossible to swim around the reef, and at first sight this seemed to be perfectly true.

For though the sea was now quite calm, the water was never still for a second; the level could rise or fall a couple of yards almost instantly over the jagged, barnacled rocks. Getting in and out of the water—especially when wearing an Aqua-Lung and carrying one or often two heavy cameras—therefore required good timing and strong nerves. You had to wait for the crest of a wave and then literally throw yourself and your gear into the sea. If you missed, you were liable to be cut to ribbons on the barnacles.

It was impossible to do any exploring really close to the reef, because the surge of the water made it far too dangerous and also stirred up such clouds of bubbles that visibility was reduced to zero. We had to aim for deeper, calmer water a couple of hundred feet away, and getting there involved fighting strong currents which always seemed to be against us, whether we were coming or going. To serve as a mobile base while we were working over the reef, we had brought along an inflatable rubber dinghy which we could anchor wherever we pleased. Our circular raft was scarcely well streamlined, and to make matters worse, the Aqua-Lung cylinders dangling beneath it caused additional drag. It usually took about twenty minutes of steady flippering to reach the spot we had aimed at, less than one hundred yards from the lighthouse.

But the struggle was worth it. Beneath us was a fantastic fairyland of caves, grottoes, coral-encrusted valleys—and fish in such numbers as I have never met anywhere else in the world. Sometimes they crowded round us so closely that we could see nothing but a solid wall of scales and had literally to push our way through it. They were inquisitive and completely unafraid; during our visit we met, in addition to the usual menagerie of small, multicolored reef dwellers, eagle rays, turtles, angelfish, jacks, tuna (up to three hundred pounds!), groupers, and sharks. Especially the latter.

One of our chief objects in coming to the reef was to get some good shark pictures; though we had encountered sharks scores of times in the past, we had never taken any really good photographs of them—they were always too shy. We had hoped that the oceangoing sharks out here on the reef would be a little less nervous than the inshore ones; this proved to be the case, and we were able to get near them without difficulty.

Although it was easy enough to get them, it was considerably more difficult to get men and sharks in the same picture. To do this, we tried an experiment which I do not recommend as a model, but which was not really as dangerous as it sounds.

Rodney shot a fish and tethered it to the seabed, gutting it so that there was plenty of blood in the water. I took up a position about ten feet away from the bait—*up*stream of the blood, needless to say—and lay flat on the seabed. Mike placed himself twenty feet to one side, so that he could get both me and the bait (we hoped the two would not be synonymous) in the same field of view.

Moreover, there were three of us, well armed with spears or heavy cameras. (A Rolleimarine case makes an excellent club at close quarters, though this is not very good for the camera.) We were in water so crystal clear that we could see a shark approaching from one hundred feet away and would have plenty of warning if it showed signs of being aggressive. Speaking for myself, my chief worry was that the sight of three such strange creatures, blowing out streams of bubbles from their Aqua-Lungs, would scare the sharks away from the neighborhood.

I need not have worried. Almost at once a fairly large shark, about eight feet in length, appeared downcurrent. It began to swim back and forth in wide arcs, hunting over the seabed as it tried to locate the source of the blood. This "quartering" is characteristic of all animals, on land or sea, that use scent to locate their prey; a bloodhound weaves to and fro in wide sweeps until it has found the trail. Then it moves directly to its goal; and so did this shark.

It took no notice of me at all, as I lay flat on the seabed only ten feet away. Just as it was about to grab the fish, another and equally large shark came in like a thunderbolt. The battle was so swift that no human eye could have followed the details; all I can remember is a blur of gray and white bodies, twisting and turning in the water. As an example of sheer ferocity and power, I have never seen anything to match this silent collision on the seabed; two dogs fighting over a single bone gives only a faint idea of its speed and violence. It was over in a second, lasting just long enough to be illuminated by the lightening stroke of Mike's flashbulb. Then the victor swept past me, gulping down the spoils, and leaving me to reflect that for the real meaning of the phrase *struggle for existence,* one must look to the sea, not the land. Even the jungle is almost peaceful by comparison.

Our experiments with sharks ended somewhat abruptly, when Rodney harpooned a fish and was charged by a large and aggressive shark before he could get it off his spear. The fish was still struggling violently, and it was doubtless these vibrations that excited the shark. It swept round Rodney in great spirals, slowly closing in with each revolution. As his gun was now unloaded, Rodney was in a very uncomfortable position; he kept spinning round in the water to face the shark, holding his empty gun out in front of him like a pike. Sometimes the shark would make a hairpin bend and double back on its tracks, and at least once it came close enough for Rodney to nudge it with his gun. It did not give up until Rodney had retreated to the dinghy and had thrown the speared fish into it. No longer able to smell its hoped-for dinner, and deciding that Rodney was too large a mouthful, the disappointed shark made off into the distance. So determined an attack is quite unusual, but is another proof of the statement that one can never tell what a shark will do next.

Winding Up

As the 1950s ended, Clarke faced one of the greatest adventures in his life:
planning the mission that led to the discovery in the early
1960s of treasure beneath the sea at Great Basses.

We started packing at dawn and managed to get everything—including ourselves—into the Oxford and the Volkswagen. We waved good-bye to Kirinda and drove uneventfully back to Colombo, stopping once or twice to show Peter Throckmorton, another companion who had joined us, some of the local sights. It seemed a pity to have come all the way from Greece and to have seen nothing of Ceylon except the seacoast, but he had to return to Athens almost at once, to make the arrangements for his next expedition.

The longest of our brief stops was at the walled town of Galle, Ceylon's chief seaport before the building of Colombo harbor in the nineteenth century. We walked on the ramparts of the old fort, looked down into the clear waters, and made plans for the future. Within a few hundred yards of us lay the wreck of the only *known* treasure ship to go down off Ceylon. This was the P. & O. mail-steamer *Malabar,* which was sunk in 1860 by a monsoon storm while actually inside the harbor. There was no loss of life, but the ship carried an extremely valuable cargo—including the suite of Lord Elgin, later viceroy of India. According to Potter's *The Treasure Diver's Guide,* some $450,000 of silver pesos also went down with her, but at least half of this was raised soon afterward. As the *Malabar* was ideally placed for salvage operations, it is most unlikely that the divers had left anything worthwhile inside her; but it would still be interesting to have a look.

There is another connection between Galle and the submarine world which must now be completely forgotten, and it is well worth recalling. Just a century ago, the first (to my knowledge) underwater color pictures were made on the coral reef near the fort, by an enterprising artist named Baron Eugene de Ransonnet. He constructed a small diving bell, just big enough to hold the upper part of his body, and weighed down with seven hundred pounds of lead. Sitting inside it with his sketchbook, the baron had himself lowered into the water, while air was supplied from a pump in an adjoining boat.

The resulting quite superb studies—almost photographically accurate—were published in *Sketches of Ceylon* (Vienna, 1867) and are still well worth reproducing. As we stood on the ramparts overlooking the spot where the baron sketched the underwater scenery of Ceylon, I could not help thinking that today we could get a better view with face mask and snorkel than he could with almost half a ton of diving bell.

One of my first acts on returning to Colombo in 1961 was to weigh the silver accumulated from the wreck. For the main lumps the figures, in pounds, were 30, 29, 29, 29, 18, 27, 24, 20. Allowing for smaller lumps, loose coins (many corroded or partly covered with coral), and the material sent to the Smithsonian, the total weight of silver we had recovered on the two expeditions was about 350 pounds. All this had come from very few hours of actual diving on an area of two or three square yards.

The night before Peter Throckmorton flew out of Ceylon, we worked into the small hours taking close-ups of all the important finds, so that he would have photographs from which he could continue his researches. We also had consultations with the director of the Colombo Museum, Dr. P. E. P. Deraniyagala, and the archaeological commissioner, Dr. C. E. Godakumbure, and started making arrangements for a permanent exhibition of the material.

We still knew nothing about the origin or nationality of the ship, but soon after our return we did make one very interesting discovery. It was the first definite piece of information we had found concerning the wreck, and it may be quite misleading. But here it is.

Among the discoveries that led to the silver trove was a bronze cannon found on the wreck and marked on the breech with the series of numbers 2 3 23 8. When we wrote to the Smithsonian and asked Mendel Peterson what this could mean, he answered at once that it was the English way of indicating the weight of a gun. The "2" would be hundredweights, the "3" quarter-hundredweights, and the "23" the odd pounds. The final "8" was probably the number of the gun. Allowing for the fact that in the peculiar British system of measures a hundredweight is 112 pounds, this would mean that the gun weighed 331 pounds.

It is not every set of kitchen scales that goes up to 331 pounds, and it was some time before we could test this theory. But finally we located a heavy-duty weighing machine, hauled the cannon into the Volkswagen, and drove it round to the Ceylon Institute of Scientific and Industrial Research.

When we had got the cannon onto the platform, we were delighted to find that it weighed 332 pounds; the few patches of coral were quite enough to account for the negligible discrepancy. And I was amused to find that the modern scientific establishment whose precision scales we were borrowing used exactly the same system as the cannon founders almost three centuries ago; it was calibrated in hundredweights, quarter-hundredweights, and pounds, so that my actual readings were 2 3 24, as against the cannon's engraved 2 3 23.

The gun, then, was definitely English. However, this proved nothing at all. In its career, a gun might serve on many ships, changing hands either by purchase or the fortunes of war. Nevertheless, we began to take more seriously the theory that the ship had belonged to the British, rather than the Dutch East India Company, and initiated research in England as well as Holland.

Our shuttle boat *Ran Muthu*'s last voyage under our flag was completely uneventful. The engine was repaired, and she sailed quietly back to Colombo and returned to the shipyard where she had been built. But by this time we had decided that seagoing cruisers were too expensive a hobby and we sold her to a local shipping company. She was a sturdy little boat and her new owners had a bargain.

A few days later, the first winds of the southwest monsoon started gusting across the city, overturning trees, scattering leaves and branches in every direction. Rain fell in torrents; the roads over which we had driven back from Kirinda were feet deep underwater—or even swept away. We—and *Ran Muthu*—had got back just in time. Our luck, though not all we might have wished for, could have been very much worse.

We regarded the weather with some satisfaction. We now believed that some reports of rival expeditions, which had caused us so much alarm, were nothing more than wild rumors or hopeful exaggerations. But even if there was some truth in them, it did not matter now—nor would it for another year.

It was easy to imagine conditions out on the reef. The breakers would be pounding against the base of the lighthouse, shooting clouds of spray far up the granite tower. Shark's Tooth Rock would be hidden beneath the foam, boiling and seething around it; no boat could possibly get near the site of the wreck without being dashed to pieces.

Part Three

THE 1960s:

KUBRICK AND

CAPE KENNEDY

Clarke receiving the Kalinga Prize for science writing from Rene Maheu,
acting director general of Unesco, in New Delhi, 1962.

COURTESY OF ARTHUR C. CLARKE

Introduction

The decade of the sixties was undoubtedly the most momentous of my life, though I did not realize it at the time. It began badly in 1962 when, after an accidental blow on the head, I became completely paralyzed. I was just able to breathe and probably missed being an iron-lung patient by a hairs-breadth. During this period I wrote *Dolphin Island,* which I regarded at the time as being my farewell to the sea, never imagining that I'd ever go diving again. However, after a few months I made what seemed to be an almost full recovery.

By 1964 I was fit enough to go to New York to work on a Time/Life project—*Man and Space.* As it turned out, this was a Very Good Move—because Stanley Kubrick was looking around for a follow-up to his very successful *Dr. Strangelove or: How I Learned to Stop Worrying and Love the Bomb.* I have described the result of this encounter in *The Lost Worlds of 2001.* Suffice it to say that for the next few years this project dominated my life, and I don't quite understand how I managed to find time to complete what may be my most important work of nonfiction, *Profiles of the Future.*

The Apollo program was now gaining momentum, and when the first men rounded the Moon in December 1968, they had already seen *2001.* I've never quite forgiven them for resisting the temptation (as they told me later) to radio back that they'd seen a Black Monolith in the center of Far Side.

In July 1969, when Apollo 11 left for the first landing on the Moon, I was with Wally Schirra and Walter Cronkite, covering the launch for CBS. In the exhilarating week that followed, none of us ever imagined that our stay on the Moon would be so brief and that, thirty years later, there would still be no definite plans for returning.

Meanwhile, my life had been further complicated by my partner Mike Wilson's discovery of a wreck full of beautiful silver rupees, six miles off the south coast of Ceylon. As I felt more mobile in the zero-g underwater environment than on land, I accompanied Mike on a return to the site, with the results chronicled in *The Treasure of the Great Reef.* We have still not discovered the origin of the wreck, though all the coins it carried were dated A.H. 1116 (A.D. 1703). I have occasional fantasies of returning to the Great Basses Reef—and hope my associates will talk me out of it.

Failures

of Nerve and

Imagination

In several essays, Clarke examines what it takes to set up business as a prophet. He says it is instructive to see what success others have made of this dangerous occupation—and it is even more instructive to see where they have failed.

With monotonous regularity, apparently competent men have laid down the law about what is technically possible or impossible—and have been proved utterly wrong, sometimes while the ink was scarcely dry from their pens. On careful analysis, it appears that these debacles fall into two classes, which I will call "failures of nerve" and "failures of imagination."

The failure of nerve seems to be the more common; it occurs when *even given all the relevant facts* the would-be prophet cannot see that they point to an inescapable conclusion. Some of these failures are so ludicrous as to be almost unbelievable and would form an interesting subject for psychological analysis. "They said it couldn't be done" is a phrase that occurs throughout the history of invention; I do not know if anyone has ever looked into the reasons *why* "they" said so, often with quite unnecessary vehemence.

It is now impossible for us to recall the mental climate that existed when the first locomotives were being built, and critics gravely asserted that suffocation lay in wait for anyone who reached the awful speed of thirty miles an hour. It is equally difficult to believe that eighty-nine years ago, the idea of the domestic electric light was pooh-poohed by all the "experts"—with the exception of a thirty-one-year-old American inventor named Thomas Alva Edison. When gas securities nose-dived in 1878 because Edison (already a formidable figure, with the phonograph and the carbon microphone to his credit) announced that he was working on the incandescent lamp, the British Parliament set up a committee to look into the matter.

The distinguished witnesses reported, to the relief of the gas companies, that Edison's ideas were "good enough for our transatlantic friends . . . but unworthy of the attention of practical and scientific men." And Sir William Preece, engineer-in-chief to the British Post Office, roundly declared that "subdivision of the electric light is an absolute *ignis fatuus.*" One feels that the fatuousness was not in the *ignis.*

The scientific absurdity being pilloried, be it noted, is not some wild-and-woolly dream like perpetual motion, but the humble little electric lightbulb, which more than three generations have taken for granted, excepted when it burns out and leaves one in the dark. Yet although in this matter, Edison saw far beyond his contemporaries, he, too, later in life was guilty of the same shortsightedness that afflicted Preece, because he opposed the introduction of the alternating current.

The most famous, and perhaps the most instructive, failures of nerve have occurred in the fields of aero- and astronautics. At the beginning of the twentieth century, scientists were almost unanimous in declaring that heavier-than-air flight was impossible, and that anyone who attempted to build airplanes was a fool.

Simon Newcomb, the great American astronomer, wrote extensively that "no possible combination of known substances, forms of machinery, and known forms of force could produce a practical machine by which men could fly long distances through the air." He made the mistake of trying to marshal the facts of aerodynamics when he did not understand the science. His failure of nerve lay in not realizing that the means of flight were already at hand.

For Newcomb received wide publicity at just about the time that the Wright brothers were mounting a gasoline engine on wings. When news of their success reached the astronomer, he was only momentarily taken aback. Flying machines might be a marginal possibility, he then conceded—but they were certainly of no practical importance, for it was quite out of the question that they could carry the extra weight of a passenger as well as a pilot.

Such refusal to face facts that now seem obvious has continued throughout the history of aviation. Another astronomer, William H. Pickering, attempted to "straighten out" the uninformed public a few years *after* the first airplanes had started to fly: "It seems safe to say that such ideas must be wholly visionary, and even if a machine could get across the Atlantic with one or two passengers the expense would be prohibitive to any but the capitalist who could own his own yacht. . . . Another popular fallacy is to expect enormous speed to be obtained."

It so happens that most of his fellow astronomers considered Pickering far too imaginative; he was prone to see vegetation—and even evidence for insect life—on the Moon. I am glad to say that by the time he died in 1938 at the ripe age of eighty, Professor Pickering had seen airplanes traveling at 400 mph and carrying considerably more than "one or two" passengers.

Closer to the present, the opening of the Space Age has produced a mass vindication (and refutation) of prophecies on a scale and at a speed never before witnessed. Having taken some part in this myself, and being no more immune than the next man to the pleasures of saying "I told you so," I would like to recall a few of the statements about spaceflight that have been made by prominent scientists in the past. It is necessary for someone to do this, and to jog the remarkably selective memories of the pessimists. The speed with which those who have once declaimed "It's impossible" can switch to "I said it could be done all the time" is really astounding.

As far as the general public is concerned, the idea of spaceflight as a serious possibility first appeared in the 1920s, largely as a result of newspaper reports of the work of the American Robert Goddard and the Hungarian-born German Hermann Oberth. (The much earlier studies of Konstantin Tsiolkovsky in Russia were then almost unknown outside his own country.) When the ideas of Goddard and Oberth, usually distorted by the press, filtered through to the scientific world, they were received with hoots of derision. For a sample of the kind of criticism the pioneers of astronautics had to face, I point to the writings by one Prof. A. W. Bickerton in 1926:

"The foolish idea of shooting at the Moon is an example of the absurd length to which vicious specialization will carry scientists working in thought-tight compartments. . . . For a projectile entirely to escape the gravitational pull of Earth, it needs a velocity of seven miles a second. The thermal energy of a gram at this speed is 15,180 calories a second. . . . The energy of our most violent explosive—nitroglycerine—is less than 1,500 calories a gram. Consequently, even had the explosive nothing to carry, it has only one-tenth of the energy necessary to escape the Earth. . . . Hence the proposition appears to be basically impossible."

Bickerton's first error lies in the sentence containing "our most violent explosive—nitroglycerine." One would have thought it obvious that energy, not violence, is what we want from a rocket fuel; and as a matter of fact nitroglycerine and similar explosives contain much less energy, weight for weight, than such mixtures as kerosene and liquid oxygen. This had been carefully pointed out by Tsiolkovsky and Goddard years before.

Bickerton's second error is much more culpable. What of it, if nitroglycerine has only a tenth of the energy necessary to escape from the Earth? That merely means that you have to use at least ten pounds of nitroglycerine to launch a single pound of payload.

For the fuel itself has not got to escape from Earth; it can all be burned quite close to our planet, and as long as it imparts its energy to the payload, this is all that matters. When *Lunik II* lifted thirty-three years after Professor Bickerton said it was impossible, most of its several hundred tons of kerosene and liquid oxygen never got very far from Russia—but the half-ton payload reached the Mare Imbrium.

Right through the 1930s and 1940s, eminent scientists continued to deride the rocket pioneers—when they bothered to notice them at all. Anyone who has access to a good college library can find, preserved for posterity in the dignified pages of the January 1941 *Philosophical Magazine,* an example that makes a worthy mate to Bickerton's.

Canadian astronomer Prof. J. W. Campbell of the University of Alberta, in *Rocket Flight to the Moon,* wrote that "this now seems less remote than television appeared one hundred years ago." The professor then "mathematically" concluded that it would take *a million tons* of takeoff weight to carry *one pound* of payload on the round-trip.

The correct figure, for today's primitive fuels and technologies, is roughly one ton per pound—a depressing ratio, but hardly as bad as that calculated by the professor. Yet his mathematics were impeccable; so what went wrong?

Merely his initial assumptions, which were hopelessly unrealistic. He chose a path for the rocket that was fantastically extravagant in energy, and he assumed the use of an acceleration so low that most of the fuel would be wasted at low altitudes, fighting the Earth's gravitational field. It was as if he had calculated the performance of an automobile—when the brakes were on. No wonder that he concluded: "While it is always dangerous to make a negative prediction, it would appear that the statement that rocket flight to the Moon does not seem so remote as television did less than one hundred years ago is overoptimistic."

Yet the correct results had been published by Tsiolkovsky, Oberth, and Goddard years before. Though the work of the first two would have been very hard to consult at the time, Goddard's paper "A Method of

Reaching Extreme Altitudes" was already a classic and had been issued by that scarcely obscure body the Smithsonian Institution. If Professor Campbell had only consulted it, he would not have misled his readers and himself.

Some of my best friends are astronomers, and I am sorry to keep throwing stones at them—but they do seem to have an appalling record as prophets. If you still doubt this, let me tell a story so ironic that you might well accuse me of making it up. But I am not that much of a cynic; the facts are on file for everyone to check.

Back in the dark ages of 1935, the founder of the British Interplanetary Society, P. E. Cleator, was rash enough to write the first book on astronautics published in England. His *Rockets Through Space* gave an (incidentally highly entertaining) account of the experiments that had been carried out by the German and American rocket pioneers, and their plans for such commonplaces of today as giant multistage boosters and satellites. Rather surprisingly, the staid scientific journal *Nature* reviewed the book on March 14, 1936:

"It must be said at once that the whole procedure sketched in the present volume presents difficulties of so fundamental a nature that we are forced to dimiss the notion as essentially impracticable, in spite of the author's insistent appeal to put aside prejudice and to recollect the supposed impossibility of heavier–than–air flight before it was actually accomplished. An analogy such as this may be misleading, and we believe it to be so in this case."

Well, though the world now knows the truth about rocketry, it is also interesting to look at the debate, which flourished both in England and America, over the efficacy of the Germans' V–2 rocket and the subsequent development of long-range missiles by the Soviet Union. But when German rockets hit London toward the end of World War II, a lot of the argument went up in smoke.

In December 1945, Dr. Vannevar Bush, the civilian general of the U.S. scientific war effort, told the Senate that the intercontinental missile being developed by the Soviets was something Americans could leave out of their thinking—it was impossible. Shortly before, Churchill's scientific adviser, Lord Cherwell, had told the House of Lords that the V–2 itself was only a propaganda rumor.

The debate against both the V–2 and intercontinental missiles came to an end in 1945, when the V–2 made its presence felt in England and the Pentagon woke up to the fact that the long-range rocket could be built now that thermonuclear bombs made it possible to build warheads five times lighter yet several hundred times more powerful than the Hiroshima bomb.

The Soviet Union had no such inhibitions. Faced with the need for a two-hundred-ton rocket, they went right ahead and built it. By the time it was perfected, it was no longer required for military purposes, for Soviet physicists had bypassed the U.S. billion-dollar tritium-bomb cul-de-sac and gone straight to the far cheaper lithium bomb. Having backed the wrong horse in rocketry, the Russians then entered a much more important event—and won the race into space.

After failure of nerve, the second kind of prophetic failure is less blameworthy, and more interesting. It arises when all the available facts are appreciated and marshaled correctly—but when the really vital facts are still undiscovered, and the possibility of their existence is not admitted. This is the failure of imagination.

One of the most celebrated failures of imagination was that by Lord Rutherford, who more than any other man laid bare the internal structure of the atom. Rutherford frequently made fun of those sensation mongers who predicted that we would one day be able to harness the energy locked up in matter. Yet only five years after his death in 1937, the first chain reaction was started in Chicago. What Rutherford, for all his wonderful insight, had failed to take into account was that a nuclear reaction might be discovered that would release more energy than that required to start it. To liberate the energy of matter, what was wanted was a nuclear "fire" analogous to chemical combustion, and the fission of uranium provided this. Once that was discovered, the harnessing of atomic energy was inevitable, though without the pressures of war it might have well taken the better part of a century.

The example of Lord Rutherford demonstrates that it is not the man who knows most about a subject, and is the acknowledged master of his field, who can give the most reliable pointers to its future. Too great a burden of knowledge can clog the wheels of imagination. I have tried to embody this fact of observation in Clarke's law:

When a distinguished but elderly scientist states that something is possible, he is almost certainly right. When he states that something is impossible, he is very probably wrong.

Perhaps the adjective *elderly* requires definition. In physics, mathematics, and astronautics it means over thirty; in the other disciplines, senile decay is sometimes postponed to the forties. There are, of course, glorious exceptions; but as every researcher just out of college knows, scientists of over fifty are good for nothing but board meetings and should at all costs be kept out of the laboratory!

Too much imagination is much rarer than too little; when it occurs, it usually involves its unfortunate professor in frustration and failure—unless he is sensible enough merely to write about his ideas and not to attempt their realization. In the first category we find all the science fiction authors, historians of the future, creators of utopias—and the two Bacons, Roger and Francis.

Friar Roger (c. 1214–92) imagined optical instruments and mechan-
ically propelled boats and flying machines—devices far beyond the existing
or even foreseeable technology of his time. It is hard to believe that these
words were written in the thirteenth century:

"Instruments may be made by which the largest ships, with only one
man guiding them, will be carried with greater velocity than if they were
full of sailors. Chariots may be constructed that will move with incredible
rapidity without the help of animals. Instruments of flying may be formed
in which a man, sitting at his ease and meditating in any subject, may beat
the air with his artificial wings after the manner of birds . . . as also machines
which will enable men to walk at the bottom of the sea."

This passage is a triumph of imagination over hard fact. Everything
in it has come true, yet at the time it was written it was more an act of faith
than of logic. It is probable that all long-range prediction, if it is to be
accurate, must be of this nature. The real future is not logically foreseeable.

One can only prepare for the unpredictable by trying to keep an open
and unprejudiced mind—a feat that is extremely difficult to achieve, even
with the best will in the world. Indeed, a completely open mind would be
an empty one, and freedom from all prejudices and preconceptions is an
unattainable ideal. Yet there is one form of mental exercise that can provide
good basic training for would-be prophets: anyone who wishes to cope
with the future should travel back in imagination a single lifetime—say to
1900—and ask himself just how many of today's technologies would be,
not merely incredible, but incomprehensible to the keenest scientific brains
of that time.

The collapse of "classical" science actually began with Roentgen's
discovery of X rays in 1895; here was the first clear indication, in a form
that everyone could appreciate, that the commonsense picture of the universe was not sensible after all. X rays—the very name reflects the baffle-
ment of scientists and laymen alike—could travel through solid matter, like
light through a sheet of glass. No one had ever imagined or predicted such
a thing; that one would be able to peer into the interior of a human body—
and thereby revolutionize medicine and surgery—was something that the
most daring prophet had never suggested.

The discovery of X rays was the first great breakthrough into the
realms where no human mind had ever ventured before. Yet it gave scarcely
a hint of still more astonishing developments to come—radioactivity, the
internal structure of the atom, relativity, the quantum theory, the uncertainty principle.

If you showed a modern diesel engine, an automobile, a steam turbine,
or a helicopter to Benjamin Franklin, Galileo, Leonardo da Vinci, and Archimedes—a list spanning two thousand years—not one of them would
have any difficulty in understanding how these machines worked. Leo-

nardo, in fact, would recognize several from his notebooks. All four men would be astonished at the materials and the workmanship, which would seem magical in its precision, but once they had gotten over that surprise they would feel quite at home—as long as they did not delve too deeply into the auxiliary control and electrical systems.

But now suppose they were confronted by a television set, a personal computer, a nuclear reactor, a radar installation. Quite apart from the complexity of these devices, the individual elements of which they are composed would be incomprehensible to any man born before this century. Whatever the degree of education or intelligence, he would not possess the mental framework that could accommodate electron beams, transistors, atomic fission, waveguides, and cathode-ray tubes.

The wholly unexpected discovery of uranium fission in 1939 made possible such absurdly simple (in principle, if not in practice) devices as the atomic bomb and the nuclear chain reactor. No scientist could have predicted them; if he had, all his colleagues would have laughed at him.

It is highly instructive, and stimulating to the imagination, to make a list of the inventions and discoveries that have not been anticipated—and those that have.

The first list contains items that have already been achieved or discovered, yet have an element of the unexpected or downright astonishing about them: X rays, nuclear energy, radio and TV, electronics, photography, sound recording, quantum mechanics, relativity, transistors, masers and lasers, superconductors and superfluids, atomic clocks, the Mossbauer effect concerning gamma radiation in crystals, determining the composition of celestial bodies, dating the past, detecting invisible planets, the ionosphere, and the Van Allen belt.

The second list contains concepts—the expected—that have been around for hundreds or thousands of years: automobiles, flying machines, steam engines, submarines, spaceships, robots, telephones, death rays, transmutation, artificial life, immortality, invisibility, levitation, teleportation, communication with the dead, observing the past and the future, and telepathy.

Now what can we expect or imagine?

We'll Never

Conquer Space

After all that has been written by explorers and visionaries, the statement that
man will never conquer space sounds ludicrous. Yet, Clarke rebuts the
naysayers here and states that man's descendants must learn again
in heartbreak and loneliness their destiny with the stars.

Our age is in many ways unique, full of events and phenomena that never occurred before and can never happen again. They distort our thinking, making us believe that what is true now will be true forever, though perhaps on a larger scale. Because we have annihilated distance on this planet, we imagine that we can do it once again. The facts are far otherwise, and we will see them more clearly if we forget the present and turn our minds toward the past.

To our ancestors, the vastness of Earth was a dominant fact controlling their thoughts and lives. In all earlier ages than ours, the world was wide indeed and no one could ever see more than a tiny fraction of its immensity. A few hundred miles—a thousand, at the most—was infinity. Great empires and cultures could flourish on the same continent, knowing nothing of one another's existence save fables and rumors faint as from a distant planet. When the pioneers and adventurers of the past left their homes in search of new lands, they said good-bye forever to the place of their birth and the companions of their youth. Only a lifetime ago, parents waved farewell to their emigrating children in the virtual certainty that they would never meet again.

And now, within one incredible generation, all this has changed. Over the seas where Odysseus wandered for a decade, the Rome-Beirut Comet whispered its way within the hour. And above that, the closer satellites span the distance between Troy and Ithaca in less than a minute.

But the new stage that is opening up for the human drama will never shrink as the old one has done. We have abolished space here on the little Earth; we can never abolish the space that yawns between the stars. Once again, as in the days when Homer sang, we are face-to-face with immensity and must accept its grandeur and terror, its inspiring possibilities and its dreadful restraints. From a world that has become too small, we are moving out into one that will be forever too large, whose frontiers will recede from us always more swiftly than we can reach out toward them.

Consider first the fairly modest solar, or planetary, distances that we are now preparing to assault. The very first Lunik made a substantial impression upon them, traveling more than 200 million miles from Earth—six times the distance to Mars. When we have harnessed nuclear energy for spaceflight, the Solar System will contract until it is little larger than the Earth today. The remotest of the planets will be perhaps no more than a week's travel from Earth, while Mars and Venus will be only a few hours away.

This achievement, which will be witnessed within a century, might appear to make even the Solar System a comfortable, homely place, with such giant planets as Saturn and Jupiter playing much the same role in our thoughts as do Africa or Asia today. (Their qualitative differences of climate, atmosphere, and gravity, fundamental though they are, do not concern us at the moment.) To some extent this may be true, yet as soon as we pass beyond the orbit of the Moon, a mere quarter million miles away, we will meet the first of the barriers that will sunder Earth from her scattered children.

To a culture that has come to take instantaneous communication for granted, as part of the very structure of civilized life, this "time barrier" may have a profound psychological impact. It will be a perpetual reminder of universal laws and limitations against which not all our technology can ever prevail, for it seems as certain as anything can be that no signal—less any material object—can ever travel faster than light.

The velocity of light is the ultimate speed limit, being part of the very structure of space and time. Within the narrow confines of the Solar System, it will not handicap us too severely, once we have accepted the delays in communication that it involves. At the worst, these will amount to eleven hours—the time that it takes a radio signal to span the orbit of Pluto, the outermost planet. Between the three inner worlds, Earth, Mars, and Venus, it will never be more than twenty minutes—not enough to interfere seriously with commerce or administration, but more than sufficient to shatter those personal links of sound or vision that can give us a sense of direct contact with friends on Earth, wherever they may be.

It is when we move out beyond the confines of the Solar System that we come face-to-face with an altogether new order of cosmic reality. Even today, many otherwise educated men—like those savages who can count to three but lump together all numbers beyond four—cannot grasp the profound distinction between *solar* and *stellar* space. The first is the space enclosing our neighboring worlds, the planets; the second is that which embraces those distant suns, the stars. *And it is literally millions of times greater.*

There is no such abrupt change of scale in terrestrial affairs. To obtain a mental picture of the distance to the nearest star, as compared with the distance to the nearest planet, you must imagine a world in which the closest object to you is only five feet away—and then there is nothing else to see until you have traveled a thousand miles.

One day—it may be soon, or it may be a thousand years from now—we shall discover a really efficient means of propelling our space vehicles. Every technical device is always developed to its limit (unless it is superseded by something better), and the ultimate speed for spaceships is the velocity of light. They will never reach that goal, but they will get very close to it. And then the nearest star will be less than five years' voyaging from Earth.

Our exploring ships will spread outward from their home in an ever-expanding sphere of space. It is a sphere that will grow at almost—but never quite—the speed of light. Five years to the triple system of Alpha Centauri, ten to that strangely matched doublet Sirius A and B, eleven to the tantalizing enigma of 61 Cygni, the first star suspected of possessing a planet. These journeys are long, but they are not impossible. Man has always accepted whatever price was necessary for his explorations and discoveries, and the price of space is time.

Even voyages that may last for centuries or millenniums will one day be attempted. Suspended animation, an undoubted possibility, may be the key to interstellar travel. Self-contained cosmic arks, which will be tiny traveling worlds in their own right, may be another solution, for they would make possible journeys of unlimited extent, lasting generation after generation. The famous time dilation predicted by the theory of relativity, whereby time appears to pass more slowly for a traveler moving at almost the speed of light, may yet be a third. And there are others.

With so many theoretical possibilities for interstellar flight, we can be sure that at least one will be realized in practice. Remember the history of the atomic bómb; there were three different ways in which it could be made, and no one knew which was best. So they were all tried—and they all worked.

Looking far into the future, therefore, we must picture a slow (little more than half a billion miles an hour!) expansion of human activities outward from the Solar System, among the suns scattered across the region of the galaxy in which we now find ourselves. These suns are on average five light-years apart; in other words, we can never get from one to the next in less than five years.

But consider the effects of the inevitable, unavoidable time lag. There could be only the most tenuous contact between the home island and its offspring. Returning messengers could report what had happened on the nearest colony—five years ago. They could never bring information more up-to-date than that, and dispatches from the more distant parts of the ocean would be from still further in the past—perhaps centuries behind the times. There would never be news from the other islands, but only history.

No oceanic Alexander or Caesar could ever establish an empire beyond his own coral reef; he would be dead before his orders reached his governors. Any form of control or administration over other islands would be utterly impossible, and all parallels from our own history would thus cease to have any meaning. It is for this reason that the popular science fiction stories of interstellar empires and intrigues become pure fantasies, with no basis of reality. Try to imagine how the War of Independence would have gone if news of Bunker Hill had not arrived in England until Disraeli was Victoria's prime minister, and his urgent instructions on how to deal with the situation had reached America during Eisenhower's second term. Stated in this way, the whole concept of interstellar administration is seen as an absurdity.

At this point, we will move the discussion on to a new level and deal with an obvious objection. Can we be sure that the velocity of light is indeed a limiting factor? So many "impassable" barriers have been shattered in the past; perhaps this one may go the way of all the others.

We will not argue the point or give the reasons scientists believe that light can never be outraced by any form of radiation or any material object. Instead, let us assume the contrary and see just where it gets us. We will even take the most optimistic possible case and imagine that the speed of transportation may eventually become infinite.

Picture a time when, by the development of techniques as far beyond our present engineering as a transistor is beyond a stone ax, we can reach anywhere we please instantaneously, with no more effort than by dialing a number. This would indeed cut the universe down to size and reduce its physical immensity to nothingness. What would be left?

Everything that really matters. For the universe has two aspects—its scale, and its overwhelming, mind-numbing complexity. Having abolished the first, we are now face-to-face with the second.

What we must now try to visualize is not size, but quantity. Most people today are familiar with the simple notation that scientists use to describe large numbers; it consists merely of counting zeros, so that a hundred becomes 10^2; a million, 10^6; a billion, 10^9; and so on. This useful trick enables us to work with quantities of any magnitude, and even defense budget totals look modest when expressed as $\$5.76 \times 10^9$ instead of $\$5,760,000,000$.

The number of other suns in our own galaxy (that is, the whirlpool of stars and cosmic dust of which the Sun is an out-of-town member, lying in one of the remoter spiral arms) is estimated at 10^{11}—or written in full, 100,000,000,000. Our present telescopes can observe something like 10^9 other galaxies. There are probably at least as many galaxies in the whole of creation as there are stars in our own galaxy, but let us confine ourselves to those we can see. They must contain a total of about about 10^{11} times 10^9 stars, or 10^{20} stars altogether.

One followed by twenty other digits is, of course, a number beyond all understanding. There is no hope of ever coming to grips with it, but there are ways of hinting at its implications.

Just now we assumed that the time might come when we could dial ourselves, by some miracle of matter transmission, effortlessly around the cosmos, as today we call a number in our local exchange. What would the cosmic telephone directory look like if its contents were restricted to suns and it made no effort to list individual planets, still less the millions of places on each planet?

And so we return to our opening statement. Space can be mapped and crossed and occupied without definable limit; but it can never be conquered. When our race has reached its ultimate achievements, and the stars themselves are scattered no more widely than the seed of Adam, even then we shall still be like ants crawling on the face of the Earth. The ants have covered the world, but have they conquered it—for what do their countless colonies know of it, or of each other?

So it will be with us as we spread out from Earth, loosening the bonds of kinship and understanding, hearing faint and belated rumors at second—or third—or thousandth hand of an ever-dwindling fraction of the entire human race. Though Earth will try to keep in touch with her children, in the end all the efforts of her archivists and historians will be defeated by time and distance, and the sheer bulk of material. For the number of distinct societies or nations, when our race is twice its present age, may be far greater than the total number of all the men who have ever lived up to the present time.

Rocket to the

Renaissance

*Though there are no more undiscovered continents on Earth, the road to the stars
has been discovered—and none too soon, according to the following essay.*

Civilization cannot exist without new frontiers; it needs them both physically and spiritually. The physical need is obvious—new lands, new resources, new materials. The spiritual need is less apparent, but in the long
run it is more important. We do not live by bread alone; we need adventure,
variety, novelty, romance. As the psychologists have shown by their
sensory-deprivation experiments, a man goes swiftly mad if he is isolated in
a silent, darkened room, cut off completely from the external world. What
is true of individuals is also true of societies; they, too, can become insane
without sufficient stimulus.

It may seem overoptimistic to contend that man's forthcoming escape
from Earth, and the crossing of interplanetary space, will trigger a new
renaissance and break the patterns into which our society, and our arts,
must otherwise freeze. Yet this is exactly what I propose to do; first, however, it is necessary to demolish some common misconceptions.

The space frontier is infinite, beyond all possibility of exhaustion; but the opportunity and the challenge it presents are both totally different from any that we have met in our world in the past. All the moons and planets in the Solar System are strange, hostile places that may never harbor more than a few thousand human inhabitants, who will be at least as carefully handpicked as the population of Los Alamos. The age of the mass colonization may be gone forever. Space has room for many things, but not for "your tired, your poor, your huddled masses yearning to breathe free." Any statue of liberty on Martian soil will have inscribed on its base, "Give me your nuclear physicists, your chemical engineers, your biologists and mathematicians." The immigrants of the twenty-first century will have much more in common with those of the seventeenth century than of the nineteenth. For the *Mayflower,* it is worth remembering, was loaded to the scuppers with eggheads.

The often-expressed idea that the planets can solve the problem of overpopulation may be a complete fallacy. Humanity is increasing at the rate of some one hundred thousand a day, and no conceivable space lift could make serious inroads in this appalling figure.

With present techniques, the combined military budgets of all nations might just about suffice to land ten men on the Moon every day. Yet even if space transportation were free, that would scarcely help matters—for there is not a single planet upon which men could live and work without elaborate mechanical aids. On all of them we shall need the paraphernalia of space suits, synthetic-air factories, pressure domes, totally enclosed hydroponic farms. One day our lunar and Martian colonies will be self-supporting, but if we are looking for living room for our surplus population, it would be far cheaper to find it in the Antarctic—or even at the bottom of the Atlantic Ocean.

No, the population battle must be fought and won here on Earth, and the longer we postpone the inevitable conflict the more horrifying the weapons that will be needed for victory. (Compulsory abortion and infanticide, and antiheterosexual legislation—with its reverse—may be some of the milder expedients.) Yet though the planets may not save us, this is a matter in which logic may not count. The weight of increasing numbers—the suffocating sense of pressure—as the walls of the ant heap crowd ever closer will help to power man's drive into space, even if no more than a millionth of humanity can ever go there.

Though the planets can give no physical relief to Earth, their intellectual and emotional contributions may be enormous. The discoveries of the first expeditions, the struggles of the pioneers to establish themselves on other worlds—these will inspire a feeling of purpose and achievement among the stay-at-homes. They will know, as they watch their television screens, that History with a capital *H* is starting again. The sense of wonder, which we have almost lost, will return to life; and so will the spirit of adventure.

It is difficult to overstate the importance of this—though it is easy to poke fun at it by making cynical remarks about escapism. Only a few people can be pioneers or discoverers, but everyone who is even half alive occasionally feels the need for adventure and excitement. If you require proof of this, look at the countless horse operas galloping across the ether. The myth of a West that never was has been created to fill the vacuum in our modern lives, and it fills it well. Sooner or later, however, one tires of myths (many of us have long since tired of this one), and then it is time to seek new territory. There is a poignant symbolism in the fact that the giant rockets now stand poised on the edge of the Pacific where the covered wagons halted only two lifetimes ago.

Already, a slow but profound reorientation of our culture is under way, as men's thoughts become polarized toward space. Even before the first living creatures left Earth's atmosphere, the process had started in the most influential area—the nursery. Space toys have been commonplace for years; so have cartoons and "take me to your leader" jokes that would have been incomprehensible in years past. Increasing awareness of the universe has even, alas, contributed to our psychopathology. A fascinating parallel could be drawn between the flying saucer cults and the witchcraft mania of the seventeenth century. The mentalities are the same, and I hereby present the notion to any would-be Ph.D. in search of a thesis.

As the exploration of the Solar System proceeds, human society will become more and more permeated with the ideas, discoveries, and experiences of astronautics. They will have their greatest effect, of course, upon the men and women who actually go out into space to establish either temporary bases or permanent colonies on the planets. Because we do not know what they will encounter, it is scarcely profitable to speculate about the societies that may evolve a hundred or a thousand years from now, upon the Moon, Mars, Titan, and the other major solid bodies of the Solar System. (We can write off the giant planets, Jupiter, Saturn, Uranus, and Neptune, which have no stable surfaces.) The outcome of our ventures in space must await the verdict of history.

As has happened so often in the past, the challenge may be too great. We may establish colonies on the planets, but they may be unable to maintain themselves at more than a marginal level of existence, with no energy left over to spark any cultural achievements. Whatever the eventual outcome, we can be reasonably certain of immediate benefits—and I am deliberately ignoring such "practical" returns as improvements in weather forecasting and communications, which may in themselves put space travel on a paying basis. The creation of wealth is certainly not to be despised, but in the long run the only human activities really worthwhile are the search for knowledge and the creation of beauty. This is beyond argument; the only point of debate is which comes first.

Only a small part of mankind will ever be thrilled to discover the electron density around the Moon, the precise composition of the Jovian atmosphere, or the strength of Mercury's magnetic field. Though the existence of whole nations may one day be determined by such facts, and others still more esoteric, these are ideas that concern the mind, and not the heart. Civilizations are respected for their intellectual achievements; they are loved—or despised—for their works of art. Can we even guess what art will come from space?

The writer cannot escape from his environment, however hard he tries. When the frontier is open we have Homer and Shakespeare—or, to choose less Olympian examples nearer to our own age, Melville, Conrad, and Mark Twain. When it is closed, the time has come for Tennessee Williams and the beatniks—and for Proust, whose horizon toward the end of his life was a cork-lined room. (If Lewis Carroll had lived today, he might not have given us Alice but Lolita.)

It is too naive to imagine that astronautics will restore the epic and the saga in anything like their original forms; spaceflight will be too well documented, and Homer started off with the great advantage of being untrammeled by too many facts. But surely the discoveries and adventures, the triumphs and inevitable tragedies that must accompany man's drive toward the stars will one day inspire a new heroic literature and bring forth latter-day equivalents of *The Golden Fleece, Gulliver's Travels, Moby-Dick, Robinson Crusoe,* or *The Ancient Mariner.*

It is perhaps too early to speculate about the impact of spaceflight on music and the visual arts. Here again one can only hope—and hope is certainly needed when one looks at the canvases upon which some contemporary painters all too accurately express their psyches. The prospect for modern music is a little more favorable; now that electronic computers have been taught to compose it, we may confidertly expect that before long some of them will learn to enjoy it, thus saving us the trouble.

Maybe these ancient art forms have come to the end of the line, and the still unimaginable experiences that await us beyond the existent gravity, for example, will certainly give rise to a strange, otherworldly architecture, fragile and delicate as a dream. And what, I wonder, will *Swan Lake* be like on Mars, when the dancers have only a third of their terrestrial weight—or on the Moon, where they will have merely a sixth?

The complete absence of gravity—a sensation that only a few human beings have experienced up until now, yet which is mysteriously familiar in dreams—will have a profound impact upon every type of human activity on Earth. It will make possible a whole constellation of new sports and games and transform many existing ones. This final prediction we can make with confidence: weightlessness will open up novel and hitherto unexpected realms of erotica. And about time, too.

All our aesthetic ideas and standards are derived from the natural world around us, and it may well turn out that many of them are peculiar to Earth. What other planet will have blue skies and seas, green grass, hills softly rounded by erosion, rivers and waterfalls, a single brilliant moon. Nowhere in space will we rest our eyes upon the familiar shapes of trees and plants, or any of the animals which share our world. Whatever life we meet will be as strange and alien as the nightmare creatures of the ocean abyss, or of the insect empire whose horrors are normally hidden from us by their microscopic scale. It is even possible that the physical environments of the other planets may turn out to be unbearably hideous; it is equally possible that they will lead us to new and more universal ideas of beauty, less limited by our earthbound upbringing.

The existence of extraterrestrial life is, of course, the greatest of the many unknowns awaiting us on the planets. We are now fairly certain that there is some form of vegetation on Mars; the seasonal color changes, coupled with recent spectroscopic evidence, give this a high degree of probability. As Mars is an old and perhaps dying world, the struggle for existence may have led to some weird results. We had better be careful when we land.

Where there is vegetation, there may be higher forms of life; given sufficient time, nature explores all possibilities. Mars has had plenty of time, so those parasites on the vegetable kingdom known as animals may have evolved there. They will be very peculiar animals, for they will have no lungs. There is not much purpose in breathing when the atmosphere is practically devoid of oxygen.

Contact with a contemporary, nonhuman civilization will be the most exciting thing that has ever happened to our race; the possibilities for good and evil are endless. Within a decade or so, some of the classic themes of science fiction may enter the realm of practical politics. It is much more likely, however, that if Mars has ever produced intelligent life, we have missed it by geological ages. Since all the planets have been in existence for at least five *billion* years, the probability of culture flourishing on two of them at the same time must be extremely small.

When our archaeologists reach Mars, they may find waiting for us a heritage as great as that which we owe to Greece and Rome. We should not, however, pin too much hope on Mars, or upon any of the worlds of this solar system. If intelligent life exists elsewhere in the universe, we may have to seek it upon the planets of other suns. They are separated from us by a gulf millions—I repeat, *millions*—of times greater than that dividing us from our next-door neighbors Mars and Venus. Until a few years ago, even the most optimistic scientists thought it impossible that we could ever span this frightful abyss, which light itself takes years to cross at a tireless 670,000,000 miles per hour. Yet now, by one of the most extraordinary and unexpected breakthroughs in the history of technology, there is a good chance that we may make contact with intelligence *outside* the solar system before we discover the humblest mosses or lichens inside it.

This breakthrough has occurred in electronics. It now appears that by far the greater part of our exploration of space will be by radio. It can put us in touch with worlds that we can never visit—even with worlds that have long since ceased to exist. The radio telescope, and not the rocket, may be the instrument that first establishes contact with intelligence beyond the Earth.

Even a decade ago, this idea would have seemed absurd. But now we have receivers of such sensitivity, and antennas of such enormous size, that we can hope to pick up radio signals from the nearer stars—if there is anyone out there to send them. The search for such signals began early in 1960 at the National Radio Astronomy Observatory, Greenback, West Virginia, and many other observatories will follow suit once they have built the necessary equipment. This is perhaps the most momentous quest upon which men have ever embarked; sooner or later, it will be successful.

From the background of cosmic noise, the hiss and crackle of exploding stars and colliding galaxies, we will someday filter out the faint, rhythmic pulses which are the voice of intelligence. At first we will know only (only!) that there are other minds than ours in the universe; later we will learn to interpret these signals. Some of them, it is fair to assume, will carry images—the equivalent of picture telegraphy, or even television. It will be fairly easy to deduce the coding and reconstruct these images. One day, perhaps not far in the future, some cathode-ray screen will show pictures from another world.

There are fascinating and endless grounds for speculation. We have known radio for barely a lifetime, and television for barely a generation; all of our techniques of electronic communication must be incredibly primitive. Yet even now, if put to it, we could send all that is best in our entire culture pulsing across the light-years.

But something must be lost in any contact between cultures; what is gained is far more important. In the ages to come we may lock minds with many strange beings and study with incredulity, delight, or horror civilizations that may be older than our Earth. Some of them will have ceased to exist during the centuries that their signals have been crossing space. The radio astronomers will thus be the true interplanetary archaeologists, reading inscriptions and examining works of art whose creators passed away before the building of the pyramids. Even this is a modest estimate; a radio wave arriving now from a star at the heart of the Milky Way must have started its journey around 25,000 B.C.

Radio prehistory—electronic archaeology—may have consequences at least as great as the classical studies of the past. The races whose messages we interpret and whose images we reconstruct will obviously be of a very high order, and the impact of their art and technology upon our own culture will be enormous.

Most of this planet's life remains to this day trapped in a meaningless cycle of birth and death. Only the creatures who dared to move from the sea to the hostile, alien land were able to develop intelligence. Now that this intelligence is about to face a still greater challenge, it may be that this beautiful Earth of ours is no more than a brief resting place between the sea of salt and the sea of stars. Now we must venture forth.

The whole structure of Western society may well be unfitted for the effort that the conquest of space demands. No nation can afford to divert its ablest men into such essentially noncreative, and occasionally parasitic, occupations as law, advertising, and banking. Nor can it afford to squander indefinitely the technical manpower it does possess. And it does not necessarily follow that the Soviet Union could do much better.

Despite the perils and problems of our times, we should be glad that we are living today. Every civilization is like a surf rider, carried forward on the crest of a wave. The wave bearing us has scarcely started its run; those who thought it was already slackening spoke centuries too soon. We are poised now in the precarious but exhilirating balance that is the essence of real living, the antithesis of mere existence. Behind us lie the reefs we have already passed; beneath us the great wave, as yet barely flecked with foam, arches its back still higher from the sea.

And ahead . . . ?

We cannot tell; we are too far out to see the unknown land. It is enough to ride the wave.

The

Obsolescence

of Man

A million years ago, an unprepossessing primate discovered that his forelimbs could be used for other purposes besides locomotion. Objects such as sticks and stones could be grasped—and, once grasped, were useful for killing, digging up roots, defending or attacking, and a hundred other jobs. In short, tools had appeared, and soon the machine could overtake man, Clarke writes.

The first users of tools were not men—a fact appreciated only in the last year or two—but prehuman anthropoids; and by their discovery they doomed themselves. For even the most primitive of tools, such as a naturally pointed stone that happens to fit the hand, provides a tremendous physical and mental stimulus to the user. He has to walk erect; he no longer needs huge canine teeth—since sharp flints can do a better job—and he must develop manual dexterity of a high order. These are the specifications of Homo sapiens; as soon as they start to be filled, all earlier models are headed for rapid obsolescence. To quote Prof. Sherwood Washburn of the University of California's anthropology department: "It was the success of the simplest tools that started the whole trend of human evolution and led to the civilizations of today."

Note that phrase—"the whole trend of human evolution." The old idea that man invented tools is therefore a misleading half-truth; it would be more accurate to say that *tools invented man.* They were very primitive tools, in the hands of creatures who were little more than apes. Yet they led to us—and to the eventual extinction of the ape-men who first wielded them.

Now the cycle is about to begin again, but neither history nor pre-history ever exactly repeats itself, and this time there will be a fascinating twist in the plot. The tools that the ape-men invented caused them to evolve into their successor, Homo sapiens. The tool we have invented in the latter part of the twentieth century is our successor. Biological evolution has given way to a far more rapid process—technological evolution. To put it bluntly and brutally, the machine is going to take over.

This, of course, is hardly an original idea. That the creations of man's brains might one day threaten and perhaps destroy him is such a tired old cliché that no self-respecting science fiction writer would dare to use it. It goes back through Karel Capek's *R.U.R.,* Samuel Butler's *Erewhon,* Mary Shelley's *Frankenstein,* and the figure of Daedalus, King Minos' one-man office of scientific research. For at least three thousand years, therefore, a vocal minority of mankind has had grave doubts about the ultimate out-come of technology. From the self-centered, human point of view, these doubts are justified. But that, I submit, will not be the only—or even the most important—point of view for much longer.

The breaching of the barrier between brain and machine is perhaps one of the greatest breakthroughs in the history of human thought, like the discovery that the Earth revolves around the Sun, or that man is part of the animal kingdom, or that $E = mc^2$. All these ideas took time to sink in and were frantically denied when first put forward. In the same way it will take a little while for men to realize that machines can not only think, but may one day think them off the face of the Earth.

At this point you may reasonably ask: "Yes—but what do you mean by *think?*" I propose to sidestep that question, using a neat device for which I am indebted to the English mathematician A. M. Turing. Turing imagined a game played by two teleprinter operators in separate rooms—this imper-sonal link being used to remove all clues given by voice, appearance, and so forth. Suppose one operator was able to ask the other any questions he wished, and the other had to make suitable replies. If, after some hours or days of conversation, the questioner could not decide whether his tele-graphic acquaintance was human or purely mechanical, then he could hard-ly deny that he/it was capable of thought. An electronic brain that passed this test would, surely, be regarded as an intelligent entity. Anyone who argued otherwise would merely prove he was less intelligent than the machine; he would be a splitter of nonexistent hairs, like the scholar who proved that the *Odyssey* was not written by Homer, but by another man of the same name.

We are still decades—but not centuries—from building such a machine, yet already we are sure that it could be done. If Turing's experiment is never carried out, it will merely be because the intelligent machines of the future will have better things to do with their time than conduct extended conversations with men. I often talk with my dog, but I don't keep it up for long.

Even machines less intelligent than men might escape from our control by sheer speed of operation. And in fact, there is every reason to suppose that machines will become much more intelligent than their builders, as well as incomparably faster.

There are still a few authorities who refuse to grant any degree of intelligence to machines, now or in the future. This attitude shows a striking parallel to that adopted by the chemists of the early nineteenth century. It was known then that all living organisms are formed from a few common elements—mostly carbon, hydrogen, oxygen, and nitrogen—but it was firmly believed that the materials of life could not be made from mere chemicals alone. No chemist could ever take carbon, hydrogen, and so forth and combine them to form any of the substances upon which life was based. There was an impassable barrier between the worlds of inorganic and organic chemistry.

Since this is not a treatise on computer design, you will not expect me to explain how to build a thinking machine. In fact, it is doubtful if any human being will ever be able to do this in detail, but one can indicate the sequence of events that will lead from H. sapiens to M. sapiens. The first two or three steps on the road have already been taken; machines now exist that can learn from experience, profiting from their mistakes and—unlike human beings—never repeating them. Machines have been built that do not sit passively waiting for instructions, but that explore the world around them in a manner that can only be called inquisitive. Others look for proofs of theorems in mathematics or logic and sometimes come up with solutions that had never occurred to their makers.

It is even possible that the first genuine thinking machines may be grown rather than constructed; already some crude but very stimulating experiments have been carried out along these lines. Several artificial organisms have been built that are capable of rewiring themselves to adapt to changing circumstances. Beyond this there is the possibility of computers that will start from relatively simple beginnings, be programmed to aim at specific goals, and search for them by constructing their own circuits, perhaps by growing networks of threads in a conducting medium. Such a growth may be no more than a mechanical analogy of what happens to every one of us in the first nine months of our existence.

All speculations about intelligent machines are inevitably conditioned—indeed, inspired—by our knowledge of the human brain. No one, of course, pretends to understand the full workings of the brain or expects that such knowledge will be available in the foreseeable future. (It is a nice philosophical point as to whether the brain can ever, even in principle, understand itself.) But we do know enough about its physical structure to draw many conclusions about the limitations of "brains"—whether organic or inorganic.

There are about 10 billion switches—or neurons—inside your skull, "wired" together in circuits of unimaginable complexity. Ten billion is such a large number that, until recently, it could be used as an argument against the achievement of mechanical intelligence. In the 1950s a famous neurophysiologist made a statement (still produced like some protective incantation by the advocates of cerebral supremacy) to the effect that an electronic model of the human brain would have to be as large as the Empire State Building and would need Niagara Falls to keep it cool when it was running.

This must now be classed with such interesting pronouncements as "No heavier than air machine will ever be able to fly." For the calculation was made in the days of the vacuum tube, the precursor of the transistor, and the transistor has now completely altered the picture. Indeed—such is the rate of technological progress today—the transistor has itself been replaced by smaller and faster devices, based upon principles of quantum physics. If the problem was merely one of space, electronic techniques today would allow us to pack a computer as complex as the human brain on only a small portion of the first floor of the Empire State Building.

The human brain surpasses the average stereo set by a thousandfold, packing its 10 billion neurons into a tenth of a cubic foot. And although smallness is not necessarily a virtue, even this may be nowhere near the limit of possible compactness.

For the cells composing our brains are slow-acting, bulky, and wasteful of energy—compared with the scarcely more than atom-sized computer elements that are theoretically possible. The mathematician John von Neumann once calculated that electronic cells could be 10 billion times more efficient than protoplasmic ones; already they are a million times swifter in operation, and speed can often be traded for size. If we take these ideas to their ultimate conclusion, it appears that a computer equivalent in power to one human brain need be not much bigger than a matchbox, and probably much, much smaller.

This slightly shattering thought becomes more reasonable when we take a critical look at flesh and blood and bone as engineering materials. All living creatures are marvelous, but let us keep our sense of proportion.

Perhaps the most wonderful thing about life is that it works at all, when it has to employ such extraordinary materials and has to tackle its problems in such roundabout ways.

Consider the eye. Suppose you were given the problem of designing a camera—for that, of course, is what the eye is—which *has to be constructed entirely of water and jelly,* without using a scrap of glass, metal, or plastic. Obviously, it can't be done.

You're quite right; the feat is impossible. The eye is an evolutionary miracle, but it's a lousy camera. You can prove this while you're reading the next sentence.

Here's a medium-length word: *photography.* Close one eye and keep the other fixed—repeat, *fixed*—on that center *g*. You may be surprised to discover that—unless you cheat by altering the direction of your gaze—you cannot see the whole word clearly. It fades out three or four letters to the right and left.

No camera ever built—even the cheapest—has as poor an optical performance as this. For color vision also, the human eye is nothing to boast about; it can operate only over a small band of the spectrum. To the worlds of the infrared and ultraviolet, visible to bees and other insects, it is completely blind.

We are not conscious of these limitations because we have grown up with them, and indeed if they were corrected, the brain would be quite unable to handle the vastly increased flood of information. Let us not make a virtue of a necessity; if our eyes had the optical performance of even the cheapest miniature camera, we would live in an unimaginably richer and more colorful world.

These defects are due to the fact that precision scientific instruments simply cannot be manufactured from living materials at this time. With the eye, the ear, the nose—indeed, all the sense organs—evolution has performed a truly incredible job against fantastic odds. But it will not be good enough for the future; indeed, it is not good enough for the present.

There are some senses that do not exist, that can probably never be provided by living structures, and that we need in a hurry. On this planet, to the best of our knowledge, no creature has ever developed organs that can detect radio waves or radioactivity. Though I would hate to lay down the law and contend that nowhere in the universe can there be organic Geiger counters or living television sets, I think it highly improbable. There are some jobs that can be done only by transistors or magnetic fields or electron beams and are therefore beyond the capability of purely organic structures.

There is another fundamental reason living machines such as you and I cannot hope to compete with nonliving ones. We are handicapped by one of the toughest engineering specifications ever issued. What sort of performance would you expect from a machine that has to grow several billionfold during the course of manufacture—and which has to be completely and continuously rebuilt, molecule by molecule, every few weeks?

Though intelligence can arise from life, it may then discard it. Perhaps at a later stage, as the mystics have suggested, it may also discard matter; but this leads us in realms of speculations that an unimaginative person like myself would prefer to avoid.

One often-stressed advantage of living creatures is that they are self-repairing and reproduce themselves with ease—indeed, with enthusiasm. This superiority over machines will be short-lived; the general principles underlying the construction of self-repairing and self-reproducing machines have already been worked out. There is, incidentally, something ironically appropriate in the fact that Turing, the brilliant mathematician who pioneered in this field and first indicated how thinking machines might be built, shot himself a few years after publishing his results. It is very hard not to draw a moral from this.

The greatest single stimulus to the evolution of mechanical—as opposed to organic—intelligence is the challenge of space. Only a vanishingly small fraction of the universe is directly accessible to mankind, in the sense that we can live there without elaborate protection or mechanical aids. If we generously assume that humanity's potential extends from sea level to a height of three miles, over the whole Earth, that gives us a total of some half billion cubic miles. At first sight this is an impressive figure, especially when you remember that the entire human race could be packaged into a one-mile cube. But it is absolutely nothing when set against Space with a capital S. Our telescopes sweep a volume at least a million million million million million million million million million million million times greater.

Though such a number is utterly beyond conception, it can be given a vivid meaning. If we reduced the known universe to the size of the Earth, then the portion in which we can live without space suits and pressure cabins is about the size of a single atom.

It is true that, one day, we are going to explore and colonize many other atoms in this Earth-sized volume, but it will be at the cost of tremendous technical efforts, for most of our energies will be devoted to protecting our frail and sensitive bodies against the extremes of temperature, pressure, or gravity found in space and on other worlds. Within very wide limits, machines are indifferent to these extremes. Even more important, they can wait patiently through the years and the centuries that will be needed for travel to the far reaches of the universe.

Creatures of flesh and blood such as ourselves can explore space and win control over infinitesimal fractions of it. But only creatures of metal and plastic can ever really conquer it, as indeed they have already started to do. The tiny brains of our Prospectors and Rangers barely hint at the mechanical intelligence that will one day be launched at the stars.

The protoplasmic computer inside your skull should now be programmed to accept the idea—at least for the sake of argument—that machines can be both more intelligent and more versatile than men, and may well be so in the very near future. So it is time to face the question: Where does that leave man?

I suspect that this is not a question of very great importance—except, of course, to man. Perhaps the Neanderthals made similar plaintive noises, around 100,000 B.C., when H. sapiens appeared on the scene, with his ugly vertical forehead and ridiculous protruding chin. Any Paleolithic philosopher who gave his colleagues the right answer would probably have ended up in the cooking pot; I am prepared to take that risk.

The short-term answer may indeed be cheerful rather than depressing. There may be a brief golden age when men will glory in the power and range of their new partners. Barring war, this age lies directly ahead of us. As Dr. Simon Remo put it: "The extension of the human intellect by electronics will become our greatest occupation within a decade." That is undoubtedly true, if we bear in mind that at a somewhat later date the word *extension* may be replaced by *extinction*.

One of the ways in which thinking machines will be able to help us is by taking over the the humbler tasks of life, leaving the human brain free to concentrate on higher things. (Not, of course, that this is any guarantee that it will do so.) For a few generations, perhaps, every man will go through life with an electronic companion, which may be no bigger than today's transistor radios. It will "grow up" with him from infancy, learning his habits, his business affairs, taking over all the minor chores like routine correspondence and income tax returns and engagements. On occasion it could even take its master's place, keeping appointments he preferred to miss, and then reporting back in as much detail as he desired. It could substitute for him over the telephone so completely that no one would be able to tell whether man or machine was speaking; a century from now, Turing's "game" may be an integral part of our social lives, with complications and possibilities that I leave to the imagination.

I do not know who first thought that machines might replace man; probably the physicist J. D. Bernal, who in 1929 published an extraordinary book of scientific predictions called *The World, the Flesh and the Devil*. In this slim and long-out-of-print volume (I sometimes wonder what the sixty-year-old Fellow of the Royal Society now thinks of his youthful indiscretion, if he even remembers it), Bernal decided that the numerous limitations of the human body could be overcome only by the use of mechanical attachments or substitutes—until, eventually, all that might be left of man's original organic body would be the brain.

This idea is already far more plausible than when Bernal advanced it, for we have seen the development of mechanical hearts, kidneys, lungs, and other organs and the wiring of electronic devices directly into the human nervous system—all from the forties through the sixties.

In a crude way—yet one that may accurately foreshadow the future—we have already extended our visual and tactile senses away from our bodies. The men who work with radio isotopes, handling them with remotely controlled mechanical fingers and observing them by television, have achieved a partial separation between brain and sense organs. They are in one place; their minds effectively in another.

Recently the word *cyborg* (cybernetic organism) has been coined to describe the machine-animal of the type we have been discussing. Drs. Manfred Clynes and Nathan Kline of Rockland State Hospital, Orangeburg, New York, who invented the name, define a cyborg in these stirring words: "An exogenously extended organizational complex functioning as a homeostatic system." To translate, this means a body that has machines hitched to it, or built into it, to take over or modify some of its functions.

I suppose I could call a man in an iron lung a cyborg, but the concept has far wider implications than this. One day we may be able to enter into temporary unions with any sufficiently sophisticated machines, thus being able not merely to control but to become a spaceship or submarine or a television network. This would give far more than purely intellectual satisfaction; the thrill that can be obtained from driving a racing car or flying an airplane may be only a pale ghost of the excitement our great-grandchildren may know, when the individual human consciousness is free to roam at will from machine to machine, through all the reaches of sea and sky and space.

But how long will this partnership last? Can the synthesis of man and machine ever be stable, or will the purely organic component become such a hindrance that it has to be discarded? If this eventually happens—and I have given good reasons for thinking that it must—we have nothing to regret, and certainly nothing to fear.

The popular idea, fostered by comic strips and the cheaper forms of science fiction, that intelligent machines must be malevolent entities hostile to man is so absurd that it is hardly worth wasting energy to refute it. I am almost tempted to argue that only *un*intelligent machines can be malevolent; anyone who has tried to start an outboard motor will probably agree. Those who picture machines as active enemies are merely projecting their own aggressive instincts, inherited from the jungle, into a world where such things do not exist. The higher the intelligence, the greater the degree of cooperativeness. If there is ever a war between men and machines, it is easy to guess who will start it.

Yet however friendly and helpful the machines of the future may be, most people will feel that it is a rather bleak prospect for humanity if it ends up as a pampered specimen in some biological museum—even if that museum is the whole planet Earth. This, however, is an attitude I find impossible to share.

No individual exists forever; why should we expect our species to be immortal? Man, said Nietzsche, is a rope stretched between the animal and the superhuman—a rope across the abyss. That will be a noble purpose to have served.

Space and the

Spirit of Man

Now that the Space Age is here, it is time to ask if predictions of a cultural revival can be justified, as the following essay points out.

That the world is now space-conscious, to an extent that would have seemed unbelievable only a few years ago, is a statement that needs no proof. But it is not yet space-minded. By this, I mean that the general public still thinks of space activities almost exclusively in terms of military strength and international prestige. These matters are, of course, vitally important; yet in the long run, if there is a long run, they will be merely the ephemeral concerns of our neurotic age. In the sane society that we have to build if we are to survive, we must forget spacemanship and concentrate on space.

Unfortunately, altogether too many educators, intellectuals, and other molders of public opinion still regard space as a terrifying vacuum, instead of a frontier with infinite possibilities. Typical of this attitude, though seldom so clearly expressed, is the following passage from Prof. Lewis Mumford's *The Transformation of Man:*

"Posthistoric man's starvation of life would reach its culminating point in interplanetary travel. . . . Under such conditions, life would again narrow down to the physiological functions of breathing, eating, and excretion. . . . By comparison, the Egyptian cult of the dead was overflowing with vitality; from a mummy in his tomb one can still gather more of the attributes of all human beings than from a spaceman."

The almost laughable falsehood of this passage was demonstrated by Cmdr. Alan Shepard's famous exclamation, "What a beautiful sight!" as his *Mercury* capsule arced over the Caribbean. I would maintain that these words are enough to settle the matter, but it must be admitted that most people would prefer more substantial evidence for the benefits of manned spaceflight.

Let me first dispose of one argument for man in space that is frequently put forward, and which only confuses the issue. It is often suggested that the complexity and unreliability of automatic space probes will make it impossible to dispense with human astronauts, even if they merely serve as troubleshooters. This is a shortsighted view; in the not-too-distant future— perhaps only fifty years from now—we will have robots as good as any flesh-and-blood explorers. The frequent and possible failures of the next decade's automatic astronauts must not blind us to the fact that they will be only clumsy, moronic toys compared with their successors half a century hence. The justification of man in space must depend not upon the deficiencies of his machines, but upon the positive advantages that he, personally, will gain from going there.

There is no point in exploring—still less colonizing—a hostile and dangerous environment unless it opens up new opportunities for experience and spiritual enrichment. Mere survival is not sufficient; there are already enough examples on this planet of societies that have been beaten down to subsistence level by the forces of nature. The questions that all protagonists of spaceflight have to ask themselves, and answer to their own satisfaction, are these: What can the other planets offer that we cannot find here on Earth? Can we do better on Mars or Venus than the Eskimos have done in the Arctic? And the Eskimos, it is worth reminding ourselves, have done very well indeed; a dispassionate observer might reasonably decide that they are the only truly civilized people on this planet.

The possible advantages of space can best be appreciated if we turn our backs upon it and return, in imagination, to the sea. Here is the perfect environment for life—the place where it originally evolved. In the sea, an all-pervading fluid medium carries oxygen and food to every organism; it need never hunt for either. The same medium neutralizes gravity, insures against temperature extremes, and prevents damage by too intense solar radiation—which must have been lethal at the Earth's surface before the ozone layer was formed.

When we consider these facts, it seems incredible that life ever left the sea, for in some ways the dry land is almost as dangerous as space. Because we are accustomed to it, we forget the price we have had to pay in our daily battle against gravity. We seldom stop to think that we are still creatures of the sea, able to leave it only because, from birth to death, we wear the water-filled space suits of our skins.

Yet until life had invaded and conquered the land, it was trapped in an evolutionary cul-de-sac—for intelligence cannot arise in the sea. The relative opacity of water, and its resistance to movement, were perhaps the chief factors limiting the mental progress of marine creatures. They had little incentive to develop keen vision (the most subtle of the senses, and the only long-range one) or manual dexterity. It will be most interesting to see if there are any exceptions to this, elsewhere in the universe.

Even if these obstacles do not prevent a low order of intelligence from arising in the sea, the road to further development is blocked by an impossible barrier. The difference between man and animals lies not in the possession of tools, but in the possession of fire. A marine culture could not escape from the Stone Age and discover the use of metals; indeed, almost all branches of science and technology would be forever barred to it.

Perhaps we would have been happier had we remained in the sea (the porpoises seem glad enough to have returned, after sampling the delights of the dry land for a few million years), but I do not think that even the most cynical philosopher has ever suggested we took the wrong road. The world beneath the waves is beautiful, but it is hopelessly limited, and the creatures who live there are crippled irremediably in mind and spirit. No fish can see the stars; but we will never be content until we have reached them.

There is one point, and a very important one, at which the evolutionary parallel breaks down. Life adapted itself to the land by unconscious, biological means, whereas the adaptation to space is conscious and deliberate, made not through biological but through engineering techniques of infinitely greater flexibility and power. At least, we think it is conscious and deliberate, but it is often hard to avoid the feeling that we are in the grip of some mysterious force or zeitgeist that is driving us out to the planets, whether we wish to go or not.

Though the analogy is obvious, it cannot be proved, at this moment of time, that expansion into space will produce a quantum jump in our development as great as that which took place when our ancestors left the sea. From the nature of things, we cannot predict the new forces, powers, and discoveries that will be disclosed to us when we reach the other planets or can set up new laboratories in space. They are as much beyond our vision today as fire or electricity would be beyond the imagination of a fish.

Yet no one can doubt that the increasing flow of knowledge and sense impressions, and the wholly new types of experience and emotion, that will result from space travel will have a profoundly stimulating effect upon the human psyche. I have already referred to our age as a neurotic one; the "sick" jokes, the decadence of art forms, the flood of anxious self-improvement books, the etiolated cadavers posing in the fashion magazines—these are minor symptoms of a malaise that has gripped at least the Western world, where it sometimes seems that we have reached fin de siècle way ahead of the calendar.

The opening of the space frontier will change all that, as the opening of any frontier must do. It has saved us, perhaps in the nick of time, by providing an outlet for dangerously stifled energies. In William James's famous phrase, it is the perfect "moral equivalent of war."

From time to time, alarm has been expressed at the danger of a "sensory deprivation" in space. Astronauts on long journeys, it has been suggested, will suffer the symptoms that afflict men who are cut off from their environment by being shut up in darkened, soundproofed rooms.

I would reverse this argument; our culture will suffer from sensory deprivation if it does not go out into space. There is striking evidence of this in what has already happened to the astronomers and physicists. As soon as they were able to rise above the atmosphere, a new and often surprising universe was opened up to them, far richer and more complex than had ever been suspected from ground observations. Even the most enthusiastic proponents of space research never imagined just how valuable satellites would actually turn out to be, and there is a profound symbolism in this.

But the facts and statistics of science, priceless as they are, tell only a part of the story. Across the seas of space lie the new raw materials of the imagination, without which all forms of art must eventually sicken and die. Strangeness, wonder, mystery, and magic—these things, which not long ago seemed lost forever, will soon return to the world. And with them, perhaps will come again an age of sagas and epics such as Homer never knew.

Though we may welcome this, we may not enjoy it, for it is never easy to live in an age of transition—indeed, of revolution. As the old Chinese curse has it: "May you live in interesting times," and the twentieth century is probably the most "interesting" period mankind has ever known. The psychological stresses and strains produced by astronautics—upon the travelers and those who stay at home—will often be unpleasant, even though the ultimate outcome will be beneficial to the race as a whole.

The American public has already experienced some emotional highs and lows that give a slight foretaste of what is to come. To date, the extremes are well represented by the explosion of the first Vanguard, and the success of the first manned suborbital shot, when the whole nation stopped its work and play to watch Cape Canaveral. But these are only pale shadows of such future triumphs as Moon landings—or the impact of a Nova-class vehicle on Miami Beach.

We now take it for granted that our planet is a tiny world in a remote corner of an infinite universe and have forgotten how this discovery shattered the calm certainties of medieval faith. Even the echoes of the second great scientific revolution are swiftly fading; today, except in a few backward regions, the theory of evolution arouses as little controversy as the statement that the Earth revolves around the Sun. Yet it is only one hundred years since the best minds of the Victorian age tore themselves asunder because they could not face the facts of biology.

Space will, sooner or later, present us with facts that are much more stubborn, and even more disconcerting. There can be little reasonable doubt that, ultimately, we will come into contact with races more intelligent than our own. That contact may be one-way, through the discovery of ruins or artifacts; it may be two-way, over radio or laser circuits; it may even be face-to-face. But it will occur, and it may be the most devastating event in the history of mankind. The rash assertion that "God made man in His own image" is ticking like a time bomb at the foundation of many faiths, and as the hierarchy of the universe is disclosed to us, we may have to recognize this chilling truth: if there are any gods whose chief concern is man, they cannot be very important gods.

Perhaps if we knew all that lay ahead of us on the road to space—a hundred or a thousand or a million years in the future—no man alive would have the courage to make the first step. But that first step—and the second—has already been taken; to turn back now would be treason to the human spirit, even though our feet must someday carry us into realms no longer human.

The eyes of all ages are upon us now, as we create the myths of the future at Cape Canaveral in Florida and Baikonur in Kazakhstan. No other generation has been given such powers, and such responsibilities. The impartial agents of our destiny stand on their launching pads, awaiting our commands. They can take us to that greater renaissance whose signs and portents we can already see, or they can make us one with the dinosaurs.

The choice is ours, it must be made soon, and it is irrevocable. If our wisdom fails to match our science, we will have no second chance. For there will be no one to carry our dreams across another Dark Age, when the dust of all our cities incarnadines the sunsets of the world.

The Uses of

the Moon

Clarke writes that it is essential that the importance of the Moon in man's future
is understood. If man does not understand, Clarke postulates, then he will have
gone there for the wrong reasons and will not know what to
do with his newfound knowledge.

Many people imagined that the whole project of lunar exploration was merely a race with the Russians—a contest in conspicuous consumption of brains and material, designed to impress the remainder of mankind. No one can deny the strong element of competition and national prestige involved, but in the long run, this was the least important aspect of the matter. If the race to the Moon were nothing more than a race, it would have made good sense to let the Russians bankrupt themselves in the strain of winning it, in the calm confidence that their efforts would collapse in recriminations and purges sometime during the 1970s. [Their efforts did collapse, and their country went broke—but it took until the nineties for that to happen.]

The Moon is a barren, airless wasteland, blasted by intolerable radiations. Yet a century from now it may be an asset more valuable than the wheatfields of Kansas or the oil wells of Oklahoma. And an asset in terms of actual hard cash—not the vast imponderables of adventure, romance, artistic inspiration, and scientific knowledge. Though, ultimately, these are the only things of real value, they can never be measured. The conquest of the Moon, however, can be justified to the cost accountants, not only to the scientists and the poets.

Let me first demolish, with considerable pleasure, one common argument for going to the Moon—the military one. Some ballistic generals have maintained that the Moon is "high ground" that could be used for reconnaissance and bombardment of the Earth. Though I hesitate to say that this is complete nonsense, it is as near to it as makes very little practical difference.

You cannot hope to see as much from 250,000 miles away as from a television satellite just above the atmosphere, and the use of the Moon as a launching site makes even less sense. For the effort required to set up one lunar military base with all its supporting facilities, at least one hundred times as many bases could be established on Earth. Also it would be far easier to intercept a missile coming from the Moon, taking many hours for the trip in full view of telescopes and radar, than one sneaking around the curve of the Earth in twenty minutes. Only if, which heaven forbid, we extend our present tribal conflicts to the other planets will the Moon become of military importance.

Before we discuss the civilized uses of our one natural satellite, let us summarize the main facts about it. They may be set down quite briefly:

The Moon is a world a quarter the diameter of Earth, its radius being just over one thousand miles. Thus its area is one-sixteenth of our planet's—more than that of Africa, and almost as much as that of both the Americas combined. Such an amount of territory is not to be despised; it will take many years (and many miles) to explore it in detail.

The amount of material in the Moon is also impressive; if you would like it in tons, the figure comes to 750,000,000,000,000,000,000,000, which is millions of times more than all the coal, iron, minerals, and ores that man has shifted in the whole of history. It is not enough mass, however, to give the Moon much of a gravitational pull; as everyone now knows, a visitor to the Moon has only a fraction (actually one-sixth) of his terrestrial weight.

The low gravity has several consequences, almost all of them good. The most important is that the Moon has been unable to retain an atmosphere; if it ever had one, it long ago escaped from the Moon's feeble clutch and leaked off into space. For all practical purposes, therefore, the lunar surface is in a perfect vacuum. In a moment, we will see why this is an advantage.

Because there is no atmosphere to weaken the Sun's rays, or to act as a reservoir of heat during the nighttime, the Moon is a world of very great temperature extremes. On our Earth, in any one spot, the thermometer seldom ranges over as much as one hundred degrees even during the course of a year. Though the temperature can exceed 100 degrees Fahrenheit in the tropics and drop to 125 degrees below in the Antarctic, these figures are quite exceptional. But every point on the Moon undergoes twice this range in a lunar day; indeed an explorer could encounter such changes within seconds, merely stepping from sunlight into shadow or vice versa.

This obviously presents problems, but the very absence of atmosphere that causes such extremes also makes it easy to deal with them—for a vacuum is one of the best possible heat insulators, a fact familiar to anyone who has ever taken hot drinks on a picnic.

No air means no weather. It is hard for us, accustomed to wind and rain, cloud and fog, hail and snow, to imagine the complete absence of all these things. None of the meteorological variations that make life interesting, unpredictable, and occasionally impossible on the surface of the planet takes place on the Moon: the only change that ever occurs is the regular, utterly unvarying cycle of day and night. Such a situation may be monotonous, but it simplifies, to an unbelievable extent, the problems facing architects, engineers, explorers, and indeed everyone who will ever conduct operations of any kind on the Moon.

The Moon turns rather slowly on its axis, so that its day (and its night) is almost thirty times longer than ours. As a result, the sharp-edged frontier between night and day, which moves at one thousand miles an hour on the Earth's equator, has a maximum speed of less than ten miles an hour on the Moon. In high lunar latitudes, a walking man could keep in perpetual daylight with little exertion. And because the Moon turns on its axis in the same time as it revolves around the Earth, it always keeps the same hemisphere turned toward us. Until the advent of *Lunik III,* this was extremely frustrating to astronomers; in another generation, as we shall see, they will be very thankful for it.

So much for the main facts; now for a few assumptions that most people would have accepted as reasonable in 1961, though they would have laughed at them before 1957.

The first is that suitably protected men can work and carry out engineering operations on the face of the Moon, either directly or by remote control through robots.

The second is that the Moon consists of the same elements as the Earth, though doubtless in different proportions and combinations. Most of our familiar minerals will be missing; there will be no coal or limestone, since these are the products of life. But there will be carbon, hydrogen, oxygen, and calcium in other forms, and we can evolve a technology to extract them from whatever sources are available. It is even possible that there may be large quantities of free (though frozen) water not too far below the Moon's surface; if this is the case, one of the chief problems of the lunar colonists will be solved.

In any event, without going into mining details, ore processing, and chemical engineering, it will be possible to obtain all the materials needed for maintaining life. The first pioneers will be content with mere survival, but at a later stage they will build up a self-supporting industry based almost entirely on lunar resources. Only instruments, specialized equipment, and men will come from Earth; the Moon will supply all the rest—and ultimately, of course, the men.

Now for the reasons why it is worth the expense, risk, and difficulty of prevailing on the inhospitable Moon. They are implicit in the question, What can the Moon offer that we cannot find on Earth?

One immediate but paradoxical answer is nothing—millions of cubic miles of it. Many of the key industries in the modern world are based on vacuum techniques; electric lighting and its offspring radio and electronics could never have begun without the vacuum tube, and the invention of the transistor has done little to diminish its importance. (The initial steps of transistor manufacture have themselves to be carried out in a vacuum.) A great many metallurgical and chemical processes, and key stages in the production of drugs like penicillin, are possible only in a partial or virtually complete vacuum, and it's impossible to make a very large one.

On the Moon there will be a hard vacuum of unlimited extent outside the door of every airlock. I do not suggest that it will be worthwhile switching much terrestrial industry to the Moon, even if the freight charges allowed it. But the whole history of science makes it certain that new processes and discoveries of fundamental importance will evolve as soon as men start to carry out operations in the lunar vacuum. Low-pressure physics and technology will proceed from rags to riches overnight; industries that today are unimagined will spring up on the Moon and ship their products back to Earth. For in that direction the freight charges will be relatively low.

And this leads to a major role that the Moon will play in the development of the Solar System: it is no exaggeration to say that this little world, so small and close at hand (the very first rocket to reach it took only thirty-five hours on the journey), will be the stepping-stone to all the planets. The reason for this is its low gravity; it requires twenty times as much energy to escape from the Earth as from the Moon. As a supply base for all interplanetary operations, therefore, the Moon has an enormous advantage over the Earth—assuming, of course, that we can find the kind of materials we need there. This is one of the reasons why the development of lunar technology and industry is so important.

From the gravitational point of view, the Moon is indeed high ground, while we on the Earth are like dwellers at the bottom of an immensely deep pit out of which we have to climb every time we wish to conduct any cosmic explorations. No wonder that we must burn one hundred tons of rocket fuel for every ton of payload we launch into space—and on a one-way trip at that. For return journeys, thousands of tons would be needed.

This is why all Earth-based plans for space travel are so hopelessly uneconomic, involving gaint boosters with tiny payloads. It is as if, in order to carry a dozen passengers across the Atlantic, we had to construct a ship weighing as much as the *Queen Elizabeth* but costing very much more. (The development costs for a large space vehicle are several billion dollars.) And, to make the whole thing completely fantastic, the vehicle can be used only once, for it will be destroyed in flight. Of the tens of thousands of tons that leave the Earth, only a small capsule will return. The rest will consist of boosters dropped in the ocean or discarded in space.

The big breakthrough toward really efficient space operations may depend upon the fortunate fact that the Moon has no atmosphere. The peculiar (by our standards—they are normal by those of the universe) conditions prevailing there permit a launching technique much more economical than rocket propulsion. This is the old idea of the "space gun," made famous by Jules Verne almost one hundred years ago.

It would probably not be a gun in the literal sense, powered by chemical explosives, but a horizontal launching track like those used on aircraft carriers, along which space vehicles could be accelerated electrically until they reached sufficient speed to escape from the Moon. It is easy to see why such a device is completely impractical on Earth, but might be of enormous value on the Moon.

To escape from the Earth, a body must reach the now familiar speed of 25,000 mph. At the fierce acceleration of ten gravities, which astronauts have already withstood for very short periods of time, it would take two minutes to attain this speed—and the launching track would have to be four hundred miles long. If the acceleration were halved to make it more endurable, the length of the track would have to be doubled. And, of course, any object traveling at such a speed in the lower atmosphere would be instantly burned up by friction. We can forget all about space guns on Earth.

The situation is completely different on the Moon. Because of the almost perfect vacuum, the lunar escape speed of a mere 5,200 mph can be achieved at the ground level without any danger from air resistance. And at an acceleration of ten gravities, the launching track need be only nineteen miles long—not four hundred, as on the Earth. It would be a massive piece of engineering, but a perfectly practical one, and it would wholly transform the economics of spaceflight.

Vehicles could leave the Moon without burning any fuel at all; all the work of takeoff would be done by fixed power plants on the ground, which would be as large and massive as required. The only fuel that a space vehicle returning to Earth need carry would be a very small amount for maneuvering and navigating. As a result, the size of the vehicle needed for a mission from Moon to Earth would be reduced tenfold; a hundred-ton spaceship could do what had previously required a thousand-tonner.

This would be a spectacular enough improvement; the next stage, however, would be the really decisive one. This is the use of a Moon-based launcher or catapult to place supplies or fuel where they are needed, in orbit around the Earth or indeed any other planet in the Solar System.

It is generally agreed that long-range spaceflight—particularly voyages beyond the Moon—will become possible only when we can refuel our vehicles in orbit. Plans have been drawn up in great detail for operations involving fleets of tanker rockets, which perhaps over a period of years could establish what are virtually filling stations in space. Such plans will, of course, be fantasically expensive, for it requires about fifty tons of rocket fuel to put a single ton of payload into orbit around the Earth, only a couple of hundred miles up.

Yet a Moon-based launcher could do the same job—from a distance of 250,000 miles!—for a twentieth of the energy and without consuming any rocket fuel whatsoever. It would launch tanks of propellants toward Earth, and suitable guidance systems would steer them into stable orbit where they would swing around endlessly until required. This would have as great an effect on the logistics of spaceflight as the dropping of supplies by air has already had upon polar exploration; indeed, the parallel is a very close one.

Though enormous amounts of power would be required to operate such lunar catapults, this will be no problem in the twenty-first century. A single hydrogen bomb, weighing only a few tons, liberates enough energy to lift 100 million tons completely away from the Moon. That energy will be available for useful purposes when our grandchildren need it; if it is not, we will have no grandchildren.

There is one other application of the lunar catapult that may be very important, though it may seem even more far-fetched at the present time. It could launch the products of the Moon's technology down to the surface of the Earth. A freight-carrying capsule, like a more refined version of nose cones and reentry vehicles, could be projected from the Moon to make an automatic landing on the Earth at any assigned spot. Once again, no rocket fuel would be needed for the trip, except a few pounds for maneuvering. All the energy of launching would be provided by the fixed power plant on the Moon; all the slowing down would be done by the Earth's atmosphere. When such a system is perfected, it may be no more expensive to ship freight from Moon to Earth than it is now to fly it from one continent to another by jet. Moreover, the launching catapult could be quite short, since it would not have to deal with fragile human passengers. If it operated at fifty gravities' acceleration, a four-mile-long track would be sufficient.

I have discussed this idea at some length for two reasons. The first is that it demonstrates how, by taking advantage of the Moon's low gravity, its airlessness, and the raw materials that certainly are there, we can conduct space exploration far more economically than by basing our operations on Earth. In fact, until some revolutionary new method of propulsion is invented, it is hard to see any other way in which space travel will be practical on the large scale.

The second reason is the slightly more personal one that, to the best of my knowledge, I was the first to develop this idea, in a 1950 issue of the *Journal of the British Interplanetary Society*. Five years earlier I had proposed the use of satellites for radio and television communications; I did not expect to see either event materialize in my lifetime, but one has already happened and now I wonder if I may see both.

The subject of communications leads us to another extremely important use of the Moon. As civilization spreads throughout the Solar System, the Moon will provide the main link between Earth and her scattered children. For though it is just as far to the other planets from the Moon as from the Earth, sheer distance is not the only factor involved. The Moon's surface is already in space, while the surface of the Earth—luckily for us—is shielded from space by a whole series of barriers through which we have to drive our signals.

The best known of these barriers—and this has been realized only during the past year—is the atmosphere itself. Thanks to the development of an extraordinary optical device called the laser, which produces an intense beam of almost perfectly parallel light, it now appears that the best agent for long-distance communications is not radio, but light. A light beam can carry millions of times as many messages as a radio wave and can be focused with infinitely greater accuracy. Indeed, a laser-produced light beam could produce a spot on the Moon only a few hundred feet across, where the beam from a searchlight would be thousands of miles in diameter. Thus colossal ranges could be obtained with very little power; calculations show that with lasers we can think of signaling to the stars, not merely to the planets.

But we cannot use light beams to send messages through the Earth's erratic atmosphere; a passing cloud could block a signal that had traveled across a billion miles of space. On the airless Moon, however, this would be no problem, for the sky is perpetually clear to waves of all frequencies, from the longest radio waves, through visible light, past the ultraviolet, and even down to the short X rays that are blocked by a few inches of air. This whole immense range of electromagnetic waves will be available for communications or any other use—perhaps such applications as the broadcasting of power, which has never been practical on Earth. There will be enough "bandwidth" or ether space for all the radio and television services we can ever imagine, no matter how densely populated the planets become and however many messages the men of the future wish to flash back and forth across the Solar System.

We can thus imagine the Moon as a central clearinghouse for interplanetary communications, aiming its tightly focused light beams to the other planets and to ships in space. Any messages that concerned Earth would be radioed across the trivial 250,000-mile gulf on those wavelengths that penetrate our atmosphere.

There are several other reasons why the Moon might almost have been designed as a base for interplanetary communications. Everyone is now familiar with the enormous radio telescopes that have been built to reach out into space and to maintain contact with such distant probes as our Pioneers and Explorers (and Rangers, Mariners, and Prospectors that will follow them). The most ambitious of these was the ill-fated six-hundred-foot giant at Sugar Grove, West Virgina—abandoned when partly built, after some scores of millions of dollars had been spent on it.

The six-hundred-foot telescope was an expensive failure because it was too heavy; the planned weight was about twenty thousand tons, but design changes later brought it up to thirty-six thousand tons. But on the Moon, both the costs and the weight of such a structure would be enormously reduced—perhaps by more than 90 percent. For thanks to the low gravity, a very much lighter construction could be used than is necessary on Earth. And the Moon's airlessness pays yet another dividend, for a terrestrial telescope has to be designed with a substantial safety factor so that it can withstand the worst that the weather can do. There is no need to worry about gales on the Moon; there is not the slightest breeze to disturb the most delicate structures.

Nor have we yet finished with the Moon's advantages from the view of those who want to send (and receive) signals across space. It turns so slowly on its axis that the problem of tracking is much simplified; *and it is a quiet place.*

Or, to be more accurate, the far side of the Moon is a quiet place—probably the quietest that now exists within millions of miles of the Earth. I am speaking, of course, in the radio sense; for the last sixty years, our planet has been pouring an ever-increasing racket into space. This has already seriously inconvenienced the radio astronomers, whose observations can be ruined by an electronic shaver one hundred miles away.

Clarke's official Royal Air Force portrait, 1943. (COURTESY OF ARTHUR C. CLARKE)

Robert Bloch (author of *Psycho*), Harlan Ellison, and Evelyn Gold with Clarke at Midwestcon in Indian Lake, Ohio, May 1952. (COURTESY OF DEAN A. GRENNEL)

Clarke in front of poster for
2001 at MGM, 1968.

(COURTESY OF ARTHUR C. CLARKE)

Clarke with Allen Ginsberg at
the Hotel Chelsea, 1968.
(COURTESY OF ARTHUR C. CLARKE)

Clarke with British author Angus Wilson,
January 1, 1970. (COURTESY OF ARTHUR C. CLARKE)

Clarke with Neil Armstrong,
June 13, 1970.

(COURTESY OF ARTHUR C. CLARKE)

Clarke with his
mother, Nora, and Aunt
Nellie at Minehead,
October 1, 1970.

(COURTESY OF ARTHUR C. CLARKE)

Clarke at the CBS anchor desk with Wally Schirra and Walter Cronkite during the Apollo 15 coverage, August 1971. (COURTESY OF ARTHUR C. CLARKE)

Clarke with Bucky Fuller and Glen Olds at the Design Science Institute, New York, 1973.

(COURTESY OF MICHAEL CRAVEN)

Clarke in Queensland,
Australia,
March 16, 1974.
(COURTESY OF ARTHUR C. CLARKE)

In 1976, the Indian government gave Clarke his first satellite dish. (COURTESY OF ARTHUR C. CLARKE)

Clarke with Stanley and Christiane Kubrick, April 19, 1976. (COURTESY OF ARTHUR C. CLARKE)

Clarke with Kathy Keeton, the founder of *Omni*, after winning the Science Fiction Achievement Award for *Fountains of Paradise*, 1979. (COURTESY OF ARTHUR C. CLARKE)

Clarke with Scott Meredith at Ballifants, July 1981. (COURTESY OF ARTHUR C. CLARKE)

Clarke with Yuri Artsutanov, the inventor of the space elevator, in Leningrad, June 16, 1982.
(COURTESY OF ARTHUR C. CLARKE)

Clarke with cosmonauts Alekei Leonov and Vitaliy Sevastyanov in Star City, Russia, June 15, 1982.
(COURTESY OF ARTHUR C. CLARKE)

Clarke with
editor Vasili
Zaharchenko at
the offices of
*Tekhnika
Molodezhi* in
Moscow, June
1982. (COURTESY OF
ARTHUR C. CLARKE)

Clarke tries unsuccessfully to involve Isaac Asimov in a more adventurous lifestyle, New York, 1984.

(COURTESY OF ARTHUR C. CLARKE)

Receiving the Vidya Jyothi award from the president of Sri Lanka, J. R. Jayewardene, 1986. (COURTESY OF ARTHUR C. CLARKE)

Clarke with telescope, May 1987. (COURTESY OF ARTHUR C. CLARKE)

Clarke with Stephen Hawking and Magnus Magnusson at the BBC, 1988. (COURTESY OF ARTHUR C. CLARKE)

Clarke with Michael Collins and Buzz Aldrin at the Fifth Planetary
Congress of the Association of Space Explorers held in Riyadh,
Saudi Arabia, November 1989. (COURTESY OF ARTHUR C. CLARKE)

Clarke with astronaut Prince Sultan
in Riyadh, Saudi Arabia, 1989.

(COURTESY OF ARTHUR C. CLARKE)

Clarke with coauthor Gentry Lee in the 1990s. (COURTESY OF ARTHUR C. CLARKE)

Clarke and Hector Ekanayake after a hundred-foot dive off of the coast of Sri Lanka, March 1992. (COURTESY OF ARTHUR C. CLARKE)

Clarke and the Ekanayakes, his extended family, in 1996. *Left to right:* Tamara, Cherene, Melinda, Clarke, Valerie, and Hector. (COURTESY OF ARTHUR C. CLARKE)

Clarke with Jean-Michel Cousteau at Dive Conference, Singapore, 1996.
(COURTESY OF ARTHUR C. CLARKE)

Clarke with Roddy McDowall in Sigiriya, Sri Lanka, 1997.

(COURTESY OF ARTHUR C. CLARKE)

Clarke, president of Sri Lanka Sirimavo Bandaranaike, and Prince Charles on a state visit to Sri Lanka, February 4, 1998.

(COURTESY OF ARTHUR C. CLARKE)

But the land first glimpsed by *Lunik III* is beyond the reach of this electronic tumult; it is shielded from the din of Earth by two thousand miles of solid rock—a far better protection than a million miles of empty space. Here, where the earthlight never shines, will be the communications centers of the future, linking together with radio and light beams all the inhabited planets. And one day, perhaps, reaching out beyond the Solar System to make contact with those other intelligences for whom the first search has already begun. That search can hardly hope for success until we have escaped from the braying of all the radio and television stations of our own planet.

In a recent discussion of space exploration plans, Prof. Harold Urey made the point that the Moon is one of the most interesting places in the Solar System—perhaps more so than Mars or Venus, even though there may be life on these planets. For the face of the Moon may have carried down through the ages, virtually untouched by time, a record of the conditions that existed billions of years ago, when the universe itself was young. On Earth all such records have long been erased by the winds and the rains and other geological forces. When we reach the Moon, it will be as if an entire library of lost volumes, a million times older than that destroyed at Alexandria, was suddenly thrown open to us.

Quite beyond price will be the skills we will acquire during the exploration—and ultimately, colonization—of this new land in the sky. I suspect that on the Moon we will learn more within a few years about unorthodox methods of food production than we could in decades on the Earth. Can we, in an almost literal sense of the phrase, turn rocks into food? We must master this art (as the plants did, eons ago) if we hope to conquer space. Perhaps most exciting of all are the possibilities opened up by low-gravity medicine and the enormous question "Will men live longer on a world where they do not wear out their hearts fighting against gravity?" Upon the answer to this will depend the future of many worlds, and of nations yet unnamed.

Much of politics, as of life, consists of the administration of the unforeseen. We can foresee only a minute fraction of the Moon's potentialities, and the Moon itself is only a tiny part of the universe. The fact that the Soviet Union made an all-out effort to get there has far deeper implications than was generally faced.

We may no longer be concerned with such trivialities. We realize that if any nation has mastery of the Moon, it will determine not merely the fate of the Earth, but the whole accessible universe.

The Playing

Fields of Space

Space travel will not be all work and no play, Clarke writes. Wherever men go, they must have relaxation, physical and mental. Though it may seem a little premature to spend much time discussing Space Age sport, the subject turns out to be highly instructive. It is also full of surprises, for the new gravitational and physical conditions beyond the Earth will not only transform many existing sports, but make possible new ones.

There will be very little space in the first spaceships; it will all be outside the walls, and everyone will be most anxious to keep it there. During their off-duty hours pioneer astronauts will have to relax with cards and chess and perhaps video games. Not until fairly large bases are established on the Moon and planets, and in satellite orbits, will space sports really come into their own.

The Moon, thanks to movies and comic strips, already has a certain cozy familiarity. As it is an utterly airless world, the first explorers will have to wear space suits when they leave the shelter of their ships. These suits will be elaborate and bulky affairs, for they must not only supply the wearer with oxygen but must also provide protection from the fierce temperature extremes that exist on the Moon—where a single step from the sunlight into shadow may bring the thermometer tumbling four hundred degrees. No one wearing a space suit will feel very athletic; but sooner or later very large areas of lunar landscape will be enclosed—either by rigid domes or flexible, air-supported structures—and provided with artificial atmospheres.

When this happens, the colonists (they will no longer be pioneers) will be able to discard their clumsy pressurized suits and will be able to move about unhampered.

They will do so with a dreamlike ease that we on Earth may well envy but will never be able to emulate. For the gravitational tug of the Moon is only one-sixth as powerful as that of our planet; a man who weighs 180 pounds on Earth will weigh just under 30 on the Moon.

All objects thrown, tossed, shot, or otherwise projected on the Moon will travel six times as far and rise six times as high as they would on Earth. A high jump on the Moon would thus be a spectacular performance, though not quite so spectacular as you might think. The present terrestrial record is just over seven feet, but this does not mean that a lunar high jumper could do six times this, or forty-two feet. When an athlete clears a seven-foot bar, he actually hoists himself less than five feet; his center of gravity, which is around waist level, is already some three feet from the ground. Allowing for this, the high-jump record for the Moon will be around thirty feet, and the whole performance will take almost ten seconds. The broad-jump record, now about twenty-seven feet, would become more than one hundred and fifty feet on the Moon.

The various objects that athletes like to hurl will cover correspondingly greater distances on the Moon. There won't be enough space, unfortunately, in our pressurized lunar cities, for throwing the discus (the terrestrial record would correspond to a lunar 1,175 feet), the javelin (1,500 feet), or the hammer (1,280 feet). Even putting the 16-pound shot—which will become a 2.5-pound shot on the Moon—will strain the available accommodation. It will travel close to 375 feet.

There are certain games, however, that would be virtually unaffected by change of gravity. These are games that depend not upon weight, but upon mass or inertia. The two characteristics are very frequently confused, though they are really quite distinct.

The weight of a body depends entirely on the gravity field it happens to be occupying. That's why the same object can weigh a pound on the Earth, a sixth of a pound on the Moon, twenty-eight pounds on the Sun, and nothing at all in an orbiting satellite. But its mass—by which we mean the opposition it gives us when we try to set it moving—is absolutely independent of gravity and is the same throughout the universe. An object that would be weightless in orbit requires the same effort there as on Earth to set it in motion.

We can make a list, therefore, of the games that can be played on the Moon and planets without any alteration to the rules or equipment. They're all games that involve rolling, sliding, or bumping, but not throwing or projecting. Two famous examples are bowling and billiards; and I'm sure that Lewis Carroll, who described the most famous croquet match in all fiction, would have loved the idea of transferring this gentle game to the Moon. It would work fine there.

It hardly seems worthwhile going to the Moon to play croquet; however, there is one lunar sport that may someday become a major tourist attraction. On the Moon, inside the air-filled domes that the future colonists will erect, a man or a woman could fly like a bird. It would be relatively easy and would probably require little more than batlike wings attached to wrists and ankles. With these, we could enjoy during waking hours an experience we have known so far only in dreams.

Muscle-powered flight opens up a whole spectrum of sports and games, from straightforward racing to an aerial equivalent of water polo. I can see the time coming, not more than thirty years from today, when the television channels will be dominated by sportscasts from the Moon. The sluggish and leaden-footed sports of Earth will seem tame compared with those that could be played on the Moon.

Everything I have said about the Moon applies, in a slightly less exaggerated degree, to Mars. This is the only planet upon whose surface we may be able to venture without elaborate protection. Mars has a thin atmosphere (though not a breathable one), and during the daytime it is sometimes comfortably warm. We may be able to manage there with simple oxygen masks, though this is by no means certain. All the other planets are much too hot, or much too cold. (Even Venus, which looked promising a few years ago, now turns out to have a temperature of about six hundred degrees. This is rather a blow as I had hoped to go skin diving there.)

The gravity of Mars is about a third of the Earth's, or twice the Moon's. The lunar records I have quoted above can thus be divided by two to indicate what we may expect to achieve on Mars. It's a pity we cannot practice there first and work our way up to lunar standards. I am afraid a lot of people will break their necks on the Moon, attempting those high jumps and coming down headfirst.

The Moon and Mars, and the major satellites of the giant planets, are fairly large bodies with respectable gravities. But besides these there are thousands of pint-sized moons and asteroids (minor planets) in the Solar System. Some very peculiar things could happen on these.

Consider little Phobos, the inner moon of Mars. It is a chunk of rock about ten miles in diameter—at least we think it is rock, though a Russian scientist has recently given plausible reasons for believing that Phobos may be an artificial satellite put up by the Martians a few million years ago. In any event, its gravity pull is tiny; a simple calculation shows that a man standing on Phobos would weigh about four ounces.

This is practically, but not quite, the same as no weight at all. A stone would take a minute to fall sixty feet, instead of the two seconds it requires on Earth. Such a state of affairs is almost impossible for us to imagine, yet this is what would happen on a world like Phobos. Anyone attempting a really high jump would be in grave danger of never coming down. Indeed, it would be possible to jump clear off the world—to reach the velocity of escape by unaided muscle power. Our own planet's velocity of escape, so far achieved by various rocket launches and various space probes, is twenty-five thousand miles an hour. But on minuscule worlds like Phobos, speeds under twenty miles an hour would be enough to send a body out into space forever.

Not many conventional games and sports are possible with such microscopic gravities. Objects hit or pushed out in the open would dwindle away, swiftly or slowly, on almost flat trajectories, and would never be seen again, except in one peculiar case. Since it is not too difficult to jump off such worlds, it is even easier to establish a satellite orbit around them. In the case of Phobos, the necessary speed is about ten miles an hour. So if you tossed a stone horizontally at exactly the right speed, it would become a moon of Phobos—a satellite of a satellite. Two or three hours later, if your aim had been perfect, it would have hit you in the back of the neck.

These examples are enough to prove that anyone who attempts to organize an interplanetary Olympics is going to be in real trouble. A few decades from now, all records listed will have to specify the planet of origin. Some athletic activities may even have to be laughed out of court; there would be little point in weight lifting on a world where a ninety-seven-ounce weakling could support five tons.

So far we have spoken of sports on worlds that have some gravity, even if it is only a thousandth of the Earth's. But what about the situation in spaceships or space stations, where the very conception of weight—but not mass—is meaningless?

Flying, of course, would be not only easy but unavoidable; it would be the only way of moving around. When we build satellites with really large interior spaces, aerobatics could become an exhilarating recreation, combining the characteristics of ballet and high diving. Zero-gravity athletics is a vast, utterly unknown territory waiting to be explored.

The behavior of liquids in the complete absence of gravity also opens up some interesting possibilities, which belong to the realm of art as much as sport. A weightless liquid forms itself into a sphere under the influence of surface tension. If set vibrating, it will oscillate through all sorts of peculiar shapes. When the oscillations become too large, it may develop a wasp waist and finally fission into smaller drops.

Some years ago, I suggested that one of the attractions of a space hotel (and we will have such things in the next century) might be a spherical swimming pool. It might even be a hollow sphere, with an air space inside where spectators might watch their friends swimming around them. I pass lightly over the technical problems—such as that of anchoring the pool in one place, or preventing it from being dispersed by the splashing of the swimmers.

There may be other factors in space besides gravity (or lack of it) that will affect physical activities and may give rise to new types of sport. It is difficult for us to conceive the possibilities. A dweller in the Sahara, unfamiliar with mountains or beaches, could hardly have imagined ski jumping and surf riding. In the same way we earthlings cannot guess what wholly novel recreations our grandchildren may invent to take advantage of peculiar conditions on the other planets.

Few people realize that a great wind continually blows outward from the Sun. It is a wind of light, and it exerts a definite pressure. This radiation pressure is, for ordinary purposes, negligible. Tiny though this force may be, it could add up to an appreciable amount over the surface of a huge, gossamer-thin sail of some reflecting material like aluminum foil or a Mylar film coating with silver. And when I say huge, I mean exactly that; the sails would have to be thousands of feet across to be of any use. Even so, by using delicate rigging, their mass need be only a few hundred pounds. Once conventional rockets had carried them up into orbit, they would be employed to tow cargoes across space.

Though the acceleration produced by such a solar sail is tiny, it would be maintained hour after hour, week after week, and could eventually build up to respectable velocities. The beauty of the system is its utter simplicity, and above all, the fact that power is free and everlasting. Even if it never has any serious applications, it suggests a beautiful and fascinating sport.

Someday, space yachtsmen will be tacking around the orbit of Mercury, racing tiny one-man vehicles not much larger than the capsules that the astronauts of the sixties rode. Billowing ahead of them will be vast, glittering surfaces, possibly miles across—flexible mirrors little thicker than soap bubbles, reefed and furled by a spiderweb of invisibly fine threads. The skippers of these fantastic little crafts would need a superb knowledge of astronautics and orbital theory, as well as skills that could not be learned in any classroom. There are many links between sea and space; here, surely, is one of the strangest. Across the centuries the spirit of the men who once sailed the windjammers around the Horn may live again as their descendants ride the eternal trade wind between the worlds.

The haunting vision of the these fragile space yachts, literally riding on sunbeams, is sufficient answer to those who think that interplanetary flight will be all cold science and massive engineering. Of course, we shall need that kind of technology to take us to the planets and to build new civilizations there, but this represents only part of life. Our picture of space is not complete if we think of it only in terms of power and knowledge; for it is also a playground whose infinite possibilities we shall not exhaust in all the ages that lie ahead.

Kalinga

Prize Speech

Following is the speech Clarke made in New Delhi when receiving the 1962 Kalinga
Prize from Rene Maheu, acting director general of Unesco. The Kalinga Prize,
awarded for science writing, is a donation of one thousand pounds annually
by the Indian industrialist and statesman B. Patnaik, and administered
by Unesco. Other winners have been Bertrand Russell, George Gamow,
Louis de Broglie, Julian Huxley, and Gerard Piel.

"I am proud to receive the Kalinga Prize, an honor which I have coveted ever since it was founded." Those words were spoken last year by my distinguished colleague and compatriot Prof. Ritchie Calder, and they express my own sentiments so perfectly that I cannot do better than repeat them.

I would also like to thank the generous donor of the prize, Mr. Patnaik, and the Unesco officials who have organized this meeting. It is my hope that, as the years pass, the great importance of this award will become universally recognized, and its fame even more widespread.

In addition to the pride I personally feel on receiving the Kalinga Prize, I would like to think that it is a tribute to the field of literature in which I have specialized—science fiction. Although at least four of the earlier prizewinners have written some science fiction, it has been only a minute and incidental portion of their output. I can claim that it is a major part of mine, for I have published just about as much fiction as nonfiction.

Many scientists, I am sorry to say, still look down on science fiction and lose no opportunity of criticizing it. For example, they often point out that 90 percent of all science fiction is rubbish—ignoring the fact that 90 percent of all fiction is rubbish. Indeed, I would claim that the percentage of competent writing in the science fiction field is probably higher than in any other. This is because much of it is a labor of love, written by enthusiasts who have considerable scientific knowledge and who are often themselves practicing scientists.

What role does science fiction actually play in the popularization of science? Though it often serves to impart information, I think its chief value is *inspirational* rather than educational. How many young people have had the wonders of the universe first opened up to them, or have been turned to a scientific career, by the novels of Verne and Wells? Many distinguished scientists have paid tribute to the influence of these great masters, and a careful survey would, I believe, reveal that science fiction is a major factor in launching many youngsters on a scientific career.

It is obvious that science fiction should be technically accurate, and there is no excuse for erroneous information when the true facts are available. Yet accuracy should not be too much of a fetish, for it is often the spirit rather than the letter that counts. Thus Verne's *From the Earth to the Moon* and *A Journey to the Center of the Earth* are still enjoyable, not only because Verne was a first-rate storyteller, but because he was imbued with the excitement of science and could communicate this to his readers. That many of his "facts" and most of his theories are now known to be incorrect is not a fatal flaw, for his books still arouse the sense of wonder.

It is this sense of wonder that motivates all true scientists, and all true artists. We encounter it in the writings of such scientific expositors as Fabre, Flammarion, Jeans, Rachel Carson, Loren Eisley, as well as many of my precursors at this function; and we meet it again in all scientific romances that are worthy of the name. Any man who can read the opening pages of Wells's *The War of the Worlds* or the closing ones of *The Time Machine* without a tingling of the blood is fit only for "treasons, stratagems, and spoils."

The cultural impact of science fiction has never been properly recognized, and the time is long overdue for an authoritative study of its history and development. Perhaps this is a project that Unesco could sponsor, for it is obvious that no single scholar will have the necessary qualifications for the task. In one field in particular—that of astronautics—the influence of science fiction has been enormous. The four greatest pioneers of space-flight—Tsiolkovsky, Oberth, Goddard, and von Braun—*all* wrote science fiction to propagate their ideas (though they did not always get it published!)

In spreading the ideas of spaceflight, science fiction has undoubtedly helped to change the world. More generally, it helps us to face the strange realities of the universe in which we live. This is well put in an article recently sent to me by a science fiction fan who also happens to be a Nobel Prize winner, Dr. Hermann J. Muller, whose discovery of the genetic effects of radiation has inadvertently inspired much recent science fiction and made *mutant* a modern bogey word. To quote Dr. Muller ("Science Fiction as an Escape," *The Humanist* 6 [1957]):

> The real world is increasingly seen to be, not the tidy little garden of our race's childhood, but the extraordinary, extravagant universe de-scribed by the eye of science. . . . If our art . . . does not explore the relations and contingencies implicit in the greater world into which we are forcing our way, and does not reflect the hopes and fears based on these appraisals, then that art is a dead pretense. . . . But man will not live without art. In a scientific age he will therefore have science fiction.

In the same paper, Dr. Muller points out another valuable service that this type of literature has performed:

> Recent science fiction must be accorded high credit for being one of the most active forces in support of equal opportunities, goodwill, and cooperation among all human beings, regardless of their racial and national origins. Its writers have been practically unanimous in their adherence to the ideal of "one free world."

That, I think, is inevitable. Anyone who reads this form of literature must quickly realize the absurdity of mankind's present tribal divisions. Science fiction encourages the cosmic viewpoint; perhaps this is why it is not pop-ular among those literary pundits who have never *quite* accepted the Co-pernican revolution, nor have grown used to the idea that man may not be the highest form of life in the universe. The sooner such people complete their education and reorient themselves to the astronomical realities, the better. And science fiction is one of the most effective tools for this urgent job.

For it is, preeminently, the literature of *change*—and change is the only thing of which we can be certain today, thanks to the continuing and accelerating scientific revolution. What we science fiction writers call "mainstream literature" usually paints a static picture of society, presenting, as it were, a snapshot of it, frozen at one moment in time. Science fiction, on the other hand, assumes that the future will be profoundly different from the past—though it does not, as is often imagined, attempt to *predict* that future in detail. Such a feat is impossible, and the occasional direct hits of Wells and other writers are the result of luck as much as judgment.

But by mapping out *possible* futures, as well as a good many impossible ones, the science fiction writer can do a great service to the community. He encourages in his readers flexibility of mind, readiness to accept and even welcome change—in one word, adaptability. Perhaps no attribute is more important in this age. The dinosaurs disappeared because they could not adapt to their changing environment. We shall disappear if we cannot adapt to an environment that now contains spaceships and thermonuclear weapons.

Sir Charles Snow ends his famous essay "Science and Government" by stressing the vital importance of the "gift of foresight." He points out that men have wisdom without possessing foresight. Perhaps we science fiction writers sometimes show foresight without wisdom; but at least we undoubtedly do have foresight, and it may rub off onto the community at large.

Before concluding, I would like to take this unique occasion of the first Kalinga presentation on Indian soil to speak about the promotion of the scientific outlook in the East. Though this task is important enough in the West, it is even more desperately urgent here. Two of the greatest evils that afflict Asia and keep millions in a state of physical, mental, and spiritual poverty are fanaticism and superstition. Science, in its cultural as well as its technological sense, is the great enemy of both; it can provide the only weapons that will overcome them and lead whole nations to a better life.

For fanaticism is incompatible with the open-minded, inquiring spirit of science—with the readiness to accept the discipline of external reality, even if it conflicts with one's personal hopes and beliefs. The motto of the fanatic is "Don't confuse me with the facts—I've made up my mind." This is the exact antithesis of the scientific outlook.

As for superstition—most of us can remember the events of February 5, 1962. On that date a natural and inevitable grouping of the planets (that has happened about twenty times since the days of the Kalinga empire!) caused needless fear to millions. How many lakhs of rupees were then expended to ward off astral influences? And most of that money was spent by families who could ill afford it.

That was a spectacular example of the evils of superstitions, but there are countless others unnoticed by the world. Recently, not far from my home in Ceylon, a villager was bitten by a snake. He could get no medical treatment because the date was inauspicious; and so he died.

Gentlemen, two years ago M. Jean Rostand, at this very function, referred to your country as "that great nation which welcomes the future without rejecting the past." That is a good policy for any nation—as long as it realizes that there are things in the past that must be rejected. Science, which, after all, is only common sense raised to the nth degree, can tell us what to preserve and what to reject. Heed its voice—if not for your own sake, then for the sake of the lovely, dark-eyed children of Asia and Africa who are born in millions every year—and die in millions the next. Their only hope of a better future lies in science combined with wisdom and foresight. I shall be happy indeed if any writings of mine have helped toward this goal.

More Than

Five Senses

Clarke asks how is it that many mammals seem to employ more than the
normal five senses? The bat, keenly tuned in to echoes from its prey,
possesses uncanny "radar" abilities. Dolphins and whales benefit
from sonar waves under the sea, and even ordinary fish
"listen" to the water around them. The following
essay explores these abilities.

Many years ago, when I was a boy in the country, I invented a rather unkind trick to play on bats. I had long been fascinated by the way in which these strange flying creatures, when they take to the air soon after dusk, are able to locate and catch insects. Even when it is almost completely dark, they will hunt confidently through the sky, suddenly changing direction and darting straight at some invisible moth or beetle.

I knew how it was done, for I had read that bats sent out streams of high-pitched sounds and listened for the echoes reflected from their prey. In the age of radar, of course, everyone is familiar with this idea, but it seemed somewhat fantastic in 1930. Anyway, I asked myself the question, Can a bat tell the difference between an insect and any other solid object in the sky?

So one evening I went out after sunset, with a handful of pebbles, and stationed myself near a tall oak where bats were always to be found at dusk. As soon as one flitted overhead, I tossed a stone into its line of flight—and sure enough, the bat did a power dive toward it. Indeed, it crashed into the stone with such a thud I expected it to be stunned.

Almost every time I repeated the experiment, the same thing happened. If the stone passed anywhere near a bat, the creature would make a sharp turn and dive straight at it. Judging by the number of collisions, it was clear that the bat's "radar" could not easily distinguish between insects and stones. This did not surprise me; after all, what sensible bat would expect to find rocks moving around the sky?

Today, we know that bats are not the only creatures who use sound to navigate or hunt their prey, often in total darkness. Marine animals such as whales and dolphins have developed the sense of "sound location" to a level that we cannot yet approach, even with our most elaborate electronic devices.

When a dolphin is swimming at night, or in dirty water where its eyes are useless, it continually utters a series of squeaks or whistles. We can hear some of these sounds, but only a small part of them; most of the noise made by the dolphins is far too high-pitched to register on human ears. But to the dolphin, these sounds are all-important; as they come echoing back from the seabed, or from schools of fish, they give a clear and accurate picture of the world through which it swims. Just as a bat can fly through a completely darkened room crisscrossed with wires, so a dolphin can swim at speed through murky water full of obstacles, avoiding them all.

Yet just as I fooled the bats, sometimes the sea can fool dolphins and whales. From time to time, great schools of these creatures run aground on shallow beaches, become stranded, and die miserably between land and water. This has long been a puzzle to naturalists, and one theory is that a gently sloping beach may not reflect any echoes back to the approaching animals; in certain conditions, it may simply absorb the sound. And so, perceiving no echoes, the poor whales and dolphins continue to swim forward, quite confident that they are heading for the open sea—and discover their mistake too late to do anything about it.

The sound-locating sense of bats, whales, and dolphins is one that we can all appreciate because we all share it to some extent. The blind man tapping his stick on the pavement and being warned of obstacles by the pattern of echoes coming back to his ears is doing just the same as the bats and the dolphins, though he cannot do it anything like so well. And in everyday life, all of us—blind or otherwise—use sound for location more often than we suspect.

I once had a dramatic proof of this when I was playing table tennis under a corrugated iron roof in a tropical rainstorm. The noise was terrific, and my game went to pieces immediately. For the first time, I realized how much I had relied upon the click of the ball upon bat or table to locate it; I shall be rather surprised to find if there are any really good deaf tennis players. Yet—almost unbelievably—I once saw a blind man acting as a referee at this game; he called every play without hesitation and never missed a fault. It was a wonderful example of what the human ear can do when it is properly trained.

All fish possess a sense organ that we only vaguely understand, because we have nothing like it. It is a thin, irregular line running from head to tail on either side of the fish, called the lateral line; apparently it detects changing pressure waves in the water, but this bald statement gives only a faint idea of its capabilities. The first time I saw it in action, I could hardly believe my eyes.

A friend with a large collection of tropical fish was showing me his hundreds of tanks, and in one of them a school of tiny fish was darting back and forth in a restless cloud. Each time it came to the end of the tank, it turned and reversed itself about half an inch from the glass—always at the same distance, just as if it had run into an invisible barrier. I was interested but not particularly impressed until my friend told me that these little fish were completely blind. Yet on every circuit of the tank they stopped and turned just an instant before they would have charged into the glass walls. How did they do, it?

It was not, as in the case of the dolphins and whales, echo location or sonar, for their feat did not depend upon sound. Every fish, when it swims through the water, produces a kind of bow wave like the one you see moving ahead of a boat—though the underwater bow wave is not an up-and-down movement, but a change of pressure. The fish's lateral line can detect this wave; when it nears an obstacle, the wave is distorted by the obstruction ahead of it, and so the fish knows that there is something approaching. It can also spot the pressure waves produced by other fish moving through the waters around it—and so can hunt for food by feeling over its whole body the currents and vibrations of its liquid world. The vital importance of the lateral line to the fish is proved by the fact that this strange organ is most highly developed in the nightmarish little monsters—all teeth and jaws—that live miles down in the oceans, where no light ever penetrates. In a world where eyes are useless, they must rely on their lateral lines to tell them when to feed—and when to flee.

A good many years ago there was a song that asked the question

"Would you rather be a fish?" Scientifically, that's not an easy question to answer, because no one knows what it would be like to have a lateral line! Perhaps you can get some faint idea of the world of the fish if you stand outdoors on a windy day, wearing no shirt, and with your eyes closed. You can feel the gusts of wind coming at you from various directions; imagine that these gusts represented objects passing through the air around you. If you run quickly, you can feel your own bow wave over your bare skin. But these tiny air currents can give only the feeblest imitation of the rich world of shifting, meaningful pressures in which the creatures of the deep pass their short and hungry lives.

Some fish have evolved a sense organ even more remarkable than the lateral line; they have developed an electric sense. They produce pulses of current, at the rate of a few hundred per second (about five times the frequency of ordinary house circuits), and set up an electric field in the water around them. The field is generated at the tail of the fish and picked up by organs near its head. If its pattern could be seen by our eyes, it would resemble the lines of force around a bar magnet, which becomes visible when iron filings are sprinkled over it.

Just as the field around a magnet is warped or distorted if another piece of iron is placed near it, so the field around the electric fish is distorted by the presence of an obstacle in the water. By sensing the changes in the field it produces, the fish can hunt its food and avoid collisions in the muddy waters where it lives

Please note that this is not an echo-sounding system like that used by the bats and dolphins, even though short pulses are involved. (It could work with DC, but the fish finds it more convenient to use AC!) The electric sense is something much more complicated and much less understandable to us than sonar, because we have nothing like it at all.

Although only a few fish have been definitely proved to possess this peculiar sense, most of them seem to have it in a partly developed form. It has been known for a long time that fish are quite sensitive to electric fields, and this is the basis of the most scientific form of fishing that has yet been discovered. By lowering metal plates into the sea and charging them to the correct voltage, fish can be compelled to swim into the nets, or even into a pipe through which they can be pumped into a ship! Unfortunately, because seawater is a good conductor and so tends to short-circuit the electric field, this method of fishing has a limited range and uses a considerable amount of electric power. It works much better in freshwater, which is a rather poor conductor.

Certain fish, as is well known, have gone beyond electric senses and have developed something still more astonishing—electric weapons. The discharges produced by electric rays and eels are so powerful that they can stun a man and can probably kill any other fish; there can be few more effective "secret weapons" in the sea. I was once about to spear an electric ray when I recognized it—just in time! The world of electric images and sensations through which these creatures move, and in which they launch their silent thunderbolts at their enemies, is certainly beyond our imagination or full understanding.

Human beings cannot detect electric fields; there has never been any reason for them to do so. Our eyes—in daylight, at least—probably do a much better job than the sonic, electric, and pressure senses that the creatures of the sea have been forced to develop. Perhaps if we lived in a world of perpetual darkness, we might have evolved such senses, or even stranger ones.

It is true that we often feel uncomfortable before a thunderstorm, when there are strong electrical fields in the air. But this sensation is almost certainly due to other causes, such as humidity and heat—not electricity. Yet nature is full of surprises; perhaps hidden somewhere in our bodies there are sense organs than can respond to electrical fields. If there is any truth in all the countless reports of thought transference (telepathy) and such mysterious abilities as water divining or dowsing, the answer may lie in some unsuspected electrical sense. I do not say that it is at all likely, but I would hate to say that it is impossible.

Whether any animals—including men—are sensitive to magnetic fields is a question that scientists have only started to ask quite recently. As far as we are concerned, the answer is almost certainly no. If you pick up a magnet, it feels exactly like any other piece of iron. Scientists working in radiation laboratories and nuclear energy establishments have often entered the enormously powerful magnetic fields of their cyclotrons, cosmotrons, and other particle accelerators. Most of them have felt nothing at all; a very few have reported slight sensations around metal fillings in their teeth.

Would a magnetic sense be of any value? To migrating birds and animals definitely, for it would give them a built-in compass, so that they could find the north when there was no other way of telling direction. It has often been suggested that homing pigeons navigate in this manner, and attempts have been made to prove this theory by attaching small magnets to pigeons before releasing them. Confused by the new field, it was argued, the poor birds would be unable to find their way. These experiments have never been very conclusive, and it is now believed that birds rely mainly on the Sun and stars for their wonderful ability to navigate thousands of miles, often over the empty sea.

Animals can do so many remarkable things that there is a great temptation to invent wonderful and mysterious senses to explain their feats. We must remember, however, that to a hypothetical intelligent being who had no eyes and knew nothing about the power of vision, our own ability to observe events at a great distance would seem a miracle. It so happens that we have developed this particular sense to such a high degree that the others have become much less important.

It might have been the other way round. In some animals, the chemical senses of taste and smell have been so enormously developed that they almost play the part of sight. If you have ever owned a dog, you will know that it spends much of its time in a world that you cannot share—a world of exciting, enjoyable, and sometimes frightening smells. A hunting dog can follow an invisible trail for miles, detecting traces of chemicals that must be present in unimaginably small amounts.

We are very seldom conscious of smells (except when they are bad ones) and must undoubtedly miss a great deal of the richness of the natural world. Many years ago, G. K. Chesterton summed this up very neatly in a humorous poem in which a dog sneered at the "noselessness of man."

Once again it may be in the sea, not on the land, that the twin senses of smell and taste are most highly developed. Fish (and perhaps dolphins) may be able to tell where they are in the ocean by sampling the water around them; every sea, and every current in the sea, may have a different taste. It is well-known that sharks are extremely sensitive to traces of blood in the water; every skin diver knows that a bleeding fish is likely to attract sharks. The many attempts to develop a shark repellent rely on the hope that there may be some substances that taste intolerable or terrifying to these powerful and dangerous creatures. Despite all that you may have read to the contrary, no one has yet found a way of discouraging a really hungry shark; the only repellent that sometimes works is a hard bang on the nose, and if matters have come to this, the situation is already pretty desperate.

The shark has two serious disadvantages; it is very slow-acting, and it often operates only in one direction. The blood from a wounded fish must take several minutes to travel any distance through water, and it will not go upcurrent. It is just as well, therefore, for the shark that it has other means of locating food. Underwater hunters have discovered, time and again, that sharks appear on the scene within seconds of a fish being speared. They must have been attracted by some sound or vibration—perhaps the desperate fluttering of the injured fish—that they can spot by means of the lateral line. This gives them their long-range detecting sense; then they close in, and sight and smell take over.

Of all our senses, the one which, we feel, puts us most closely into direct touch with the real universe is the sense of touch. I have deliberately left that statement in its clumsy form to show how natural choice of words emphasizes this very point: "the sense which, we *feel,* puts us most closely into *direct touch."* The long-distance senses of sight and hearing can easily be misled; if we want to be quite sure that an object is really what it seems to be, we reach out and touch it.

Some animals can reach very much farther than we can; they have turned touch into a medium—or even a long-range sense. Cats and many deep-sea fish have done this by the simple trick of growing whiskers or feelers, but they are all beaten by the spider, which sits at the center of a great web hundreds of times larger than its own body, ready to detect anything blundering into it. The spider has, in effect, built an artificial world so that its sense of touch can be extended over a vast area. Looking at it from this point of view makes us realize what a wonderful achievement a spider's web really is; it is much more than a trap—it is a communications network. There is nothing else like it until we get to man and his telephone systems.

You may care to amuse yourself by inventing some unlikely—yet scientifically possible—organs of sense. And to put you in the right frame of mind for the task, I would like to mention a famous painting, called *The Blind Girl,* by the Victorian artist Sir John Millais (1829–96). It shows a beautiful English landscape with a thunderstorm in the distance and a glorious rainbow arched across it. The whole composition is in that detailed, photographically exact style so unpopular today because it requires hard work and a brilliant technique.

In the foreground sits a blind girl, unaware of all the beauty around her. A butterfly has alighted on her shawl, and her little companion—perhaps her sister—looks at it with wonder. To the blind girl, both butterfly and rainbow might not exist.

It is a touching picture, and it still moves me although I have not seen it for twenty years. And it teaches an even deeper lesson than the one that the artist intended.

We think that we see and hear and touch and taste and smell the world around us well enough to know it as it really is. Yet compared with bats and dolphins, we are as good as deaf; to dogs we must appear to suffer from a permanent cold in the nose; and our eyes can see only one narrow band of the spectrum of light. Of electrical, magnetic, or radioactive senses, we have no trace.

The universe has existed for billions of years, and the human race is very young. There may be creatures among the stars who have evolved all the senses that we can imagine, and many more. They would pity us as we pity Millais's blind girl.

Many years ago an American poet, whose name I have forgotten and would be delighted to rediscover, summed up this thought perfectly, in four lines that express all that I have been trying to say in several thousand words.

> *A being who hears me tapping*
> *The five-sensed cane of mind*
> *Amid such greater glories*
> *That I am worse than blind.*

Read the verse carefully, and ponder its meaning. Once you understand it, the world will never again be quite the same to you.

Son of

Dr. Strangelove

The first steps on the rather long road to 2001: A Space Odyssey *were taken in
March 1964, when Stanley Kubrick wrote to Clarke, saying that he wanted
to do the proverbial "really good" science fiction movie. His main interests,
he explained, lay in two broad areas: "(1) The reasons for believing in
the existence of intelligent extraterrestrial life. (2) The impact
(and perhaps even lack of impact in some quarters) such discovery
would have on Earth in the near future."*

As this subject had been my main preoccupation (apart from time out for
World War II and the Great Barrier Reef) for the previous thirty years, this
letter naturally aroused my interest. The only movie of Stanley Kubrick's I
had seen was *Lolita,* which I had greatly enjoyed, but rumors of *Dr. Strangelove* had been reaching me in increasing numbers. Here, obviously, was a
director of unusual quality, who wasn't afraid of tackling far-out subjects.
It would certainly be worthwhile having a talk with him; however, I refused
to let myself get too excited, knowing from earlier experience that the
mortality rate of movie projects is about 99 percent.

I examined my published fiction for film-worthy ideas and very quickly settled on a short story called "The Sentinel," written over the 1948 Christmas holiday for a BBC contest. (It didn't place.) This story developed a concept that has since been taken quite seriously by the scientists concerned with the problem of extraterrestrials, or ETs for short.

During the last decade, there has been a quiet revolution in scientific thinking about ETs; the view now is that planets are at least as common as stars—of which there are some 100 billion in our local Milky Way galaxy alone. Moreover, it is believed that life will arise automatically and inevitably where conditions are favorable; so there may be civilizations all around us that achieved space travel before the human race existed, and then passed on to heights that we cannot remotely comprehend. . . .

But if so, why haven't they visited us? In "The Sentinel," I proposed one answer (which I now more than half believe myself). We may indeed have had visitors in the past—perhaps millions of years ago, when the great reptiles ruled the Earth. As they surveyed the terrestrial scene, the strangers would guess that one day intelligence could arise on this planet; so they might leave behind them a robot monitor, to watch and to report. But they would not leave their sentinel on Earth itself, where in a few thousand years it would be destroyed or buried. They would place it on the almost unchanging Moon.

And they would have a second reason for doing this. To quote from the original story:

> They would be interested in our civilization only if we proved our fitness to survive—by crossing space and so escaping from the Earth, our cradle. That is the challenge that all intelligent races must meet, sooner or later. It is a double challenge, for it depends in turn upon the conquest of atomic energy, and the last choice between life and death. Once we had passed that crisis, it was only a matter of time before we found the beacon and forced it open. . . . Now we have broken the glass of the fire alarm, and have nothing to do but to wait.

This, then, was the idea that I suggested in my reply to Stanley Kubrick as the takeoff point for a movie. The finding—and triggering—of an intelligence detector, buried on the Moon aeons ago, would give all the excuse we needed for the exploration of the universe.

By a fortunate coincidence, I was due in New York almost immediately, to complete work on the Time-Life Science Library's *Man and Space,* the main text of which I had written in Colombo. On my way through London I had the first chance of seeing *Dr. Strangelove* and was happy to find that it lived up to the reviews. Its impressive technical virtuosity certainly augured well for still more ambitious projects.

My first meeting with Stanley Kubrick took place at Trader Vic's in the Plaza Hotel. The date—April 22, 1964—coincided with the opening of the ill-starred New York World's Fair, which, might or might not be regarded as an unfavorable omen. Stanley arrived on time and turned out to be a rather quiet, average-height New Yorker (to be specific, Bronxite) with none of the idiosyncrasies one associates with major Hollywood movie directors. He had a night-person pallor, and one of our minor problems was that he functions best in the small hours of the morning, whereas I believe that no sane person is awake after 10 P.M. and no law-abiding one after midnight. He never tried with me his usual tactic of phoning at 4 A.M. to discuss an important idea. But this courtesy did not stop him from being absolutely inflexible once he had decided on some course of action. Tears, hysterics, flattery, sulks, threats of lawsuits, will not deflect him one millimeter.

Another characteristic that struck me at once was that of pure intelligence; Kubrick grasps new ideas, however complex, almost instantly. He also appears to be interested in practically everything; the fact that he never came near entering college, and had a less-than-distinguished high school career, is a sad comment on the American educational system.

On our first day together, we talked for eight solid hours about science fiction, *Dr. Strangelove,* flying saucers, politics, the space program, Senator Goldwater—and, of course, the projected next movie.

For the next month, we met and talked an average of five hours a day—at Stanley's apartment, in restaurants, movie houses, and art galleries. Besides talking endlessly, we had a look at the competition. In my opinion there were a number of good—or at least interesting—science fiction movies prior to the mid-1960s. They include the Pal-Heinlein *Destination Moon, The War of the Worlds, The Day the Earth Stood Still, The Thing,* and *Forbidden Planet.* However, my affection for the genre perhaps caused me to make greater allowances than Stanley, who was highly critical of everything we screened. After I had pressed him to view H. G. Wells's 1936 classic, *Things to Come,* he exclaimed in anguish, "What are you trying to do to me? I'll never see anything you recommend again!"

Eventually, the shape of the movie began to emerge from the fog of words. It would be based on "The Sentinel" and five of my other short stories of space exploration; our private title for the project was *How the Solar System Was Won.* What we had in mind was a kind of semidocumentary about the first pioneering days of the new frontier; though we soon left that concept far behind, it still seems quite a good idea. Later, I had the quaint experience of buying back—at a nominal fee—my unused stories from Stanley.

Stanley worried about possibilities no one else would think of. He always acted on the assumption that if something can go wrong, it will;

ditto if it can't. There was a time as the *Mariner 4* space probe approached Mars when he kept worrying about alternate story lines—just in case signs of life were discovered on the Red Planet. But I refused to cross that bridge until we came to it. If there *were* Martians, we could work them in somehow—and the publicity for the movie would be simply wonderful.

Once the contract had been signed, the actual writing took place in a manner that must be unusual and may be unprecedented. Stanley hates movie scripts; like D. W. Griffith, I think he would prefer to work without one, if it were possible. But he had to have something to show MGM what it was buying; so he proposed that we sit down and first write the story as a complete novel. Though I had never collaborated with anyone before in this way, the idea suited me fine.

Stanley installed me, with electric typewriter, in his Central Park West office, but after one day I retreated to my natural environment in the Hotel Chelsea, where I could draw inspiration from the company of Arthur Miller, Allen Ginsberg, Andy Warhol, and William Burroughs—not to mention the restless shades of Dylan Thomas and Brendan Behan. Every other day Stanley and I would get together and compare notes; during this period we went down endless blind alleys and threw away tens of thousands of words. The scope of the story steadily expanded, both in time and space.

During this period, the project had various changes of title: it was first announced as *Journey Beyond the Stars*—which I always disliked because there have been so many movie *Voyages* and *Journeys* that confusion would be inevitable. Indeed, *Fantastic Voyage* was coming up shortly, and Salvador Dalí had been disporting himself in a Fifth Avenue window to promote it. When I mentioned this to Stanley, he said, "Don't worry—we've already booked a window for you." Perhaps luckily, I never took him up on this.

The merging of our streams of thought was so effective that, after this lapse of time, I am no longer sure who originated what ideas; we finally agreed that Stanley should have prime billing for the screenplay, while only my name would appear on the novel. Only the germ of the "Sentinel" concept is now left; the story as it exists today is entirely new—in fact, Stanley was still making major changes at a very late stage in the actual shooting.

The first version of the novel was finished on December 24, 1964; I never imagined that two Christmases later we would still be polishing the *2001* manuscript, amid mounting screams from publishers and agents.

But the first version, incomplete and undeveloped though it was, allowed Stanley to set up the deal. Through 1965, he gathered around him the armies of artists, technicians, actors, accountants, and secretaries without whom no movie can be made; in this case, there were endless additional complications, as we also needed scientific advisers, engineers, genuine space hardware, and whole libraries of reference material. Everything was accumulated during the year at MGM's Borehamwood Studios, some fifteen miles north of London; the largest set of all, however, had to be built just six miles south of the city, at Shepperton-on-Thames.

Seventy years earlier, in the twelfth chapter of his brilliant novel *The War of the Worlds,* H. G. Wells's Martians had destroyed Shepperton with their heat ray. But, in the 1960s man obtained his first close-ups of Mars, via *Mariner 4.* As I watched our film astronauts making their way over the lunar surface toward the ominously looming bulk of the Sentinel, Stanley directed them through the radios in their space suits. I remembered that within five years, at the most, men would *really* walk on the Moon.

Fiction and fact were indeed becoming hard to disentangle. I hope that in *2001: A Space Odyssey,* Stanley and I have added to the confusion, but in a constructive and responsible fashion. For what we were trying to create is a realistic myth—and we may well have to wait until the year 2001 itself to see how successful we have been.

Possible,

That's All!

The galactic novels of Clarke's esteemed friend Isaac Asimov gave him such
pleasure, Clarke relates, that he is reluctant to challenge some of
Dr. Asimov's statements in his article "Impossible, That's All" in
the February 1967 issue of the Magazine of Fantasy & Science
Fiction. *Clarke's rebuttal appeared in the October*
1968 issue of the same magazine.

The possibility, or otherwise, of speeds greater than that of light cannot be disposed of quite as cavalierly as Isaac Asimov does in his article. First of all, even the restricted, or special, theory of relativity does not deny the existence of such speeds. It only says that speeds *equal* to that of light are impossible—which is quite another matter.

The naive layman who has never been exposed to quantum physics may well argue that to get from below the speed of light to above it one has to pass *through* it. But this is not necessarily the case; we might be able to jump over it, thus avoiding the mathematical disasters that the well-known Lorentz equations predict when one's velocity is *exactly* equal to that of light. Above this critical speed, the equations can scarcely be expected to apply, though if certain interesting assumptions are made, they may still do.

I am indebted to Dr. Gerald Feinberg of Columbia University for this idea. His paper "On the Possibility of Superphotic Speed Particles" points out that since sudden jumps from one state to another are characteristic of quantum systems, it might be possible to hop over the "light barrier" without going through it. If anyone thinks that this is ridiculous, I would remind him that quantum-effect devices doing similar tricks are now on the market—witness the tunnel diode. Anything that can rack up sales of hundreds of thousands of dollars should be taken very seriously indeed.

Even if there is no way through the light barrier, Dr. Feinberg suggests that there may be another universe on the other side of it, composed entirely of particles that cannot travel slower than the speed of light. (Anyone who can visualize just what is meant by that phrase "on the other side of" is a much better man than I am.) However, as such particles—assuming that they still obey the Lorentz equations—would possess imaginary mass or negative energy, we might never be able to detect them or use them for any practical purpose like interstellar signaling. As far as we are concerned, they might as well not exist.

This last point does not worry me unduly. Similar harsh things were once said about the neutrino, yet it is now quite easy to detect this improbable object, if you are prepared to baby-sit a few hundred tons of equipment for several months two miles down in an abandoned gold mine. Anyway, mere trifles like negative energy and imaginary mass should not deter any mathematical physicist worthy of his salt. Odder concepts are being bandied around all the time in the quark-infested precincts of Brookhaven and CERN.

Perhaps at this point I should exorcise a phantom which, rather wisely, Dr. Asimov refrained from invoking. There are many things that do travel faster than light, but they are not exactly "things." They are only appearances, which do not involve the transfer of energy, matter, or information.

One example—familiar to thousands of radar technicians—is the movement of radio waves along the rectangular copper pipes known as waveguides. The electromagnetic patterns traveling through a waveguide can only move faster than light—never at less than this speed! But they cannot carry signals; the changes of pattern that alone can do this move more slowly than light, and by precisely the same ratio as the others exceed it. (I.e., the product of the two speeds equals the square of the speed of light.)

If this sounds complicated, let me give an example that I hope will clarify the situation. Suppose we had a waveguide one light-year long and fed radio signals into it. Under no circumstances could anything emerge from the other end in less than a year; in fact, it might be ten years before the message arrived, moving at only a tenth of the speed of light. But once the waves had got through, they would have established a pattern that swept along the guide at ten times the speed of light. Again I must emphasize that this pattern would require a change of some kind of transmitter, which would take ten years to make the one-light-year trip.

If you have ever watched storm waves hitting a breakwater, you may have seen a similar phenomenon. When the line of waves hits the obstacle at an acute angle, a veritable waterspout appears at the intersection and moves along the breakwater at a speed that is always greater than that of the oncoming waves and can have any value up to infinity (when the lines are parallel, and the whole seafront erupts at once). But no matter how ingenious you are, there is no way in which you could contrive to use this waterspout to carry signals—or objects—along the coast. Though it contains a lot of energy, it does not involve any movement of that energy. The same is true of the hyperspeed patterns in a waveguide.

If you wish to investigate this subject in more detail, I refer you to the article by s.f. old-timer Milton A. Rothman, "Things That Go Faster than Light" in *Scientific American* for July 1960. The essential fact of the existence of such speeds ("phase velocities") in no way invalidates the theory of relativity.

The point I wish to make, however, is that despite its formidable success in many *local* applications, relativity may not be the last word about the universe. Indeed, it would be quite unprecedented if it were.

The general theory—which deals with gravity and accelerated motions, unlike the special theory, which is concerned only with unaccelerated motion—may already be in deep trouble. One of the world's leading astrophysicists (he may have changed his mind now, so I will not identify him beyond saying that his name begins with *Z*) once shook me by remarking casually, as we were on the way up to Mount Palomar, that he regarded all the three "proofs" of the general theory as disproved. And only this week I read that Professor Dicke has detected a flattening of the Sun's poles, which accounts for the orbital peculiarities of Mercury, long regarded as the most convincing evidence for the theory.

If Dicke is right, the fact that Einstein's calculations gave the correct result for the precession of Mercury will be pure coincidence. And then we shall have an astronomical scandal at both ends of the Solar System: for Lowell's prediction of Pluto's orbit also seems completely fortuitous. Pluto is far too small to have produced the perturbations that led to its discovery. (Has anyone yet written a story suggesting that it is the satellite of a much larger but invisible planet?)

At least one mathematical physicist—Prof. J. A. Wheeler—has constructed a theory of space-time that involves what he has picturesquely called "worm holes." These have all the classic attributes of the space warp (a convenient shortcut taken by so many writers of interstellar fiction); you disappear at A and reappear at B, without ever visiting any point in between. Unfortunately, in Wheeler's theory the average speed from A to B, even via a worm hole, still works out at less than the speed of light. This seems very unenterprising and I hope the professor does a little more homework.

Another interesting, and unusual, attempt to demolish the light barrier was made in the last chapter of the book *Islands in Space,* by Danridge M. Cole and Donald W. Cox. They pointed out that all the tests of the relativity equations had been carried out by particles accelerated by external forces, not self-propelled systems like rockets. It was unwise, they argued, to assume that the same laws applied in this case.

And here I must admit to a little embarrassment. I had forgotten, until I referred to my copy, that the preface to *Islands in Space* ends with a couple of limericks making a crack at me for saying (in *Profiles of the Future*) that the velocity of light could never be exceeded.

The Mind of

the Machine

*Clarke recounts here that the twentieth century has been the one in which all of
man's ancient dreams appear to be coming true. The conquest of the air, the
transmutation of matter, journeys to the Moon, even the elixir of life. Among them,
he says, the one most fraught with peril is the machine that can think.*

In some form or another, the idea of artificial intelligence goes back at least
three thousand years. Before he turned his attention to aeronautical engineering, Daedalus—King Minos' one-man office of scientific research—
constructed a metal man to guard the coast of Crete. Talos, however, was
only a physical and not an intellectual giant; perhaps a better prototype of
the thinking machine is the brazen head generally linked with the name of
Friar Bacon, though the legend precedes him by some centuries. This head
was able to answer any question given to it, relating to past, present, or
future; as is customary with oracles, there was no guarantee that the inquirer
would be pleased with what he heard.

Over these tales there usually hangs the aura of doom or horror associated with such names as Prometheus, Faust, and—above all—Frankenstein, though that unfortunate scientist's creation was not a mechanical one. Perhaps the finest work in this genre is that little classic of Ambrose Bierce's, "Moxon's Master," which opens with the words "Are you serious? Do you really believe that a machine thinks?"

Critics of this viewpoint (who are probably now in the minority) may argue that the brain is in some fundamental way different from any nonliving device. But even if this is true, it does not follow that its functions cannot be duplicated, or even surpassed, by a nonorganic machine. Airplanes fly better than birds, though they are built of very different materials.

For obvious psychological reasons, there are people who will never accept the possibility of artificial intelligence and would deny its existence even if they encountered it. As I write these words, there is a chess game in progress between computers in California and Moscow; both are playing so badly that there is clearly no human cheating on either side. Yet no one really doubts that eventually the world champion will be a computer; and when that happens, the diehards will retort: "Oh, well, chess doesn't involve *real* thinking."

Though one can sympathize with this attitude, to resent the concept of a rational machine is itself irrational. We no longer become upset because machines are stronger or swifter or more dexterous than human beings, though it took us several painful centuries to adapt to this state of affairs. How our outlook has changed is well shown by the ballad of John Henry; today, we should regard anyone who challenged a steam hammer as merely crazy—not heroic. I doubt if contests between calculating prodigies and electronic computers will ever provide inspiration for future folk songs, though I am happy to donate that theme to Tom Lehrer.

It is, of course, the advent of the modern computer that has brought the subject of thinking machines out of the realm of fantasy into the forefront of scientific research. One could not have a plainer answer to the question that Ambrose Bierce posed three-quarters of a century ago than this question from MacGowan and Ordway's recent book, *Intelligence in the Universe:* "It can be asserted without reservation that a general purpose digital computer can think in every sense of the word. This is true no matter what definition of thinking is specified; the only requirement is that the definition of thinking be explicit."

That last phrase is, of course, the joker, for there must be almost as many definitions of thinking as there are thinkers; in the ultimate analysis they probably boil down to "Thinking is what I do." One neat way of avoiding this problem is a famous test proposed by the British mathematician Alan Turing, even before the digital computer existed. Turing visualized a "conversation" over a teleprinter circuit with an unseen entity "X." If, after some hours of talk, one could not decide whether there was a man or machine at the other end of the line, it would have to be admitted that X was thinking.

For the Turing test to be applied properly, the conversation should not be restricted to a single narrow field, but should be allowed to range over the whole arena of human affairs. ("Read any good books lately?" "Do you think. . . . will be nominated?" "Has your wife found out yet?" etc., etc.) We are certainly nowhere near building a machine that can fool many of the people for much of the time; sooner or later, today's models give themselves away by irrelevant answers that show only too clearly that their replies are indeed "mechanical" and that they have no real understanding of what is going on. As Oliver Selfridge of MIT has remarked sourly, "Even among those who believe that computers can think, there are few these days, except for a rabid fringe, who hold that they are actually thinking."

Very few, if any, studies of the social impact of computers have yet faced up to the problems posed by the ominous phrase "assuming that we would want to." This is understandable; the electronic revolution has been so swift that those involved in it have barely had time to think about the present, let alone the day after tomorrow. Moreover, the fact that today's computers are very obviously not "intellectually superior" has given a false sense of security—like that felt by the 1900 buggy-whip manufacturer every time he saw a broken-down automobile by the wayside. This comfortable illusion is fostered by the endless stories—part of the transient folklore of our age—about stupid computers that had to be replaced by good old-fashioned human beings, after they had sent out bills for $1,000,000,004.95, or threatening legal action if outstanding debts of $0.00 were not settled immediately. The fact that these gaffes are almost invariably due to oversights by human programmers is seldom mentioned.

Though we have to live and work with (and against) today's mechanical morons, their deficiencies should not blind us to the future. In particular, it should be realized that as soon as the borders of electronic intelligence are passed, there will be a kind of chain reaction, because the machines will rapidly improve themselves. In a very few generations— computer generations which by this time may last only a few months— there will be a mental explosion; the merely intelligent machine will swiftly give way to the ultraintelligent machine.

One scientist who has given much thought to this matter is Dr. John Irving Good, of Trinity College, Oxford—author of papers with such challenging titles as "Can an Android Feel Pain?" (This term for artificial man, incidentally, is older than generally believed. I had always assumed that it was a product of the modern science fiction magazines and was astonished to come across "The Brazen Android" in an *Atlantic Monthly* for 1891.) Dr. Good has written: "If we build an ultraintelligent machine, we will be playing with fire. We have played with fire before, and it helped to keep the other animals at bay."

Well, yes—but when the ultraintelligent machine arrives, we may be the "other animals"—and look what has happened to them.

It is Dr. Good's belief that the very survival of our civilization may depend upon the building of such instrumentalities, because if they are indeed more intelligent than we are, they can answer all our questions and solve all our problems. As he puts it in one elegant phrase, "The first ultraintelligent machine is the last invention that man need make."

Need is the operative word here. Perhaps 99 percent of all the men who have ever lived have known only need; they have been driven by necessity and have not been allowed the luxury of choice. In the future, this will no longer be true. It may be the greatest virtue of the ultraintelligent machine that it will force us to think about the purpose and meaning of human existence. It will compel us to make some far-reaching and perhaps painful decisions, just as thermonuclear weapons have made us face the realities of war and aggression, after five thousand years of pious jabber.

These long-range philosophical implications of machine intelligence obviously far transcend today's more immediate worries about automation and unemployment. Somewhat ironically, these fears are both well grounded and premature. Although automation has already been blamed for the loss of many jobs, the evidence indicates that so far it has created many more opportunities for work than it has destroyed. (True, this is small consolation for the particular semiskilled worker who has just been replaced by a couple of milligrams of microelectronics.) *Fortune* magazine, in a hopeful attempt at self-fulfilling prophecy, has declaimed: "The computer will doubtless go down in history not as the explosion that blew unemployment through the roof, but as the technological triumph that enabled the U.S. economy to maintain the secular growth on which its greatness depends." I suspect that this statement may be true for some decades to come; but I also suspect that historians (human and otherwise) of the late twenty-first century would regard that "doubtless" with wry amusement.

For the plain fact is that long before that date, the talents and capabilities of the average—and superior—man will be as unsalable in the marketplace as his muscle power. Only a few specialized and distinctly non-white-collar jobs will remain the prerogative of nonmechanical labor; one cannot easily picture a robot handyman, gardener, construction worker, fisherman. . . . These are professions that require mobility, dexterity, alertness, and general adaptability—for no two tasks are precisely the same— but not a high degree of intelligence or data-processing power. And even these relatively few occupations will probably be invaded by a rival and frequently superior labor force from the animal kingdom, for one of the long-range technological benefits of the space program (though no one has said much about it yet, for fear of upsetting the trade unions) will be a supply of educable anthropoids filling the gap between man and the great apes.

It must be clearly understood, therefore, that the main problem of the future—and a future that may be witnessed by many who are alive today—will be the construction of social systems based on the principle not of full employment but rather of full *un*employment. Some writers have suggested that the only way to solve this problem is to pay people to be consumers; Fred Pohl, in his amusing short story "The Midas Plague," described a society in which you would be in real trouble unless you used up your full quota of goods poured out by the automatic factories. If this proves to be the pattern of the future, then today's welfare states represent only the most feeble and faltering steps toward it. The uproars about Medicare will seem completely incomprehensible to a generation that assumes every man's right to a constantly rising income.

I leave others to work out the practical details of an economic system in which it is antisocial, and possibly illegal, not to wear out a suit every week, or to eat three six-course meals a day, or to throw out last month's car. Though I do not take this picture seriously, it should serve as a reminder that tomorrow's world may differ from ours so radically that such terms as *labor, capital, communism, private enterprise, state control* will have changed their meaning completely—if indeed they are still in use. At the very least, we may expect a society that no longer regards work as meritorious, or leisure as one of the devil's more ingenious devices. Even today, there is not much left of the old Puritan ethic; automation will drive the last nails into its coffin.

The need for a change of outlook has been well put by the British science writer Nigel Calder in his remarkable book *The Environment Game:* "Work was an invention that can be dated with the invention of agriculture. . . . Now, with the beginning of automation, we have to anticipate a time when we must disinvent work and rid our minds of the inculcated habit."

The disinvention of work: What would Horatio Alger have thought of that concept? Calder's thesis (too complex to do more than summarize here) is that man is now coming to the end of his brief ten-thousand-year agricultural episode; for a period of a hundred times longer than that he was a hunter, and any hunter will indignantly deny that his occupation is "work." We now have to abandon agriculture for more efficient technologies—first because it has patently failed to feed the exploding populations, second because it has compelled five hundred generations of men to live abnormal—in fact, artificial—lives of repetitive, boring toil. Hence many of our present psychological problems, to quote Calder again: "If men were intended to work the soil, they would have longer arms."

"If men were intended to . . ." is of course a game that everyone can play. Yet now, with the ultraintelligent machines lying just below our horizon, it is time that we played the game in earnest, while we still have some control over the rules. In a few more years, it will be much too late.

The astronomer Fred Hoyle once remarked to me that it was pointless for the world to hold more people than one could get to know in a single lifetime. Even if one were president of United Earth, that would set the figure somewhere between ten thousand and one hundred thousand; with a very generous allowance for duplication, wastage, special talents, and so forth, there really seems no requirement for what has been called the global village of the future to hold more than a million people scattered over the face of the planet.

And if such a figure appears unrealistic—since we are already past the 3 billion mark and heading for at least twice as many by the end of the century—it should be pointed out that once the universally agreed upon goal of population control is attained, any desired target can be reached in a remarkably short time. If we really tried (with a little help from the biology labs), we could reach a trillion within a century—four generations. It might be more difficult to go in the other direction for fundamental psychological reasons, but it could be done. If the ultraintelligent machines decide that more than a million human beings constitute an epidemic, they might order euthanasia for anyone with an IQ of less than 150, but I hope that such drastic measures will not be necessary.

Whether the population levels off, a few centuries from today, at a million, a billion, or a trillion human beings is of much less importance than the ways in which they will occupy their time. Since all the immemorial forms of "getting and spending" will have been rendered obsolete by the machines, it would appear that boredom will replace war and hunger as the greatest enemy of mankind.

One answer to this would be the uninhibited, hedonistic society of Aldous Huxley's *Brave New World;* there is nothing wrong with this, so long as it is not the only answer. (Huxley's unfortunate streak of asceticism prevented him from appreciating this point.) Certainly much more time than at present will be devoted to sports, entertainment, the arts, and everything embraced by the vague term *culture.*

In some of these fields, the background presence of superior nonhuman mentalities would have a stultifying effect, but in others the machines could act as peacemakers. Does anyone really imagine that when all the grand masters are electronic, no one will play chess? The humans will simply set up new categories and play better chess among themselves. All sports and games (unless they become ossified) have to undergo technological revolutions from time to time; recent examples are the introduction of fiberglass in pole-vaulting, archery, and boating. Personally, I can hardly wait for the advent of Marvin Minsky's promised robot table-tennis player.

These matters are not trivial; games are a necessary substitute for our hunting impulses, and if the ultraintelligent machines give us new and better outlets, that is all to the good. We shall need every one of them to occupy us in the centuries ahead.

The ultraintelligent machines will certainly make possible new forms of art, and far more elaborate developments of the old ones, by introducing the dimensions of time and probability. Even today, a painting or a piece of sculpture that stands still is regarded as slightly passé. Although the trouble with most "kinetic art" is that it only lives up to the first half of its name, something is bound to emerge from present explorations on the frontier between order and chaos.

The insertion of an intelligent machine into the loop between a work of art and the person appreciating it opens up some fascinating possibilities. It would allow feedback in both directions; by this I mean that the viewer would react to the work of art; then the work would react to the viewer's reactions, then . . . and so on, for as many stages as was felt desirable. This sort of to-and-fro process is already hinted at, in a very crude way, with today's primitive "teaching machines"; and those modern novelists who deliberately scramble their text are perhaps also groping in this direction. A dramatic work of the future, reproduced by an intelligent machine sensitive to the varying emotional states of the audience, would never have the same form, or even the same plotline, twice in succession. It would be full of surprises even to its human creator—or collaborator.

What sort of art intelligent machines would create for their own amusement, and whether we would be able to appreciate it, are questions that can hardly be answered today. The painters of the Lascaux Caves could not have imagined (though they would have enjoyed) the scores of art forms that have been invented in the twenty thousand years since they created their masterpieces. Though in some respects we can do no better, we can do much more—more than any Paleolithic Picasso could possibly have dreamed. And our machines may begin to build on the foundations we have laid.

Yet perhaps not. It has often been suggested that art is a compensation for the deficiencies of the real world; as our knowledge, our power, and above all *our maturity* increase, we will have less and less need for it. If this is true, the ultraintelligent machines would have no use for it at all.

Even if art turns out to be a dead end, there still remains science— the eternal quest for knowledge, which has brought man to the point where he may create his own successor. It is unfortunate that, to most people, *science* now means incomprehensible mathematical complexities; that it could be the most exciting and entertaining of all occupations is something that they find impossible to believe. Yet the fact remains that, before they are ruined by what is laughingly called education, all normal children have an absorbing interest in and curiosity about the universe, which if properly developed could keep them happy for as many centuries as they wish to live.

Education is ultimately the key to survival in the coming world of ultraintelligent machines. The truly educated man (I have been lucky enough to meet two in my lifetime) can never be bored. The problem that has to be tackled within the next fifty years is to bring the entire human race, without exception, up to the level of semiliteracy of the average college graduate. This represents what may be called the minimum survival level; only if we reach it will we have a sporting chance of seeing the year 2200.

Perhaps we can now glimpse one viable future for the human race, when it is no longer the dominant species on this planet. As he was in the beginning, man will again be a fairly rare animal, and probably a nomadic one. There will be a few towns in places of unusual beauty or historical interest, but even these may be temporary or seasonal. Most homes will be completely self-contained and mobile, so that they can be moved to any spot on Earth within twenty-four hours.

The land areas of the planet will have largely reverted to wilderness; they will be much richer in life-forms (and much more dangerous) than today. All adolescents will spend part of their youth in this vast biological reserve, so that they never suffer from that estrangement from nature that is one of the curses of our civilization.

And somewhere in the background—perhaps in the depths of the sea, perhaps orbiting beyond the ionosphere—will be the culture of the ultraintelligent machines, going their own unfathomable way. The societies of man and machine will interact continuously but lightly; there will be no areas of conflict, and few emergencies, except geological ones (and those would be fully foreseeable). In one sense, for which we may be thankful, history will have come to an end.

All the knowledge possessed by the machines will be available to mankind, though much of it may not be understandable. There is no reason why this should give our descendants an inferiority complex; a few steps into the New York Public Library or the British Museum can do that just as well, even today. Our prime goal will no longer be to discover but to understand and to enjoy.

Would the coexistence of man and machine be stable? I see no reason why it should not be, at least for many centuries. A remote analogy of this kind of dual culture—one society encapsulated in another—may be found among the Amish of Pennsylvania. Here is a self-contained agricultural society, which has deliberately rejected much of the surrounding values and technology, yet is exceedingly prosperous and biologically successful. The Amish, and similar groups, are well worth careful study; they may show us how to get along with a more complex society that perhaps we cannot comprehend, even if we wish to.

For in the long run, our mechanical offspring will pass on to goals that will be wholly incomprehensible to us; it has been suggested that when this time comes, they will head on out into galactic space looking for new frontiers, leaving us once more the masters (perhaps reluctant ones) of the Solar System, and not at all happy at having to run our own affairs.

God and

Einstein

Though Clarke has, elsewhere in this book, discussed going beyond the speed of light, here he examines Einstein and God.

For some years I have been worried by the following astrotheological paradox. It is hard to believe that no one else has ever thought of it, yet I have never seen it discussed anywhere.

One of the most firmly established facts of modern physics and the basis of Einstein's theory of relativity is that the velocity of light is the speed limit of the material universe. No object, no signal, no *influence,* can travel any faster than this. Please don't ask why this should be; the universe just happens to be built that way. Or so it seems at the moment.

But light takes not millions, but *billions* of years to cross even the part of creation we can observe with our telescopes. So, if God obeys the laws He apparently established, at any given time He can have control over only an infinitesimal fraction of the universe. All hell might (literally?) be breaking loose ten light-years away, which is a mere stone's throw in interstellar space, and the bad news would take at least ten years to reach Him. And then it would be another ten years, at least, before He could get there to do anything about it. . . .

You may answer that this is terribly naive—that God is already "everywhere." Perhaps so, but that really comes to the same thing as saying that His thoughts and His influence can travel at an infinite velocity. And in this case, the Einstein speed limit is not absolute; it *can* be broken.

The implications of this are profound. From the human viewpoint it is no longer absurd—though it may be presumptuous—to hope that we may one day have knowledge of the most distant parts of the universe. The snail's pace of the velocity of light need not be an eternal limitation, and the remotest galaxies may one day lie within our reach.

But perhaps, on the other hand, God Himself is limited by the same laws that govern the movements of electrons and protons, stars and spaceships. And that may be the cause of all our troubles.

He's coming just as quickly as He can, but there's nothing that even He can do about that maddening 186,000 miles a second.

It's anybody's guess whether He'll be here in time.

THE 1970s:

TOMORROW'S

WORLDS

Clarke in Sri Lanka.

I n t r o d u c t i o n

Perhaps the most important event of the seventies for me was when, early in 1974, I moved into the large Colombo house that has been my "home" for almost a third of my life—far longer than any other place. Yet I still had to spend at least six months in every year out of Ceylon, otherwise the local tax laws would have ruined me.

Making a virtue of a necessity, I used the time away from the island lecturing and visiting interesting places—in one case, with near disastrous results. In 1973 I was snorkeling in fairly deep water in the Virgin Islands with Dr. George Mueller, then associate administrator for Manned Space Flight, and other NASA friends. Coming up from a deep dive, I was suddenly seized by vertigo and could not tell which direction was up. Luckily, I was able to hit the quick-release button of my weight belt and surfaced just beside a couple in a rubber dingy. This was the narrowest escape I ever had in almost half a century of diving.

During this period, despite my travels, I produced several books, particularly the novels *Imperial Earth* and *The Fountains of Paradise* (of which more anon). Very enjoyable was a long-distance collaboration with the dean of all astronomical artists, Chesley Bonestell, on *Beyond Jupiter,* which tried to anticipate what would happen when the planned grand tour of the planets took place later in the decade. In fact the discoveries of the Voyager and Galileo space probes far outdid our imaginations.

Another memorable event was when in 1975 the Indian government presented me with my first satellite dish—a monstrous twenty-footer, with which I was able to receive programs broadcast from experimental satellite ATS6. Needless to say, the arrival of the first TV broadcasts from space brought a host of visitors to my now rather conspicuous residence.

Then, just as the decade ended, I opened the local newspaper to read a headline: "Clarke to Become Chancellor." My immediate reaction was "I wonder who this Clarke can be—I didn't know there was another on the island." Reading on, I discovered that President J. R. Jayewardene, without bothering to ask me, had appointed me chancellor of our mini MIT, the University of Moratuwa. When I next met him I said, "Mr. President, is this a life sentence, or can I get time off for good behavior?" He replied, "Well . . . I might grant a remission." But he never did, and I am still, twenty years later, proud to sign myself as chancellor, even though I am no longer physically able to deliver a full-length convocation address.

And now more on *The Fountains of Paradise.* Soon after this novel was

published, I made a recording—Caedmon TC 1606—on a twelve-inch
record. (Remember them?) The sleeve notes were written by my old friend
Buckminster Fuller, whom I'd recently flown around the places that had
inspired the story. He even supplied a sketch of Ceylon with the "space
elevator" reaching up from it to stationary orbit. Well—and this is one of
the most extraordinary, if not eerie, coincidences I've ever encountered:
When I wrote the novel, the only material from which a space elevator
could be constructed was diamond. Now a third form of carbon has been
discovered—C60—which was instantly christened Buckminsterfullerene
because its molecules are identical in shape with Bucky's famous geodesic
domes. This material, in its tubular configuration, is the strongest known
substance and would make the space elevator possible! I only wish that
Bucky had lived long enough to enjoy this astonishing—or, I should say,
serendipitous—coincidence.

O N E

Satellites

and Saris

*On August 20, 1971, after several years of byzantine negotiations, the final
agreement setting up the world satellite communications system (Intelsat)
were signed at the State Department in Washington. At the invitation of Secretary
of State William Rogers and Amb. Abbott Washburn, Clarke was asked to speak
at the ceremony. He referred to the forthcoming Indian educational satellite
experiment, then still almost four years in the future. Later,
comments on the subject were read into the* Congressional Record
*on January 27, 1972, by William Anderson, skipper of the atomic
submarine* Nautilus, *which had reached the North Pole in 1958.*

Mr. Secretary, your excellencies, distinguished guests . . .

Whenever I peer into my cloudy crystal ball and try to visualize the
future of communications satellites, I remember an incident that occurred
in England almost one hundred years ago.

The very alarming news had just been received from the United States that a certain Mr. Bell had invented the telephone. This, of course, was very disturbing. So, as we British do in an emergency, we called a parliamentary commission. It listened to the evidence of expert witnesses, who gave the reassuring news that nothing further would be heard of this impractical Yankee invention.

Among the witnesses called was the chief engineer of the British Post Office. Someone on the commission said to him, "We understand that the Americans have invented a machine that can transmit human speech. Do you think that this—*telephone*—will be of any use in Great Britain?" The chief engineer thereupon replied, "No, sir. The Americans have need of the telephone, but we do not. *We* have plenty of messenger boys."

This very able man totally failed to see the possibilities of the telephone, and who can blame him? Could *anyone,* back in 1880, have imagined that the time would come when every home would have a telephone, and business and social life would depend upon it almost completely?

I submit, ladies and gentlemen, that the eventual impact of the communications satellite upon the whole human race will be at least as great as that of the telephone upon the so-called developed societies. In fact, as far as real communications are concerned, there are as yet no developed societies; we are all in the semaphore and smoke-signal stage. And we are now about to witness an interesting situation in which many countries—particularly in Asia and Africa—are going to leapfrog a whole era of communications technology and go straight into the Space Age. They will never know the vast networks of cables and microwave links that this continent has built up at such enormous cost. Satellites can do far more, at far less expense.

Intelsat, of course, is concerned primarily with point-to-point communications involving large ground stations, often only one per country. It provides the first reliable, high-quality wide band with links between all nations that wish to join, and the importance of this cannot be overestimated. Yet it is only a beginning, and I would like to look a little further into the future. . . .

Two years from now, NASA will launch the first satellite—ATS-6—which will have sufficient power for its signals to be picked up by an ordinary domestic television set, plus about $200 worth of additional equipment. In 1974 [Note: the program actually began in 1975] this satellite will be stationed over India, and if all goes well, the first experiment in the use of space communications for mass education will begin.

I have just come from India, where I have been making a television film, *The Promise of Space.* We erected, in a village outside Delhi, the prototype antenna: a simple umbrella-shaped, wire-mesh affair, three meters across. Anyone can put it together in a few hours; it needs only one per village to start a social and economic revolution.

The *engineering* problems of bringing education, literacy, improved hygiene, and agricultural techniques to every human being on this planet have not been solved. The cost would be on the order of a dollar per person—*per year*. The benefits in health, happiness, and wealth would be immeasurable.

But, of course, the technical problem is the easy one. Do we have the imagination—and the statesmanship—to use this new tool for the benefit of all mankind? Or will it be used merely to peddle detergents and propaganda?

I am an optimist; anyone interested in the future has to be, otherwise he would simply shoot himself. . . . I believe that communications satellites can unite mankind. Let me remind you that this great country was virtually created one hundred years ago by two inventions. Without them, the United States was impossible; with them, it was inevitable. Those inventions were, of course, the railroad and the electric telegraph.

Today we are seeing, on a global scale, an almost exact parallel to that situation. What the railroads and the telegraph did here a century ago, the jets and the communication satellites are doing now to all the world.

I hope you will remember this analogy in the years ahead. For today, whether you intend it or not, whether you *wish* it or not, you have signed far more than yet another intergovernmental agreement.

You have just signed the first draft of the Articles of Federation of the United States of Earth.

For thousands of years, men have sought their future in the starry sky. Now this old superstition has at last come true, for our destinies do indeed depend on celestial bodies—those that we have created ourselves.

Since the midsixties, the highly unadvertised reconnaissance satellites have been quietly preserving the peace of the world, the weather satellites have guarded millions against the furies of nature, and the communications satellites have acted as message carriers for half the human race. Yet these are merely the first modest applications of space technology to human affairs; its real impact is still to come. And, ironically, the first country to receive the benefits of space directly at the home and village level will be India.

In 1975, there will be a new Star of India; though it will not be visible to the naked eye, its influence will be greater than that of any zodiacal signs. It will be the satellite ATS-6 (Applications Technology Satellite 6), the latest in a very successful series launched by the National Aeronautics and Space Administration. For one year, ATS-6 will be loaned to the Indian government by the United States and will be "parked" thirty-six thousand kilometers above the equator and will make one revolution every twenty-four hours, remaining poised over the same spot on the turning Earth; in effect, India will have a television tower from which programs can be received with almost equal strength over the entire country.

Since the launch of the historic Telstar in 1962, there have been several generations of communications satellites. The latest, Intelsat IV, can carry a dozen television programs or up to nine thousand telephone conversations across the oceans of the world. But all these satellites have one thing in common: their signals are so feeble that they can be received only by large earth stations, equipped with antennas twenty or more meters across, and costing several million dollars. Most countries can afford only one such station, and indeed that is all that they need to connect their television, telephone, or other services—where these exist—to the outside world.

ATS-6, built by the Fairchild Corporation, represents the next step in the evolution of communications satellites. Its signals will be powerful enough to be picked up, not merely by multimillion-dollar earth stations, but by simple receivers, costing two or three hundred dollars, which all but the poorest communities can afford. This level of cost would open up the entire developing world to every type of electronic communication, not only television. The emerging societies of Africa, Asia, and South America could thus bypass much of today's ground-based technology and leap straight into the Space Age. Many of them have already done something similar in the field of transportation, going from oxcart to airplane with only a passing nod at cars and trains.

It can be difficult, for those who come from developed (or overdeveloped) countries, and who accept libraries, telephones, cinemas, radio, television, as part of their daily lives, to see any sense in spreading these boons to places which do not yet enjoy them. Because they frequently suffer from the modern scourge of information pollution, they cannot imagine the deadly opposite: information starvation. For any Westerner, however well-meaning, to tell an Indian villager that he would be better off without access to the world's news, knowledge, and entertainment is an impertinence. . . .

Those who actually live in the East and know its problems are in the best position to appreciate what cheap and high-quality communications could do to improve standards of living and reduce social inequalities. Illiteracy, ignorance, and superstition are not merely the results of poverty; they are part of its cause, forming a self-perpetuating system which has lasted for centuries, and which cannot be changed without fundamental advances in education. India is now beginning a Satellite Instructional Television Experiment (SITE) as a bold attempt to harness the technology of space for this task; if it succeeds, the implications for all developing nations will be enormous.

SITE's first order of business will be instruction in family planning, upon which the future of India (and all other countries) now depends. Puppet shows are already being produced to put across the basic concepts; Punch and Judy may find this idea faintly hilarious. However, there is probably no better way of reaching audiences who are unable to read, but who are familiar with the traveling puppeteers who for generations have brought the sagas of Rama and Sita and Hanuman into the villages.

Some officials have stated, perhaps optimistically, that the only way in which India can check its population explosion is by mass propaganda from satellite—which alone can project the unique authority and impact of the television set into every village in the land. If this is true, we have a situation which should indeed give pause to those who have criticized the billions spent on space.

The emerging countries of what is called the third world may need rockets and satellites much more desperately than the advanced nations that built them. Swords into plowshares is an obsolete metaphor; we can now turn missiles into blackboards.

Next to family planning, India's greatest need is increased agricultural productivity. This involves spreading information about animal husbandry, new seeds, fertilizers, pesticides, and so forth; the ubiquitous transistor radio has already played an important role here. In certain parts of the country, the famous "miracle rice" strains, which have unexpectedly given the whole of Asia a few priceless years in which to avert famine, are known as "radio paddy," because of the medium through which farmers were introduced to the new crops. But radio . . . cannot match the effectiveness of television; and of course there are many types of information that can be fully conveyed only by images. Merely *telling* a farmer how to improve his herds or harvest is seldom effective. But seeing is believing, if he can compare the pictures on the screen with the scrawny cattle and the dispirited crops around him.

Although the SITE project sounds very well on paper, only experience will show if it works. . . . It is the software . . . that will determine the success or failure of the experiment. In 1967 a pilot project was started in eighty villages around New Delhi, which were equipped with television receivers tuned to the local station. . . . It was found that an average of four hundred villagers gathered at each of the evening "tele-clubs," to watch programs on weed control, fertilizers, packaging, high-yield seeds—plus five minutes of song and dance to sweeten the educational pill. . . .

Surveys have been carried out to assess the effectiveness of these programs. In the area of agricultural knowledge, television viewers have shown substantial gains over nonviewers. To quote from the report . . . of the Indian National Committee for Space Research: "The information given . . . was more comprehensive and clearer compared to that of the other mass media. Yet another reason cited for the utility of television was its appeal to the illiterate and small farmers *to whom information somehow just does not trickle"* [emphasis added].

In February 1971 . . . I visited one of these TV-equipped villages, Sultanpur, a prosperous and progressive community just outside of Delhi. . . . Dr. Vikram Sarabhai, chairman of the Atomic Energy Commission, had kindly lent us a prototype of the three-meter-wide, chicken-wire receiving dish that will collect signals from ATS-6. . . . While the village children watched, the pie-shaped pieces of the reflector were assembled, a job that can be performed by unskilled labor in a couple of hours. When it was finished, we had something that looked like a large aluminum sunshade or umbrella with a collecting antenna in place of the handle. As the . . . assembly was . . . lifted onto the roof of the highest building, it looked as if a small flying saucer had swooped down upon Sultanpur.

With the Delhi transmitter standing in for the still unlaunched satellite, we were able to show a preview of—one hopes—almost any Indian village of the 1980s. The program we actually had on the screen . . . was a lecture-demonstration in elementary mechanics, which could not have been of overwhelming interest to most of the audience; nevertheless, it seemed to absorb viewers whose ages ranged from under ten to over seventy. Yet it was . . . over six hundred kilometers away at Ahmedabad that I really began to appreciate what could be done through even the most elementary education at the village level.

Near Ahmedabad is the big parabolic dish, sixteen meters in diameter, of the experimental satellite-communications ground station, through which the programs will be beamed up to the hovering satellite. Also in this area is AMUL, the largest dairy cooperative in the world, to which more than *a quarter of a million* farmers belong. After we had finished filming at the big dish, our camera team drove out to the AMUL headquarters, and we accompanied the chief veterinary officer on his rounds.

At our first stop, we ran into a moving little drama that we could never have contrived deliberately, and which summed up half the problems of India in a single episode. A buffalo was dying, watched over by a tearful old lady who now saw most of her worldly wealth about to disappear. If she had called the vet a few days before—there was a telephone in the village for this very purpose—he could easily have saved the calf. But she had tried charms and magic first; they are not *always* ineffective, but antibiotics are rather more reliable. . . .

I will not quickly forget the haggard, tear-streaked face of that old lady . . . yet her example could be multiplied a million times. The loss of real wealth throughout India because of ignorance and superstition must be staggering. If it saved only a few calves per year, or increased productivity only a few percentage points, the television set in the village square would quickly pay for itself. The very capable people who run AMUL realize this;

they are so impressed by the possibilities of television education that they plan to build their own station to broadcast to their quarter of a million farmers. They have the money, and they cannot wait for the satellite, though it will reach an audience two thousand times larger, for over *500 million* people will lie within the range of ATS-6.

There is a less obvious, yet perhaps even more important way in which the prosperity and sometimes the very existence of the Indian villagers will one day depend upon space technology. The life of the subcontinent is dominated by the monsoon, which brings 80 percent of the annual rainfall between June and September. The date of the onset . . . however, can vary by several weeks, with disastrous results to the farmer if he mistimes the planting of his crops.

Now, for the first time, the all-seeing eye of the meteorological satellites . . . gives real hope of dramatic improvements in weather forecasting. But forecasts will be no use unless they get to the farmers in their half a million scattered villages, and to quote from a recent Indian report: "This cannot be achieved by . . . telegrams and wireless broadcasts. Only a space communications system employing TV will be . . . able to provide the farmer with something like a personal briefing. . . . Such a nationwide rural TV broadcasting system can be expected to effect an increased agricultural production of at least 10 percent through the prevention of losses—a savings of $1,600 million per annum."

Even if this figure is wildly optimistic, it appears that the costs of such a system would be negligible compared to its benefits.

And those who are unimpressed by mere dollars should also consider the human aspect—as demonstrated by the great Bangladesh cyclone of 1971. *That* was tracked by the weather satellites, but the warning network that might have saved several hundred thousand lives did not exist. Such tragedies will be impossible in a world of efficient space communications.

Yet it is the quality, not the quantity, of life that really matters. People need information news, mental stimulus, entertainment. For the first time in five thousand years, a technology now exists that can halt and perhaps even reverse the flow from the country to the city. The social implications of this are profound; already, the Canadian government has discovered that it has to launch a satellite so that it can develop the Arctic. Men accustomed to the amenities of civilization simply will not live in places where they cannot phone their families or watch their favorite television show. The communications satellite can put an end to cultural deprivation caused by geography. It is strange to think that, in the long run, the cure for Calcutta (not to mention London, New York, Tokyo) may lie thirty-six thousand kilometers out in space.

The SITE project will run for one year and will broadcast to about five thousand television sets. . . . This figure may not seem impressive . . .

but it requires only one receiver to a village to start a social, economic, and educational revolution. If the experiment is . . . a success . . . then the next step would be for India to have a full-time communications satellite of her own.

One of the most magical moments of Satyajit Ray's exquisite *Pather Panchali* is when the little boy Apu hears for the first time the aeolian music of the telegraph wires on the windy plain. Soon those singing wires will have gone forever; but a new generation of Apus will be watching, wide-eyed, when the science of a later age draws down pictures from the sky—and opens up for all the children of India a window on the world.

Mars and the

Mind of Man

On November 12, 1971, the space probe Mariner 9 *arrived at Mars, went into orbit,
and commenced taking a series of photographs that over the next few months
revolutionized our knowledge of the planet. The day before, a panel discussion was
arranged at Cal Tech and included Ray Bradbury, Carl Sagan, Bruce Murray
(a director of the Jet Propulsion Laboratory), and Clarke. When,
a year later, they had had an opportunity to examine the results
of the mission, they wrote afterthoughts that resulted in "Mars and
the Mind of Man." The passage that follows contains Clarke's
unrehearsed remarks at Cal Tech. The afterthoughts give his
carefully considered views a year later.*

I want to go along with Ray Bradbury's views on the importance of Edgar
Rice Burroughs. It was Burroughs who turned me on, and I think he is a
much underrated writer. The man who can create Tarzan, the best-known
character in the whole of fiction, should not be taken too lightly!

Of course, there's not much left of his Mars, and his science was always rather dubious. I can still remember even as a boy feeling there was something a little peculiar about cliffs of solid gold, studded with gems. I think it might be an interesting exercise for a geology student to see how that phenomenon could be brought about.

Another writer I'd like to pay tribute to, partly because he lived such a tragically short time, was Stanley G. Weinbaum, whose *Martian Odyssey* came out around 1935. And then, of course, the other great influence on me was our Boston Brahmin. Whatever we can say about Percival Lowell's observational abilities [Lowell founded his famous observatory at Flagstaff, Arizona, and brought the so-called canal controversy to the boil in the early part of this century by claiming that Mars was covered with a spiderweb of fine lines, which he believed to represent a vast irrigation system], we can't deny his propagandistic power, and I think he deserves credit at least for keeping the idea of planetary astronomy alive and active during a period when it might have been neglected.

Anyway, I was very moved the other day when I visited the Lowell Observatory for the first time and actually looked through his twenty-six-inch telescope. He is buried right beside it; his tomb is in the shape of the observatory itself. I was distressed to find that his papers had been rather neglected. As a result, I have initiated a series of events which may now result in them being classified and edited. Whatever nonsense he wrote, I hope that one day we will name something on Mars after him.

And then, of course, you mentioned H. G. Wells. He certainly did a lot for Mars. Movie director George Pal has also done a lot for Mars, not to mention Los Angeles, in *The War of the Worlds*. He gleefully destroyed City Hall and a few other places around here.

We are now in a very interesting historic moment with regard to Mars. I'm not going to make any definite predictions, because it would be very foolish to go out on a limb, but whatever happens, whatever discoveries are made in the next few days or weeks or months, the frontier of our knowledge is moving inevitably outward.

It has already embraced the Moon. We still have a great deal to learn about the Moon and there will be many surprises even there, I'm sure. But the frontier is moving on and our viewpoint is changing with it. We're discovering, and this is a big surprise, that the Moon, and I believe Mars, and twilight parts of Mercury and especially space itself, are essentially benign environments—to our technology, not necessarily to organic life. Certainly benign as compared with the Antarctic or the oceanic abyss, where we have already been. This is an idea that the public still hasn't got yet, but it's a fact.

I think the biological frontier may very well move past Mars out to Jupiter, where I think the action is. Carl Sagan has just gone on record as saying that Jupiter may be a more hospitable home for life than any other place, *including Earth itself*. It would be very exciting if this turns out to be true.

AFTERTHOUGHTS (1973)

Reading the transcript of this discussion is a curious experience, because it already seems to belong to another age—the prehistory of Martian studies. All of us knew, that November evening while *Mariner 9* approached its moment of destiny, how important this mission might be, but I doubt if any of us would have dared to predict the full extent of its success. True, *Mariner*'s cameras revealed no Martians carrying banners with the strange device BRADBURY WAS RIGHT (or even rival groups with NO—CLARKE WAS RIGHT). But what they did show was exciting enough. At last, we are zeroing in on the real Mars.

For most of this century, Mars has been haunted by the ghost of Percival Lowell, the man with the tessellated eyeballs. *Mariners 4, 6,* and 7 started to exorcise that ghost; *Mariner 9* completed the job. The famous "canals" are gone forever. Why they ever appeared in the first place could be material for a valuable study of psychology and physiological optics.

Now that we have good-quality photographs of Mars, someone should compare Lowell's drawings with the reality to try to find just what happened up there at Flagstaff at the turn of the century. How was it possible for a man to sustain a self-consistent and extremely detailed optical illusion (if that's what it was) over a period of more than twenty years? How did he convince others? What correlation, if any, was there between the ability of other astronomers to see the canals and their position on the Lowell Observatory payroll?

Recent work on the nature of vision has shown that the eye is capable of feats which, a priori, one would have said were completely impossible. "Eidetic imagery" is an example which may be very relevant here. Dr. Bela Julesz of Bell Labs discusses a case, in "Foundations of Cyclopean Perception," where a subject was able to store an apparently random pattern of 10,000 picture elements—a 100×100 matrix of dots—and fuse a stereoscopic image! These experiments suggest that the eye-brain system has an astonishing capacity to store detailed images. Could Lowell have built up, over a period of years, a largely mental picture of Mars from the fleeting patterns glimpsed through his telescope? The mind has an extraordinary ability to "see" things that are hoped for, assembling any chance visual clues that may come to hand. When you are expecting to meet a friend in a crowd, how often do you see him before he really appears!

If Lowell's Mars was indeed a largely subjective one, it also had to be dynamic. It must have changed continually with rotation, distance, seasons, to match the changing appearance of the real Mars. Certainly a fantastic feat of creative imagination, of the greatest interest to psychologists.

And although I'm speculating, could there be some connection between his superbly maintained and brilliantly proselytized delusion (remember, plenty of other observers "saw" the canals) and the hundreds of intelligent, sober, and altogether reliable citizens who have honestly "seen" brightly moving lights in the sky and all the other familiar UFO phenomena?

However, back to the real Mars. It now appears that, by one of those ironies not uncommon in science, the earlier Mariner results caused the pendulum to swing too far to the other extreme—away from the hopelessly romantic view. From 1965 to 1972 Mars was a cosmic fossil like the moon—no, not even a fossil, because it could never have known life. The depressing image of a cratered, desiccated wilderness was about as far removed from the Lowell-Burroughs fantasy as it was possible to get.

There were some, undoubtedly, who accepted the new "revelation" with considerable relief—even glee. Now there would be no further fear of that dreaded cry in the night, "The Martians are coming! The Martians are coming!" We were comfortably alone in the Solar System, if not the universe.

Well, perhaps we are, but it seems more and more unlikely. The new Mars that has suddenly emerged from the photos, a world of immense canyons and volcanoes and erosion patterns—and, dare one say, dried-up seabeds?—is a much more active and exciting place than we would have ventured to hope, only a few years ago.

It is not really a coincidence that, while *Mariner 9* was being built, the first positive evidence for the chemical evolution of complex organic molecules beyond the Earth was being discovered. The basic building blocks of life were being found in *meteorites,* of all places, perhaps as hostile an environment as could be imagined. In view of this, and the obvious signs of past water activity shown in the Mariner photos, the biologists will have some explaining to do—if there is *no* life on Mars.

Meanwhile, we science fiction writers had best be cautious for a few years—perhaps until soft-landers start to do some detailed reporting in the midseventies. For myself, I'm already a little embarrassed to see that *The Sands of Mars* (1951) contains the sentence "There are no mountains on Mars" (in italics). Well, it took over twenty years to shoot *that* one down, so it had a good run for its money. And on the plus side, we now have some perfectly beautiful photos of Martian sand dunes, so at least my title was completely valid. The sands of Mars have survived very much better than the oceans of Venus.

There are some not very bright and/or badly educated people who complain that scientific research destroys the wonder and magic of nature. More, one can imagine the indignant reaction of Tennyson or Shelley to this nonsense, and surely it is better to know the truth than to dabble in delusions, however charming they may be. Almost invariably, the truth turns out to be far more strange and wonderful than the wildest fantasy. The great biologist J. B. S. Haldane put it very well when he said, "The universe is not only queerer than we imagine—it is queerer than we *can* imagine."

I feel sure that *Mariner 9*—and its successors—will provide many further proofs of this statement. We have already learned an instructive lesson from the Moon, which is becoming more complicated and more interesting with every expedition. The same thing will happen with Mars. Whether we find life or not, we will discover things which we could *never* have imagined. And these will provide material for the deeper and richer fantasies of the future, just as the earlier observations inspired the fantasies of the past.

And the beauty of it is—we can have it both ways! When we are actually living on Mars, we will be reading the latest works of the lucky science fiction writers who are starting their careers now, at the beginning of the fourth golden age. Yet at the same time, they will still be able to enjoy, from their new perspective, the best of Wells and Burroughs.

And, I hope, of Bradbury and Clarke.

The Sea of

Sinbad

*Among Clarke's illustrious acquaintances during his Hotel Chelsea days were
Arthur Miller, Andy Warhol, Allen Ginsberg, Norman Mailer, Virgil Thompson . . .
and a young American photographer, Peter Castellano, who eventually brought
his equipment to Sri Lanka to do photo stories for British Airways and
The Observer of London. Clarke wrote the following two essays
to accompany them.*

The island of Ceylon is a small universe; it contains as many variations of
culture, scenery, and climate as some countries a dozen times its size. What
you get from it depends on what you bring; if you never stray from your
hotel bar or the dusty streets of westernized Colombo, you could perish of
fulminating boredom in a week, and it would serve you right. But if you
are interested in people, history, nature, and art—the things that *really* matter—you may find, as I have, that a lifetime is not enough.

Of course, not everybody feels that way; the writer Eric Linklater
once left in disgust, rashly announcing to the world that "Ceylon stinks."

The Ceylonese have carefully mispronounced his name ever since. Other distinguished visitors have been more flattering; Marco Polo declared that Ceylon is "undoubtedly the finest island of its size in all the world," and one could dig up similar quotations throughout the ages. The most beautiful tribute ever paid, however, must be that by a papal legate six centuries ago: "From Ceylon to Paradise, according to native tradition, is forty miles; there may be heard the sound of the fountains of Paradise."

You certainly won't hear the fountains if you arrive when the monsoon breaks, and the rain comes down in the best Somerset Maugham tradition. The finest time of the year for the western (Colombo) side of the island is December through March; there may be occasional showers, but most days are sunny and temperate, with the thermometer around eighty degrees Fahrenheit. Between April and July it becomes humid and generally unpleasant in the west, but on the other side of Ceylon, behind a barrier of mountains up to two thousand meters high, the weather will be fine. Thanks to these mountains, there are two climates in a country less than three hundred kilometers across. By careful choice of time and place (and travel agent!) it is possible to enjoy Ceylon in any month of the year.

One of the factors that adds immeasurably to the convenience of visiting Westerners is that—in contrast to many exotic countries—English is spoken everywhere, and practically all street signs are in Roman characters (as well as in Sinhalese and Tamil). All educated Ceylonese are bilingual, since the country was a British colony until 1948. It must be admitted that, as a result of a drive to make Sinhalese the national language, the standard of English among the younger generation has plummeted during the last few years. However, attempts are now being made to correct this, and it would take a fairly determined traveler to find a place where not a single person understood that famous Victorian phrase-book appeal "Help! My postilion has been struck by lightning."

The British influence, though waning, remains powerful; one of its most obvious signs is the large number of double-decker London Transport buses, still painted red, which come to Ceylon to die. Their corpses may frequently be seen by the wayside, mourned by passengers patiently waiting for a replacement.

Slightly older, and rather more interesting, ruins will be found in the sacred cities of Anuradhapura and Polonnaruwa, which were the capitals of the Sinhalese kingdoms for fifteen centuries, until about A.D. 1250. Dominating Anuradhapura are three great shrines, or dagobas, bell-shaped piles of solid masonry almost rivaling the pyramids in size. And here may also be found the famous 2,260-year-old bo tree, reputed to be a sapling from the original under which the Buddha attained enlightenment.

The finest statues of the Buddha himself will be found at Polonnaruwa, which became the island's capital when Anuradhapura was

abandoned (after repeated invasions from India) around A.D. 1100. A huge
reclining figure, fifteen meters long, represents the Buddha at the moment
of death; standing at its head is a smaller—but still impressive—statue with
arms crossed and a look of downcast meditation or sadness. This is usually
taken to be the Buddha's chief disciple, Ananda, grieving over the departure
of his master; but most modern scholars believe it to be another manifes-
tation of the Enlightened One himself.

When Polonnaruwa was abandoned in turn, the Sinhalese kings retreated to the hill capital of Kandy, which still flourishes; during World War
II, it was Lord Mountbatten's headquarters. Today its chief claim to fame
is the Temple of the Tooth, from which every year caparisoned elephants
accompanied by dancers, bands, and torchbearers march in a spectacular
procession honoring the sacred relic of the Buddha, which, however, never
leaves the temple precincts.

Fifty kilometers south of Kandy is an equally holy spot, the 2,200-
meter-high mountain Adam's Peak—an equatorial Matterhorn, completely
clothed with trees. Perched on its summit is a tiny temple, and every year
thousands of pilgrims make the ascent up what must be the longest stairway
in the world. The climb is not difficult, but can you imagine a stairway
several kilometers long? I managed it once, and my legs were paralyzed for
the next three days. But it was worth it, for at dawn I saw the spectacle for
which the peak is famous. As the sun rose, the perfectly triangular shadow
of the mountain was cast on the clouds below, stretching for perhaps fifty
kilometers into the west. It lasted for a magical ten or fifteen minutes, then
faded out to cries of "Sadu, sadu" (Holy, holy) from the pilgrims.

Of all Ceylon's archaeological wonders, however, the most remarkable—and certainly the most useful—is the enormous irrigation system,
which, for over two thousand years, has brought prosperity to the rice
farmers in regions where it may not rain for six months at a time. Frequently
ruined, abandoned, and rebuilt, this legacy of the ancient engineers is one
of the island's most precious possessions. Some of its artificial lakes are ten
or twenty kilometers in circumference and abound with birds and wildlife.

Ceylon was once a hunter's paradise; Victorian "sportsmen" boasted
of killing hundreds of elephants, bears, leopards, and birds beyond com-
putation. There are now less than two thousand wild elephants on the
island, most of them rigidly protected in great game sanctuaries. Wilpattu,
the largest of these, is only a few hours' drive from Colombo, yet it seems
so remote from man and all his ways that the visitor can easily imagine he
is back in the primeval forest. No wonder Ceylon has become popular with
makers of jungle-type movies; though Tarzan has not yet arrived, Elephant
Bill and Mowgli are on the way, and who can forget Elizabeth Taylor facing
the stampede of tuskers in the blazing finale of *Elephant Walk*? But the best
film made in the country is still *The Bridge on the River Kwai* (which was

actually set in Burma). If you want a vicarious trip to Ceylon, check out David Lean's masterpiece at your video shop.

The mountains, the jungles, the ruined cities, the lakes, the game reservations, the misty upland tea estates, will all have their advocates; but to me there is nothing that can compare with the beaches and the reefs. To see these at their best, you have to get well away from the towns. Drive south of Colombo for fifty kilometers, or fly across the island to the sun-drenched loneliness of the east coast, and you may find yourself the only human being on a crescent of soft, white sand, enclosing a bay of the purest blue water, fringed by palm trees of astonishing height and slenderness.

The Indian Ocean is one of the last great unexplored regions of this planet. It is a curious experience to stand at the southernmost tip of Ceylon, Dondra Head, and to know that there is nothing but empty sea all the way south to the icy ramparts of Antarctica. After the swift sunset (for here you are only six hundred kilometers from the equator) the last lighthouse of the northern hemisphere will start to flash its warning above you, and far to the east you may see the flicker of its companion on the deadly Great Basses Reef. A long chain of barely submerged rocks, ten kilometers out to sea, this has been trapping ships ever since men started to sail the Indian Ocean. A few years ago, my colleagues discovered the wreck of a large armed merchantman, carrying at least a ton of beautiful silver rupees. We nearly went bankrupt salvaging it, and I wouldn't wish sunken treasure on my worst enemy.

A better investment in time and effort would be hunting for emeralds and sapphires in the island's gem pits, famous since the days of Sinbad. His "Valley of Gems" was located in Ceylon, but the elephant-catching rocs which nested in the mountains above it are fortunately extinct.

Today's jumbos fly under their own power and land at the very hand-some Bandaranaike airport thirty kilometers north of Colombo. When you disembark here, you may encounter for the first time the phrase *ayu bowan*—which means both "greetings" *and* "good-bye."

With the adoption of a new constitution in 1972, Ceylon officially changed its name back to the ancient Sinhalese form. So—"Ayu bowan, Ceylon; Sri Lanka, ayu bowan."

Either way, the country will still be just as beautiful.

In 1966, the island of Ceylon cut itself adrift from the rest of the world. It abandoned the seven-day week and reverted to the traditional Buddhist calendar, based on the phases of the Moon. Thus each lunar quarter became a holiday (Poya Day), and the day before it a half-holiday.

The result was instant chaos. It was useless planning to meet anyone on, for example, Monday week, since Monday (or any other day) might be Poya, and all offices and shops would be closed. This was merely an inconvenience as far as the country's internal affairs were concerned, but it caused utter confusion in all dealings with the outside world. Once every six weeks or so, Poya Day fell on a Sunday and the Ceylonese were briefly in step with the rest of humanity. But most of the time, the tea brokers of Mincing Lane could telephone their Colombo offices only three or four days out of every week, and they were never quite sure *which* days those would be. . . .

This attempt to put the calendar back a couple of thousand years was a move by Prime Minister Senanayake's mildly left-of-center government to gain the approval of the priests, who have considerable influence in a country that is primarily Buddhist. But the opposition, the United Front party—led by Mrs. Bandaranaike—was just as keen on the idea.

Despite its cost to the country, and the grumblings of the businessmen, no politician dared to attack this exercise in nostalgia; yet before it had completed its fifth year, it was quietly abolished and Monday, Tuesday, Wednesday . . . reentered the Ceylonese vocabulary. The sudden ending of the Poya calendar was a small but significant by-product of a national trag-edy which proved that there is no way back into the past, and the real or imagined qualities of one's ancestors have little relevance to the problems of modern life.

On the night of April 4, 1971, when the little island of 16 million appeared reasonably contented under a government that had been elected by a large majority, a force of several thousand well-trained insurgents at-tempted to take over the country. Attacks were launched upon many pro-vincial police stations, and the rebels were thus able to seize arms and obtain temporary control in some rural areas.

Although there had been some earlier signs of trouble, the govern-ment was taken almost completely by surprise. It appealed for help, which promptly arrived from a remarkable variety of sources, including the United Kingdom, the United States, the Soviet Union, India, Pakistan, the United Arab Republic, Yugoslavia, *and* China.

After a few weeks of often bitter fighting, the insurrection was put down and some fifteen thousand rebels and suspects were rounded up in prison camps. By the end of 1972, the country was back to normal, though the human and economic cost of the tragedy would never be fully assessed, and many of the factors that had provoked it still remain. The ordinary tourist, jetting into Colombo's elegant Bandaranaike airport, would see no evidence of the traumatic experience. For politicians and governments and social systems come and go; the land and the sun and the sea remain in exquisite proportion.

From the earliest times, Ceylon seems to have captured the hearts of travelers, preserved in the glamorous names they gave it: Serendip, Taprobane, the Resplendent Isle, Land Without Sorrow.

And to the English, of course, Ceylon has a special place—not merely because it was a colony for a century and a half, but because for most of that time it was a principal source of tea, without which the United Kingdom would have come to a screeching halt. Early in 1940, some genius at the Ministry of Food realized this, and all the tea chests in the London docks were dispersed throughout the land—just ahead of the blitz. I spent my last civilian summer battling with innumerable bills of lading, trying to find exactly where the stuff had gone. It was then that I came across, for the first time, the names of the great estates and tea-growing districts—Uva, Dimbulla, Nuwara, Eliya, Bandarawela. Little did I know that sixteen years later these exotic places would be the background of my own everyday life.

Until a few months ago, I was unable to account for this strange attraction to a country I had never seen, and had scarcely thought about, until my late thirties. On the frequent occasions when I was asked why I liked Ceylon, I had plenty of answers, but they were never really convincing. There was always something missing; what it was, I discovered only recently—and on the other side of the globe from Ceylon.

Of course, some of the island's charm was blatantly obvious. To anyone who had survived a score and a half of English winters, the idea of spending Christmas sunbathing under the palms, or swimming in warm water of the purest blue, was irresistible.

There is another reason why I like Ceylon; it is the right size—slightly smaller than Ireland. There is no point that cannot be reached from any other in a day's driving, over roads that are usually adequate and often excellent, which is more than can be said of most of the drivers and their vehicles. I vividly recall a Ceylon Transport Board bus that once shed its entire transmission fifty meters ahead of my car. As one hundred kilograms of metal bounced closer and closer, I unselfishly prayed that the world of letters would not sustain a major loss. I happened to have Gore Vidal riding beside me.

Geography and climate do not make a country, though they determine what kind of a country it will be. There are islands in the Pacific perhaps more lovely and more temperate, but they have no culture, no sense of the past—nothing to engage the intellect. Ceylon offers far more than the empty, mindless beauty that lured Gauguin to destruction; it has 2,500 years of *written* history, and the ruins of cities that were once among the greatest in the world.

Over the millennia, successive waves of invaders have poured into Ceylon from the north; they brought with them their language, their technology, their art, and above all, their religion. When Buddhism declined in India, it took root in Ceylon; and here, so its adherents claim, it is still to be found in its purest form. Certainly it survived almost five centuries of persecution and indifference, under three successive foreign regimes. The Portuguese, Dutch, and British occupations each lasted about 150 years.

The Portuguese came with sword and cross; the Dutch with ledger and lawbook; the British with roads and railways. Each invasion had shattering but not wholly destructive effects on the indigenous lifestyles, and much of Ceylon's fascination today arises from this extraordinary mixture of cultures.

All these are set in a background of pure Sinhalese, who constitute over 70 percent of the population, and theirs is now the official language—to the considerable disadvantage of the other main racial group, the Tamils.

The population now stands at about 16 million, and most of them owe their lives to Dr. Hermann Muller, who discovered the useful properties of DDT. Until the 1940s, malaria was endemic in Ceylon. Despite a high birthrate, for centuries the population has been stabilized—and enervated—by the mosquito. In what is now the textbook case of insect control by DDT, the carrier and the disease were virtually eliminated in the late 1940s.

As a result, the population doubled in thirty years, with the resultant problems of unemployment, inadequate social services, food shortages. A country so rich in natural resources could support many more people than it does today, and at a much higher standard of living. But the *rate* of increase does not allow time for the necessary planning.

And it is hard to plan for the future, when the sun beats down from a cloudless sky, the waves whisper softly up the beach, the terraced fields of ripening paddy seduce the eye with their soft greens and golds. The men of the cold north believe that the tropics are hostile to civilization because the struggle for existence can too easily be won. There is much truth in this, but there are also times when sun and drought can provoke as great a response as storm and snow. This happened, in Ceylon, before the beginning of the Christian era, when a series of tremendous irrigation works transformed the island's dry zone into what must have been a fertile paradise. Some of the artificial lakes created then are many kilometers in circumference; there are thousands of these unfortunately named "tanks," linked by intricate networks of canals. Only a stable, well-organized, and technically advanced society could have undertaken such massive projects; such a society was seldom allowed to exist in peace for long, and successive invasions destroyed much of the work of the ancient engineers.

In modern times, many of the old tanks have been restored, and new ones built; these little inland seas, surrounded by mountains, are often so tranquil that they act as perfect mirrors for the passing clouds. Yet there can be few places on earth where the past seems more alive, more linked to the present by generations of toil and skill, all directed to the same end. For the modern electrically operated sluicegates may be set in timeworn stones that were carved before Caesar came to Britain. At the remoter tanks, the harassed survivors of Ceylon's once countless wild elephants may be seen coming to the water's edge to drink; in another decade, alas, they will be gone, through it is to be hoped a thousand or two will flourish indefinitely in the great game reservations where the only shooting allowed is by telephoto lens. Here, four or five hours' drive from Colombo's airport, is another world, even more ancient than that of the great engineer-kings.

The wildlife of the land has been famous for centuries, but only in this generation have we grown to know the wildlife of the sea—which is even more prolific, and far more colorful. Much of the island is surrounded by coral reefs, the new playgrounds of our age. It was these that first drew me to Ceylon and gave me the keenest moments of delight (and fear) I have ever known; only much later did I discover the lovely land that they guard against waves thundering across thousands of kilometers of open sea.

Ceylon is the last outpost of the Northern Hemisphere, jutting into the still largely unexplored emptiness of the Indian Ocean; it may be the base for the great undersea expeditions of the century to come. And the southern coast of the island, I realize now, is the source of the magic that holds me here.

I was born by the sea and spent much of my childhood on the beach of the great curving Bay of Bristol in the west of England; no wonder that I have always been haunted by Houseman's lines of summer days spent on the "smooth between sea and land."

But I lost the sea when I was about ten years old and thereafter saw it only on holidays and brief visits. School, the civil service, the war, college, and a new career separated me further and further from the games of childhood.

Even a year on the Great Barrier Reef did not unlock the doors of memory. Not until I came to Ceylon did I fall in love with an exquisite arc of beach on the island's south coast and decide to establish a home there.

It takes a long time to see the obvious, and in this case perhaps there was some excuse. After all, there was little apparent similarity between the gray English sea and turquoise Indian Ocean; between boardinghouses, Butlin holiday camp, railway station—and an unbroken wall of closely packed coconut palms.

One day, after a lecture in the American Midwest, a young lady asked me just *why* I liked Ceylon. I was about to switch on the sound track I had played a hundred times before, when suddenly I saw those two beaches, both so far away. Do not ask me why it happened then; but in that moment of double vision, I knew the truth.

The drab, chill northern beach on which I had so often shivered through an English summer was merely the pale reflection of an ultimate and long-unsuspected beauty. Like the three princes of Serendip, I had found far more than I was seeking—in Serendip itself.

Ten thousand kilometers from the place where I was born, I had come home.

Willy and

Chesley

One of the most beautiful and influential books ever written about space travel was
The Conquest of Space, *written in 1949. It consisted of paintings of astronomical*
scenes by Chesley Bonestell (an architectural draftsman, who attracted great
attention in the forties when he produced his famous views of the Moon,
Saturn, and Jupiter for Life *magazine), with text by Willy Ley, one of*
the founders of the German Rocket Society (which included the teenager
Wernher von Braun). Though Willy was trained as a zoologist,
he wrote hundreds of articles on almost every branch of science,
and his work Rockets, *begun in 1944, and its revisions still*
provide an invaluable history of the subject. To millions,
they were a foretaste of the coming Space Age. Unfortunately, after
writing about space travel for half a century, Willy died just a
month before the first man reached the Moon.

My very last meeting with Willy Ley could hardly have occurred in more
appropriate surroundings. I was descending the stairs of a New York subway
when I spotted his figure some considerable distance ahead of me and was

about to hail him when I had a sudden qualm; suppose it's really someone else and I make a fool of myself? Then I remembered the clincher: This was the Forty-second Street and Fifth Avenue entrance, and the New York Public Library was immediately overhead. Anyone around here who looked like Willy *would* be Willy—and so it turned out.

Willy was more widely read, and owned a larger number of books, than anyone else I have ever met. His initial and lifelong interest was zoology—itself a sufficiently enormous subject—but the early European speculations about the possibility of spaceflight diverted him into rocket research and astronomy, as he recounted in his classic *Rockets, Missiles and Men in Space* in 1968. Eventually, he became—to use the title of another of his works—equally at home "on Earth and in the sky," though he was certainly best known for his writing and lecturing on astronautics.

Nowadays, it is easy to forget that it required considerable courage to preach the possibility of space travel in the 1930s. In those days there were many experts (not a few of them now happily riding the space bandwagon) ready and eager to prove that all thoughts of escaping from Earth were scientific nonsense. Willy Ley helped to educate the generation that turned the fantasy into hardware.

Although his most important work was undoubtedly *Rockets, Missiles and Men in Space,* perhaps the book which was of the greatest inspirational value and aesthetic appeal was *The Conquest of Space.* In this handsome volume Willy matched his text to Chesley Bonestell's beautiful paintings of the Moon and the planets. When *Conquest* appeared in 1949, few of those who gazed entranced at the vistas of lunar landscapes, or the Earth as seen from space, could have imagined that within a mere twenty years the vision of the scientist would have matched that of the artist. However, the subject of Chesley Bonestell's masterpiece—the crescent Saturn in the sky of its giant moon Titan—remains a spectacle which human eyes may not witness until the twenty-first century.

Yet Willy sometimes backed the wrong horse. In the first edition of *Rockets,* he implied that the German "secret weapon" was *not* a rocket, arguing that the performance required for long ranges demanded liquid propellants, which would not be practical for military applications. He was unlucky in his timing, for *Rockets* and the V-2 arrived at about the same moment. My colleague A. V. Cleaver, then director and general manager of the Rolls-Royce Rocket Division, has recorded meeting Willy in New York in the early 1940s and being "astonished to find that, for some reason, he had decided that the rumors were a lot of nonsense. He spent much time and effort assuring me that his ex-countrymen were most unlikely to have developed such a weapon. . . . I argued weakly against these conclusions, but being very conscious of wartime security . . . forbore to tell him that I had personally heard the 'rumors' arriving." But Willy's point about the military disadvantages of liquid-propelled rockets was well taken. As soon as possible, they were dropped in favor of solids; the Atlases and Titans have been replaced by Minutemen and Poseidons.

Though Willy was now an expert in many fields, and in both appearance and accent was the typical German scholar, he was no pedant, and his countless articles were not only entertaining but frequently witty. Few people could demolish a crank with a more dexterous rapier thrust; perhaps the fact that in the early days he must often have been taken for a crank himself gave him the technique for dealing with the genuine variety.

His writing was not, however, entirely restricted to fact; he was also the author of two interesting short stories. One, "Fog," is an account of the general confusion during a political revolution, based on his own experiences in Germany. Better known is "At the Perihelion," published under the name Robert Willey, which, as might be expected, is a soundly based tale of spaceflight. It introduced the idea of generating centrifugal "gravity" by setting two halves of a spaceship spinning at the opposite ends of a long cable. This technique was first tested on the Gemini 12 mission in November 1966.

Only once did I catch Willy out in a matter of scientific fact. In the preface to one edition of *Rockets,* he chided me for deserting the Northern Hemisphere to go to live in Ceylon. Having been under exactly the same misapprehension myself at one time, I gleefully told him to have another look at the map.

As countless aspiring writers and inquisitive science fans can testify, Willy was a kindhearted and helpful person. During World War II, he earned my eternal gratitude by keeping me supplied with otherwise unobtainable issues of the American science fiction magazines—particularly the handsome, large-size *Astounding Stories* of the lamented golden age.

I was on my way to the Apollo 11 launch when I heard the news of Willy's sudden death by heart attack. My immediate reaction—like that of all his friends—was one not only of sorrow but of something approaching anger. For forty-five years Willy had devoted the greater part of his life to the conquest of space. Although he had seen the total vindication of all his ideas, he had missed the final triumph by just four weeks.

Many have expressed the hope that one of the lunar craters will be named after Willy Ley, when the committee of the International Astronomical Union charged with this task gets to work on Farside. But in any event, his books will ensure that his name is not forgotten. I know that I shall be consulting *Rockets* for the rest of my life; and whenever I do, I shall remember Willy.

———

When I pored over the Jovian and Saturnian vistas depicted in *The Conquest of Space,* back in those distant days when no rocket had ascended much more than a hundred kilometers from Earth, I never dreamed that the time would come when I myself would be collaborating with the artist. Yet in 1951, I did pay a somewhat novel tribute to Chesley Bonestell, when I wrote a story called "Jupiter Five" in *Reach for Tomorrow.* This involved an expedition by "*Life* Interplanetary" to Saturn and Jupiter for the specific purpose of photographing the reality and comparing it with Chesley's century-old paintings.

Incredibly, just twenty years later, I was involved with Chesley on a project that would do precisely this. In 1971, NASA was about to launch the space probes *Pioneer 10* and *11,* which, if all went well, would radio back our first views of the mightiest of all planets. *Beyond Jupiter: The Worlds of Tomorrow,* our joint description of the project in illustrations and text, was beautifully produced by Little, Brown in 1972 and dedicated, "To Willy, who is now on the Moon." It is quite uncanny to compare Chesley's painting of Jupiter from close quarters with the marvelous color images later radioed back from the Pioneers, which are still heading on toward the stars, sending back information to the giant radio telescopes which follow their progress into the abyss.

They carry, as everyone knows, the famous plaques designed by Carl Sagan and his associates—the Space Age equivalents of those messages corked in a bottle and dropped into the sea. I was not aware of this (it was, in fact, a last-minute modification) when I wrote the concluding paragraphs of *Beyond Jupiter,* describing the ultimate fate of our first voyagers to the stars:

> As our space-faring powers develop, we may overtake them with the vehicles of a later age and bring them back to our museums, as relics of the early days before men ventured beyond Mars. And if we do not find them, others may.
>
> We should therefore build them well, for one day they may be the only evidence that the human race ever existed. All the works of man on his own world are ephemeral, seen from the viewpoint of geological time. The winds and rains which have destroyed mountains will make short work of the pyramids, those recent experiments in immortality. The most enduring monuments we have yet created stand on the Moon, or circle the Sun; but even these will not last forever.
>
> For when the Sun dies, it will not end with a whimper. In its final paroxysm, it will melt the inner planets to slag and set the frozen outer giants erupting in geysers wider than the continents of Earth.

Nothing will be left, on or even near the world where he was born, of man and his works.

But hundreds—thousands—of light-years outward from Earth, some of the most exquisite masterpieces of his hand and brain will still be drifting down the corridors of stars. The energies that powered them will have been dead for aeons, and no trace will remain of the patterns of logic that once pulsed through the crystal labyrinths of their minds.

Yet they will still be recognizable, while the universe endures, as the work of beings who wondered about it long ago and sought to fathom its secrets.

The Snows

of Olympus

Early in 1973, Playboy *magazine asked Clarke to write a short essay on any subject*
he pleased. He chose one of the most impressive discoveries made by Mariner 9,
Mount Olympus. It was located at a spot the old mapmakers,
peering at the tiny telescopic image of Mars, had given the astonishingly
prescient name Nix Olympica, the "Snows of Olympus." Here is the
result, which Clarke considers one of his best pieces of nonfiction.

In 1972, for only the third time in history, mankind discovered a new
world. It happened first in 1492. The impact of that discovery was immediate, its ultimate benefits incalculable. It created a new civilization and
revivified an old one. The second date, not quite so famous, is 1610. In the
spring of that year, Galileo turned his primitive "optic tube" toward the
Moon and saw with his own eyes that Earth was not unique. Floating out
there a quarter of a million miles away was another world of mountains
and valleys and great shining plains—empty, virginal—awaiting, like Michelangelo's Adam, the touch of life. And 362 years later, life came, riding
on a pillar of fire.

With the end of the Apollo program, there will now be a short pause until much cheaper transportation systems are developed. Then we will return and the history of the Moon will begin. But it may be quickly overshadowed by a greater drama, on a far more impressive stage. The third new world was not found by sailing ship or by telescope; yet, like the two earlier discoveries, it was a shocking surprise that resulted in the overthrow of long-cherished beliefs. No one knew that such a place existed, and when the evidence started to accumulate, early in 1972, many scientists were literally unable to believe their eyes.

This new world has nearly twice the diameter of the Moon and is almost four times as large as both Americas. And it has the most spectacular scenery yet discovered anywhere in the universe. Think of the Grand Canyon, the greatest natural wonder of the United States. Then quadruple its depth and multiply its width five times, to an incredible seventy-five miles. Finally, imagine it spanning the whole continent, from Los Angeles to New York. Such is the scale of the canyon that is carved along its equator.

Yet even this is not the planet's most awesome feature, for it is dominated by volcanoes that dwarf any on Earth. The mightiest, Nix Olympica—the Snows of Olympus—is almost three times the height of Everest and more than three hundred miles across. Those volcanoes are slumbering now, but not long ago they were blasting into the thin atmosphere all the chemicals of life, including water; there are dried-up riverbeds that give clear indication of recent flash floods—the first evidence ever found for running water outside our Earth. It even appears that this may be a young world, geologically speaking; if life has not already begun there, that will be yet another surprise.

By now, you may have guessed the identity of this new world. It is Mars—the *real* Mars, not the imaginary one in which we believed until *Mariner 9* swept aside the illusions of decades. It will be years before we absorb all the lessons of this, the most successful robot space mission ever flown; but already it seems that Mars, not the Moon, will be our main order of extraterrestrial business in the century to come.

This news may be received with less than enthusiasm at the very moment when NASA's budget is being cut to the bone and voices everywhere are calling for an attack on the evils and injustices of our world. But Columbus did more for Europe by sailing westward than whole generations of men who stayed behind. True, we must rebuild our cities and our societies and bind up the wounds we have inflicted upon Mother Earth. But to do this, we will need all the marvelous new tools of space—the weather and communications and resources satellites that are about to transform the economy of mankind. Even with their aid, it will be a difficult and often discouraging task, with little glamour to fire the imagination.

Yet, "where there is no vision, the people perish." Men need the mystery and romance of new horizons almost as badly as they need food and shelter. In the difficult years ahead, we should remember that the Snows of Olympus lie silent beneath the stars, waiting for our grandchildren.

SIX

Writing

to Sell

"So many publishers and authors have asked me to comment on books," Clarke writes, "or to write prefaces, that I am now forced to turn down all such requests no matter how good the cause." This form is sent out by his secretary in answer to 95 percent of Clarke's mail.

Scott Meredith, my agent, jolly well knows this, for his office sent out skillions of form letters in the hope of heading off the mailman at the pass. So what the heck am I doing here?

I'll tell you exactly how it happened. It began with a phone call to Scott from my dear friend Isaac Asimov. "Scott," he said desperately, "I'm only one hundred fifty books ahead of Arthur—he's catching up. If you can slow him down just a bit, I'll give you the Lower Slobovian Second Serial Rights of *Asimov's Guide to Cricket*—without TV residuals, of course."

"Throw in that illustrated braille *Kama Sutra* I know you're working on," Scott replied instantly, "and you have a deal."

"Done," said Isaac, whereupon Scott merely threatened to give my address to 589 people who want to know the real and secret message in *2001: A Space Odyssey,* and here I am. . . .

It was, for heaven's sake, just over a quarter of a century ago that Scott and I first met in person as I stepped off the ocean liner onto the sacred concrete of Manhattan. At that time he looked about eighteen and wasn't too much older than that and was thus only the second or third best literary agent in the United States; he was also still in shock over the fact that he'd cabled to say that my book *The Exploration of Space* had just become a Book-of-the-Month Club selection, and I'd replied by asking innocently, "What is the Book-of-the-Month Club?" And it is quite a shock to me, so many years later, to look at the copyright page and realize that he had already written, and Harper had already published, the first edition of the volume you, gentle would-be writer, hold in your hands. . . .

Yet another shock, while we are about it. On going through what I laughingly refer to as my records, I've just discovered that it was in the late forties that I sent Scott my first submission—the short story that later become the opening for *Childhood's End.*

It sometimes seems to me that every writer I know was represented by Scott and acquired his basic skills as a staffer at Scott's agency.

I'm exaggerating, of course; I know that Norman Mailer and Ernest K. Gann and Jessica Mitford and Ellery Queen and Carl Sagan and Taylor Caldwell and hundreds of other Scott Meredith clients never worked for him. But he certainly seems to have had a hand in the beginnings and development of an apparently endless list of best-seller authors. There was Harry Kemelman, for whose first novel Scott was only able to get an advance of $1,000, but whose contracts in the seventies approached half a million dollars. There's Hank Searls, whose first sale via Scott was a short story to a magazine for $50, and whose novel *Overboard* in the 1970s was a best-seller, a selection of the Literary Guild, Reader's Digest Book Club, and four other book clubs, with the paperback rights selling for hundreds of thousands of dollars, a best-seller in a dozen other countries as well, and a motion picture as well. There's also a young Englishman named Clarke who came to Scott in 1947 with that short story that Scott sold, as I recall it, for $70, and whose other novels, handled by Scott, were sold for—well, a bit more than that. And the list goes on and on. Scott moved from 580 Fifth Avenue to larger quarters at 845 Third Avenue in the 1970s. [After his death, some of his clients moved on to Scovil Chichak Galen, where I am now represented by Russell Galen, who, of course, had worked with Scott.] But Scott's "scarred and lacerated spot on the wall," against which he beat his head from time to time, was carefully framed and followed him from premises to premises. When he wasn't using it, he pretended it was a Jackson Pollock on loan from the Museum of Modern Art.

In all seriousness, I do not believe that any writer, however experienced he may fondly consider himself to be, could fail to benefit from Scott Meredith's book *Writing to Sell,* Harper and Row, 1977. The fact that it is also highly entertaining doesn't do any harm; several times I found myself laughing out loud. It says a good deal for Scott's sense of humor that it has survived so many decades of contact with authors . . . not to mention editors.

It would be impertinent of me to add anything to Scott's hard-won advice—so, I'll be impertinent.

It's always seemed to me that the biggest single problem an author has to face is, When should I give up? Scott would say, "Never!" and as he describes in "Inspiration, Perspiration, Desperation," he has a special padded cell, in which he occasionally locks up authors to prove this theory. But it really isn't as simple as that.

Agreed, authors have an amazing capacity for inventing excuses not to work. (Mine is a beautiful little monkey who cries piteously if not loved every hour, on the hour.) But there are times when no amount of staring at the typewriter and sweating blood will produce anything except frustration; you must learn to recognize those times.

Then you have two choices: you can switch to writing *something completely different* or, if that doesn't work, you must quit altogether. It was, I believe, Hemingway who said, "Writing is not a full-time occupation." That's true in more ways than one. You must live before you can write. And you must live *while* you are writing. But you mustn't kid yourself and make excuses to stop work by confusing laziness with that old standby "lack of inspiration." There is a lot of truth in the hard saying "A professional can write even when he doesn't feel like it. An amateur can't—even when he *does.*"

The other big problem is to know when a job is finished. Sorry about all these quotations, but this is the most important one yet: "No work of art is ever finished; it is only abandoned."

Old-time pros may snort indignantly at this; racing deadlines, they couldn't afford the time even for a second draft. Lester del Rey—another thoroughbred from the Scott Meredith stables, another staff editor at the agency who became a Scott Meredith client—could sit at the typewriter and produce, *within two hours,* a pretty good six-thousand-word story. That's professionalism, which I can only view with incredulous awe.

On another occasion, Lester sat down after breakfast and mailed off a twenty-thousand-word novelette the same night. He admits that it might have been better if he'd left it until the next morning . . . but would it have been all that *much* better? You can go on tinkering and revising and polishing forever; sometimes it is as hard to *stop* work on a piece as it was to start in the first place. But unless you're a poet turning out one slim volume every twenty years, you must learn to recognize the point of diminishing returns and send your 99.99 percent completed manuscript out into the cruel, hard world. There's nothing wrong with amateurism (in the best sense of that misused word); but all serious artists are interested in money. And that means all *great* artists, too; take a look someday at Beethoven's correspondence with the London Philharmonic.

Now, to give you inspiration, here are a few ideas for books that I feel *somebody* ought to write. They cover a pretty wide range—biography, history, crime, science fiction, medicine—with a couple of guaranteed best-sellers thrown in:

Jonathan Livingstone Sea-Slug

The inspirational saga of one of nature's humblest organisms, an adventurous Abominable Sea Slug (Mucus horribilis.) Jonathan, born—or rather, fissioned—in a sewer outlet off Flushing, feels a dim impulse for higher things and conceives the brave ambition of slithering upward to the glorious world above the waves. Unfortunately, despite the amazing camouflage that makes him almost indistinguishable from his surroundings, Jonathan is eaten by an even more revolting creature, the Squamous Scavenger Fish (Scatophagus vomitous), before completing his odyssey.

Screaming Flesh

The memoirs of a famous surgeon, pioneer of navel transplants, who is knighted by a grateful government when his revolutionary "brain bypass" operation transforms the fortunes of a political party. The author recalls, with brutal frankness and in loving detail, carefree days as a young medical student in the Terminal Accident Ward at St. Sepulchre's. Though there are lighter moments—such as the hilarious episode of the electrified bedpans—sensitive readers may not get past chapter 2, which gives the book its striking title.

Sunset on the Boulevard

A moving account of the work carried out by a rescue mission among the poor of Bel-Air and Beverly Hills in the years following the Final Depression. There are harrowing stories of the destitute drinking the last drops of water in their swimming pools; maddened by hunger, trying to open Andy Warhol soup cans—and finally succumbing to fatal sunburn when the protective smog vanishes.

Defender of the Doomed

The aptly entitled autobiography of the famous criminal lawyer whose courtroom exploits—hopefully—are never likely to be equaled. The author tells how he fought to save no less than ninety-eight clients from the gas chamber or electric chair—and lost them all. Now, for the first time, we learn exactly how he did it.

If you can't find something in this to trigger your imagination, perhaps you'd better stick to running your father's pre-stressed liverwurst business.

Just one minor point. My agent and I would each like 10 percent of the loot.

And 0 percent of the lawsuits.

Part Five

THE 1980s:

STAY OF

EXECUTION

Clarke with director Peter Hyams, Ray Bradbury, and Gene Roddenberry
(left to right) at the Hollywood premiere of *2010*, December 1984.

Introduction

The decade of the eighties, though it contained many highlights, is one which I have tried to forget. It's only with the aid of Neil McAleer's invaluable biography that I am able to reconstruct it.

In February 1980 my mother died at her home in Somerset; she had broken her hip in a fall while visiting me in Sri Lanka and had never fully recovered. On the day of her funeral I had to be in India filming the total eclipse of the sun, that opened my TV series *Arthur C. Clarke's Mysterious World*.

Later in 1981 I had the pleasure of flying Robert and Virginia Heinlein over the three places that had the greatest influence on my life—Adam's Peak, Sigiriya, and the Great Basses. That was my last view of the dangerous reef where I had my most memorable diving experience.

In 1982 my lifestyle was changed abruptly with the arrival of my first computer (5 MB of memory—who'd ever need so much?). *2010: Odyssey Two* was "Archie's" first production, and I was never to touch a typewriter again.

Other high points of this rather hectic decade were receiving the Marconi Fellowship Award; my first visit to Russia and Star Village, where I was entertained by Cosmonaut Aleksei Leonov; and the U.N. Conference on the Exploration & Peaceful Uses of Outer Space (UNISPACE '82) in Vienna. I used this opportunity to deplore the militarization of space—over a year before President Reagan gave his notorious "Star Wars" speech—and was delighted when Rep. George Brown of California placed my address in the *Congressional Record*.

Other highlights were speaking from the famous U.N. podium during World Telecommunications Year; presenting Pope John Paul II with my "scientific autobiography," *Ascent to Orbit*; and most memorable of all, meeting Prince Charles and Princess Diana at the London premiere of *2010*.

For most of my life, 1984 had been an ominous future year; in my case, that turned out to be 1986. For some time I had experienced difficulty in walking, so I checked into a London hospital, where I was told I was suffering from motor neuron disease (aka Lou Gehrig's disease) and probably had less than two years to live. Determined to disprove this diagnosis, I returned to Sri Lanka for a strict regimen of physiotheraphy. After almost a year, I seemed to be actually improving, and it was suggested that I get another medical opinion. A new friend, presidential adviser Dr. George

Keyworth (ironically, the drafter of the above-mentioned "Star Wars" speech!), arranged for me to visit Johns Hopkins Medical Center.

On my way through London I took part in a discussion with Carl Sagan and Stephen Hawking. It was a moving experience because every time I looked at Stephen in his elaborate life-support system, I wondered if that was a forecast of my own future.

However, when I got to Johns Hopkins, a team lead by Prof. Dan Drachman gave me the good news that I was actually suffering from the recently discovered postpolio syndrome, which afflicts many polio survivors some decades after the initial illness. So there was a chance that I might see the year 2001 after all!

The Steam-Powered

Word Processor

Clarke describes the following as a "heavily researched piece of spurious scholarship taken quite seriously by some European readers." However, he says, it went over quite well in the Colonies, being printed in the January 1986 Analog *and later appearing in the Science Fiction Writers of America* Nebula *Awards Volume in 1987.*

This article won me a magnum of champagne, which the *Sunday Times* columnist Godfrey Smith offered as a prize for the best alternative to the clumsy phrase *word processor*. I submitted *word loom*, which seems to have taken off like the proverbial lead balloon.

Needless to say I never saw the champagne. It was gleefully consumed by my siblings, the directors of the Rocket Publishing Company, at their annual meeting.

Very little biographical material exists relating to the remarkable career of the now almost forgotten engineering genius the Reverend Charles

Cabbage (1815–188?), onetime vicar of St. Simians in the Parish of Far Tottering, Sussex. After several years of exhaustive research, however, I have discovered some new facts which, it seems to me, should be brought to a wider public.

I would like to express my thanks to Miss Drusilla Wollstonecraft Cabbage and the good ladies of the Far Tottering Historical Society, whose urgent wishes to disassociate themselves from many of my conclusions I fully understand.

As early as 1715 *The Spectator* refers to the Cabbage (or Cubage) family as a cadet branch of the de Coverleys (bar sinister, regrettably, though Sir Roger himself is not implicated). They quickly acquired great wealth, like many members of the British aristocracy, by judicious investment in the slave trade. By 1800, the Cabbages were the richest family in Sussex (some said in England), but as Charles was the youngest of eleven children, he was forced to enter the church and appeared unlikely to inherit much of the Cabbage wealth.

Before his thirtieth year, however, the incumbent of Far Tottering experienced a remarkable change of fortune, owing to the untimely demise of all his ten siblings in a series of tragic accidents. This turn of events, which contemporary writers were fond of calling "The Curse of the Cabbages," was closely connected with the vicar's unique collection of medieval weapons, oriental poisons, and venomous reptiles. Naturally, these unfortunate mishaps gave rise to much malicious gossip and may be the reason why the Reverend Cabbage preferred to retain the protection of Holy Orders, at least until his abrupt departure from England. (All these events have led Ealing Studious to deny that Alec Guinness's *Kind Hearts and Coronets* had anything to do with the reverend. It is known, however, that at one time Peter Cushing was being considered for the role of Cabbage.)

It may well be asked why a man of great wealth and minimal public duties should devote the most productive years of his life to building a machine of incredible complexity, whose purpose and operations only he could understand. Fortunately, the recent discovery of the Faraday-Cabbage correspondence in the archives of the Royal Institution now throws some light on this matter. Reading between the lines, it appears that the reverend gentleman resented the weekly chore of producing a two-hour sermon on basically the same themes 104 times a year. (He was also incumbent of Tottering-in-the-Marsh, population seventy-three.) In a moment of inspiration that must have occurred around 1851—possibly after a visit to the Great Exhibition, that marvelous showpiece of confident Victorian know-how—he conceived a machine that would automatically reassemble masses of text in any desired order. Thus he could create any number of sermons from the same basic material.

This crude initial concept was later greatly refined.

Although—as we shall see—the Reverend Cabbage was never able to complete the final version of his "word loom," he clearly envisaged a machine that would operate not only upon individual paragraphs but single lines of text. (The next stage—words and letters—he never attempted, though he mentions the possibility in his correspondence with Faraday and recognized it as an ultimate objective.)

Once he had conceived the word loom, the inventive cleric immediately set out to build it. His unusual (some would say deplorable) mechanical ability had already been amply demonstrated through the ingenious mantraps that protected his vast estates, and which had eliminated at least two other claimants to the family fortune.

At this point, the Reverend Cabbage made a mistake that may well have changed the course of technology—if not history. With the advantage of hindsight, it now seems obvious to us that his problems could only have been solved by the use of electricity. The Wheatstone telegraph had already been operating for years, and he was in correspondence with the genius who had discovered the basic laws of electromagnetism. How strange that he ignored the answer that was staring him in the face!

We must remember, however, that the gentle Faraday was now entering the decade of senility preceding his death in 1867. Much of the surviving correspondence concerns his eccentric faith (the now extinct religion of "Sandemanism," with which Cabbage could have little patience).

Moreover, the vicar was in daily (or at least weekly) contact with a very advanced technology with over a thousand years of development behind it. The Far Tottering church was blessed with an excellent twenty-one-stop organ manufactured by the same Henry Willis whose 1875 masterpiece at North London's Alexandra Palace was proclaimed by Marcel Dupre as the finest concert organ in Europe. Cabbage was himself no mean performer on this instrument and had a complete understanding of its intricate mechanism. He was convinced that an assembly of pneumatic tubes, valves, and pumps could control all the operations of his projected word loom.

It was an understandable but fatal mistake. Cabbage had overlooked the fact that the sluggish velocity of sound—a miserable 330 meters a second—would reduce the machine's operating speed to a completely impracticable level. At best, the final version might have attained an information-handling rate of 0.1 baud—so that the preparation of a single sermon would have required about ten weeks!

It was some years before the Reverend Cabbage realized this fundamental limitation; at first he believed that by merely increasing the available power he could speed up his machine indefinitely. The final version absorbed the entire output of a large steam-driven threshing machine—the clumsy ancestor of today's farm tractors and combine harvesters.

At this point, it may be as well to summarize what little is known about the actual mechanics of the word loom. For this, we must rely on garbled accounts in the *Far Tottering Gazette* (no complete runs of which exist for the essential years 1860–80) and occasional notes and sketches in the Reverend Cabbage's surviving correspondence. Ironically, considerable portions of the final machine were in existence as late as 1942. They were destroyed when one of the Luftwaffe's stray incendiary bombs reduced the ancestral home of Tottering Towers to a pile of ashes. A small portion—two or three gearwheels and what appears to be a pneumatic valve—are still in the possession of the local Historical Society. These pathetic relics reminded me irresistibly of another great technological might-have-been, the famous Anticythera Computer.

The machine's "memory" was based—indeed, there was no practical alternative at the time—on the punched cards of a modified Jacquard loom; Cabbage was fond of saying that he would weave thoughts as Jacquard wove tapestries. Each line of output consisted of twenty (later thirty) characters, displayed to the operator by letter wheels rotating behind small windows.

The principles of the machine's card operating system have not come down to us, and it appears—not surprisingly—that Cabbage's greatest problem involved the location, removal, and updating of the individual cards. Once text had been finalized, it was cast in type-metal; the amazing clergyman had built a primitive Linotype at least a decade before Mergenthaler's 1886 patent!

Before the machine could be used, Cabbage was faced with the laborious task of punching not only the Bible but the whole of Cruden's Concordance onto Jacquard cards. He arranged for this to be done, at negligible expense, by the aged ladies of the Far Tottering Home for Relics of Decayed Gentlefolk—now the local disco and break-dancing club. This was another astonishing first, anticipating by a dozen years Hollerith's famed mechanization of the 1890 United States census.

But at this point, disaster struck. Hearing, yet again, strange rumors from the parish of Far Tottering, no less a personage than the archbishop of Canterbury descended upon the now obsessed vicar. Understandably appalled by discovering that the church organ had been unable to perform its original function for at least five years, Canterbury issued an ultimatum. Either the word loom must go—or the Reverend Cabbage must resign. (Preferably both: there were also hints of exorcism and reconsecration.)

This dilemma seems to have produced an emotional crisis in the already unbalanced clergyman. He attempted one final test of his enormous and unwieldy machine, which now occupied the entire western transept of St. Simian's. Over the protests of the local farmers (for it was now harvest time) the huge steam engine, its brassware gleaming, was trundled up to the church, and the belt-drive connected (the stained-glass windows having long ago been removed to make this possible).

The reverend took his seat at the now unrecognizable console (I cannot forbear wondering if he booted the system with a foot pedal) and started to type. The letter wheels rotated before his eyes as the sentences were slowly spelled out, one line at a time. In the vestry, the crucibles of molten lead awaited the commands that would be laboriously brought to them on puffs of air. . . .

"Faster, faster!" called the impatient vicar as the workmen shoveled coal into the smoke-belching monster in the churchyard. The long belt, snaking through the narrow window, flapped furiously up and down, pumping horsepower upon horsepower into the straining mechanism of the loom.

The result was inevitable. Somewhere, in the depths of the immense apparatus, something broke. Within seconds, the ill-fated machine tore itself into fragments. The vicar, according to eyewitnesses, was very lucky to escape with his life.

The next development was both abrupt and totally unexpected. Abandoning church, wife, and thirteen children, the Reverend Cabbage eloped to Australia with his chief assistant, the village blacksmith.

To the class-conscious Victorians, such an association with a mere workman was beyond excuse (even an under footman would have been more acceptable!).

The very name of Charles Cabbage was banished from polite society, and his ultimate fate is unknown, though there are reports that he later became chaplain of Botany Bay. The legend that he died in the outback when a sheep-shearing machine he had invented ran amok is surely apocryphal.

How D. H. Lawrence ever heard of this affair is still a mystery. As is now well known, he had originally planned to make the protagonist of his most famous novel not Lady Chatterley but her husband; however, discretion prevailed, and the Cabbage connection was revealed only when Lawrence foolishly mentioned it, in confidence, to Frank Harris, who promptly published it in the *Saturday Review*. Lawrence never spoke to Harris again; but then, no one ever did.

The rare-book section of the British Museum possesses the only known copy of the Reverend Cabbage's sermons in Steam, long claimed by the family to have been manufactured by the word loom. Unfortunately, even a casual inspection reveals that this is not the case; with the exception of the last page (223–24), the volume was clearly printed on a normal flatbed press.

Page 223–24, however, is an obvious insert. The impression is very uneven, and the text is replete with spelling mistakes and typographical errors.

Is this indeed the only surviving production of perhaps the most re-markable—and misguided—technological effort of the Victorian age? Or is it a deliberate fake, created to give the impression that the word loom actually operated at least once—however poorly?

We shall never know the truth, but as an Englishman, I am proud of the fact that one of today's most important inventions was first conceived in the British Isles. Had matters turned out slightly differently, Charles Cab-bage might now have been as famous as James Watt, George Stevenson—or even Isambard Kingdom Brunel.

Afterword:

''Maelstrom II''

There cannot be many science fiction novels that end with a forty-page appendix
full of mathematical equations and electric-circuit diagrams, Clarke writes,
adding "Don't worry, this isn't one of them; but just such a book inspired
it, half a century ago. And with any luck, during the next
half century it will cease to be fiction."

It must have been in 1937 or 1938 when I was treasurer of the five-year-old British Interplanetary Society (annual budget to start the conquest of space about $200) that the BIS was sent a book with a rather odd title, by an author with an even odder name. Akkas Pseudoman's *Zero to Eighty* must now be quite a rarity.

The snappy subtitle says it all:

Being my lifetime doings, reflections, and inventions
also
my journey round the Moon.

Quite an "also"; I can hear the author's modest cough.

He was not, of course, really Mr. Pseudoman, as the preface made clear. This was signed "E. F. Northrup" and explained that the book had been written to show that the Moon may be reached by means of known technologies, without "invoking any imaginary physical features or laws of nature."

Dr. E. F. Northrup was a distinguished electrical engineer and the inventor of the induction furnace which bears his name. His novel, which is obviously a wish-fulfillment fantasy, describes a journey to the Moon (and around it) in a vehicle fired from the Earth by a giant gun, as in Jules Verne's classic *From the Earth to the Moon*. Northrup, however, tried to avoid the obvious flaws in Verne's naive proposal, which would quickly have converted Ardan et al. into small blobs of protoplasm inside a sphere of molten metal.

Northrup used an electric gun, two hundred kilometers long, most of it horizontal but with the final section curving up Mount Popocatépetl, so that the projectile would be at an altitude of more than five kilometers when it reached the required escape velocity of 11.2 kilometers per second. In this way, air resistance losses would be minimized, but a small amount of rocket power was available for any necessary corrections.

Well—it makes more sense than Verne's moongun, but not by much. Even with two hundred kilometers of launch track, the unfortunate passengers would have to withstand 30 g, for more than half a minute. And the cost of the magnets, power stations, transmission lines, etc., would run into billions; rockets would be cheaper, as well as far more practical.

A few years after reading Dr. Northrup's book (which is still full of interesting ideas, including a remarkably sympathetic treatment of Russian technology) it occurred to me that he had put his electric launcher on the wrong world: it made no sense on Earth—but was ideal for the Moon.

First, there's no atmosphere to heat up the vehicle or destroy its momentum, so the launching track could be laid out horizontally. Once it's given escape velocity, the payload would slowly rise up from the surface and head out into space.

Second, lunar escape velocity is only one-fifth of Earth's and can therefore be attained with a correspondingly shorter launch track—and a twenty-fifth of the energy. Not a bad way to export goods from the Moon. Suitably protected human passengers could be handled by larger systems, if there was enough traffic to justify them.

I wrote up this idea, with the necessary calculations, in a paper, "Electromagnetic Launching as a Major Contribution to Space-Flight," published in the November 1950 *BIS Journal*. And because a good idea should be exploited, I used it in fiction on two occasions: in *Islands in the Sky* and in the short story "Maelstrom II," originally published in *Playboy,* April 1965.

Some twenty years after the publication of "Electromagnetic Launching," the concept was taken much further by Gerald O'Neill, who made it a key element of his "space colonization" projects (who is justifiably annoyed by the Star Warriors' preemption of his title). He showed that the large space habitats he envisaged could be most economically constructed from materials mined and prefabricated on the Moon, and then shot into orbit by electromagnetic catapults to which he gave the name "mass drivers." (I've challenged him to produce any propulsion device that doesn't fit this description.)

The other scientific element in "Maelstrom II" has a much longer history; it's the branch of celestial mechanics known as perturbation theory. I've been able to get considerable mileage out of it since my applied maths instructor, Dr. George C. McVittie, introduced me to the subject at King's College in the late forties. (However, I'd come across it—without realizing—in dear old *Wonder Stories* almost two decades earlier.) Here's a challenge to you: spot the flaw in the following scenario. . . .

The first expedition has landed on Phobos, the inner moon of Mars. Gravity there is only about a thousandth of Earth's, so the astronauts have a great time seeing how high they can jump. One of them overdoes it and exceeds the tiny satellite's escape velocity of about thirty kilometers an hour. He dwindles away into the sky, toward the mottled red Marsscape; his companions realize that they'll have to take off and catch him before he crashes into the planet only six thousand kilometers below.

A dramatic situation which opens Laurence Manning's 1932 serial, *The Wreck of the Asteroid.* Manning, one of the most thoughtful science fiction writers of the thirties, was an early member of the American Rocket Society and was very careful with his science. But this time, I'm afraid, he was talking nonsense; his high jumper would have been perfectly safe.

Look at the situation from the point of view of Mars. If he's simply standing on Phobos, he's orbiting the planet at almost eight thousand kilometers an hour. As space suits are massive affairs, not designed for athletic events, I doubt if the careless astronaut could achieve that critical thirty kilometers an hour. Even if he did, it would be less than a half percent of the velocity he already has, relative to Mars. Whichever way he jumped, therefore, it will make virtually no difference to his existing situation; he'll still be traveling in almost the same orbit as before. He'd recede a few kilometers away from Phobos—and be right back where he started, just one revolution later! (Of course, he could run out of oxygen in the meantime—the trip around Mars will take seven and a half hours. So maybe his friends should go after him—at their leisure.)

This is perhaps the simplest example of "perturbation theory," and I developed it a good deal further in "Jupiter V" in 1956. This story was based on what seemed a cute idea in the early fifties. A decade earlier, *Life* magazine had published space-artist Chesley Bonestell's famous paintings of the outer planets. Wouldn't it be nice, I thought, if sometime in the twenty-first century, *Life* sent one of its photographers out there to bring back the real thing and compare it with Chesley's hundred-year-old visions?

Well, little did I imagine that, in 1976, the Voyager space probe would do just this—and that, happily, Chesley would still be around to see the result. Many of his carefully researched paintings were right on target— though he couldn't have anticipated such stunning surprises as the volcanoes of Io, or the multiplex rings of Saturn.

Mother Nature

Got There First

This piece was intended for Omni *magazine's flippant "Last Word" column, but to Clarke's astonishment it was turned down. However, it was promptly snapped up by* Analog *(née* Astounding*), which has long had a vested interest in such matters, and appeared in the August 1990 issue.*

Man never invents anything that nature hasn't tried out millions of years earlier.

Look at flying; would we ever have thought of such a crazy idea if the birds, the bats, and the bees hadn't first shown us that it was possible?

Air-conditioning? How do you imagine the termites keep their vast and complicated "cities" cool in the tropics? They also invented agriculture around the time of the dinosaurs (give or take a few million years) and operate highly efficient fungus farms.

Fluorescent lighting? Yes, fireflies are an obvious example, but not a very impressive one. Deep-sea fish put on infinitely more spectacular displays. The Mariana Trench on a Saturday night puts Times Square to shame.

Even rotary motion—until recently regarded as a human exclusive—has now turned up in the microworld. Believe It or Not, there are bacteria with propellers.

And look at that wonderful twentieth-century invention the jet. Squids were jet-setting across the oceans long before the Concordes and 747s got into the act. Not quite so fast, of course; but what's the big hurry?

While we're on the subject of squids—admittedly not the most wildly popular conversational gambit—here's another one for Ripley. No television screen can beat the display a squid can put on over its entire body, changing pattern and color in the twinkling of an eye. To match this amazing feat, a human being would have to produce instant tattoos.

Rockets? Well, there's a beetle that mixes violently reactive chemicals in its guts to produce a blast of scalding steam. Though it does this to discourage predators, I wouldn't be at all surprised if some relative in the Borneo jungles has taken the next step and developed JATO—jet-assisted takeoff.

Perhaps the worst shock to our pride was the discovery that World War II's secret weapons, radar and sonar, had been anticipated by bats, dolphins, whales—and doubtless other creatures we haven't discovered yet. (As the old riddle goes, "Which is the most cunning of the animals? That which no man hath yet seen.")

If you think we're still ahead because bats and the like use sound waves while we sophisticates use radio, don't push your luck. Electric eels are an amazing (I resisted *shocking*) example of what evolution can do when it really sets its mind to it.

Yet now it seems that nature's equivalents of Con Edison do more than merely stun their prey; some fish use the electric fields they generate for navigation and obstacle detection. Which is the next best thing to genuine radar.

It was while contemplating such wonders of the animal kingdom that I formulated this working hypothesis: "If it can be done, nature's done it already." After all, she's had long enough to experiment, so we Johnny-come-latelies shouldn't feel too embarrassed.

This led to a further thought. Are there any examples in nature of technological breakthroughs that still elude us? Well, fasten your seat belts. . . .

For years science fiction writers have dreamed of a "space drive" which could get us to the planets in a more civilized manner without all that noise and fuss. Every few years some hopeful inventor claims to have demonstrated such a device, usually depending on gyroscopes or vibrating masses to produce a unidirectional thrust. They have all come to grief as action and reaction cancel exactly (Newton's third law, where every action produces an opposite reaction).

What we want is a totally closed system that can produce motion without ejecting any matter: all it needs is some source of internal energy to produce a thrust.

Impossible, conventional physics says. Well, I have news for you. Once again, Mother Nature has beaten us to it—and shown us the real way to the stars.

And I do think it's ironic that the American rocketeers of the late forties, at White Sands, New Mexico, had the proof right under their noses. They only had to go across the border to find—the Mexican jumping bean! (And remember—you read it here first!)

The Mexican jumping bean is the seed of a shrub in which a small moth (*Carpocapsa saltitans* for you purists) lays its egg. On hatching, the larva feeds on the pulp and hollows out the interior of the bean. If placed on a warm surface, it promptly launches itself in search of cooler surroundings. To the uninitiated, the effect is quite startling; contrary to the *Encyclopaedia Britannica*, "the familiar jumping movement" is not at all familiar.

Seriously—is the nonreactive "space drive" beloved of science fiction writers a genuine possibility? I have always argued—except in my own fiction, where anything goes—that there is no way of violating Newton's third law.

Yet several competent scientists and engineers believe that—like Newton's law of gravitation!—the third law is only a (very accurate) first-order approximation. My old friend Harry Stine—whom I met at the White Sands Proving Ground and who may well have shown me my first Mexican jumping bean on that occasion!—tells me that as long ago as 1962, he coauthored a paper on "non-Newtonian space propulsion," which was presented to the American Physical Society. Perhaps because of its revolutionary implications, in the 1962 spring *Bulletin* it was given the inoffensive title "Some aspects of certain transient mechanical systems."

Harry described such an "impulse drive" in his 1980 novel, *Star Driver*—which, he says in a recent letter, "will certainly muddy the patent situation." He goes on to add: "I'll probably never see an 'impulse drive,' but I know how to build one . . . and it won't be that difficult. . . . It does, however, require a paradigm shift. And maybe the time is coming."

Maybe it is. In February 1994, one of the world's leading scientific journals, *Physics Review,* printed what may be a revolutionary paper by Bernhard Haisch, Alfonso Rueda, and Harold E. Puthoff: "Inertia as a Zero-Point Field Lorentz Force." Based on earlier work by Sakharov, this purports to explain the origin of inertia—the intrinsic "sluggishness" of all matter when any attempt is made to change its velocity (Newton again—the first law). The paper argues that gravitational and inertial forces are due to the enormous energies ("zero point fluctuations") that pervade all space, even the so-called vacuum itself. To quote Puthoff: "There is a possibility (at least in principle) that inertial and gravitation masses can . . . be affected."

In English? It is theoretically possible to have a completely self-contained propulsion system. Many years ago the famous science fiction writer E. E. Smith coined a name for it: "the inertialless drive." With such a drive the smallest force would produce an almost infinite acceleration: you could reach the velocity of light from a standing start in a matter of seconds without feeling a thing. (However, some spoilsport suggested that because the chemical reactions which power the human body depend on the inertia of its constituent molecules, you'd be dead anyway. A good point: back to the drawing board?)

In any case, even if Puthoff et al. are correct, we'll still be stuck with the noisy, inefficient, polluting, and downright dangerous rocket for decades to come. After all, it took something like forty years to get from $E = mc^2$ to atomic energy, with most arguing that it was fantasy right to the very end.

But isn't it exciting to think that sometime in the next century, the road to the stars will really be opened up—and the conquest of space will begin in earnest. . . .

Message

to Comsat,

February 18, 1988

Sri Lanka is the perfect spot for astronomical observations. Clarke chose to
live on this remote spot, then called Ceylon, in 1956.

This is Arthur Clarke, sending greetings to you from practically on the equator. It gives me a nice feeling to know that the Indian Ocean satellite that keeps me in touch with the world is right overhead.

And so, by an interesting coincidence, is the most stable point in the Earth's gravitational field. Exhausted geostationary satellites also end up there, milling round and round above Sri Lanka in a celestial Sargasso Sea when they've run out of gas.

Now I'm going to say something that may upset quite a few people, especially the bean counters who calculate our phone bills. I can see someone yelling, "Cut him off!" about a minute from now.

For I want to remind you of something that happened in England just 150 years ago. In those days, sending a letter from one part of the country to another was enormously complicated and expensive. Why? Because an army of clarkes—sorry, clerks—calculated the exact amount you had to pay on every piece of mail, according to the distance it traveled! Just think of the manpower and paperwork that must have been involved!

Then along came a genius named Rowland Hill, who did what we'd now call "a systems analysis." He discovered (surprise! surprise!) that the cost of sending a letter was almost independent of distance—virtually all the labor went into the handling at the beginning and the end of the journey.

So, Mr. Hill made a revolutionary suggestion. He said: "Let's have a flat rate, irrespective of distance. People can pay for letters in advance, simply by purchasing a stamp. I calculate that it need cost only one penny, and even if we don't break even at first, the explosion in correspondence will soon give us a profit. And the benefits to commerce and society will be immeasurable."

Rowland Hill was one of the creators of the world we know, and needless to say, he was shouted down at first. But he persisted, and the Penny Post started in 1840.

Surprisingly, it only took five years for Mr. Hill to win his battle with the bureaucrats. They did things a lot faster in those days. Not so many committees to deal with.

I'm sure you'll see what I'm driving at, and because I can never resist an opportunity for a commercial, I'd like to end by reading to you a paragraph from my latest last book [*2061: Odyssey Three*]:

". . . In the beginning, the Earth had possessed the single supercontinent of Pangaea, which over the aeons had split asunder. So had the human species, into innumerable tribes and nations; now it was merging together, as the old linguistic and cultural divisions began to blur. With the historic abolition of long-distance charges on December 31, 2000, every telephone call became a local one, and the human race greeted the new millennium by transforming itself into one huge, gossiping family."

Graduation Address:

International Space

University

*Through the courtesy of Comsat, on August 20, 1988, Clarke addressed via
satellite the first graduating class from his home in Sri Lanka.*

You are the first representatives of a true space-faring species, whose world
is not bounded by the confines of one planet. So let me share with you a
passage I wrote in 1969 for the history of the Apollo 11 mission, first on
the Moon:

> Half a billion years ago, the Moon summoned life out of its first home,
> the sea, and led it onto the empty land. For, as it drew the tides across
> the barren continents of primeval Earth, their daily rhythm exposed
> to the sun and air the creatures of the shallows. Most perished—but
> some adapted to the new and hostile environment. The conquest of
> land had begun.
>
> We shall never know when this happened, on the shores of what
> vanished sea. There were no eyes or cameras to record the event.

Now, the Moon calls again—and this time, life responds with a roar that shakes the sky. When a Saturn V soars spaceward on four thousand tons of thrust, it signifies more than a triumph of technology. It opens the next chapter of evolution.

Yes, we will return to the Moon. But when we go there again, it will be in vehicles that will make the Saturn V—for all its staggering complexity and its 150 million horsepower—look like a clumsy, inefficient dinosaur of the early Space Age. And this time, we will stay.

Space travel is a technological mutation that really should not have arrived until the twenty-first century. But thanks to the ambition and genius of Wernher von Braun and Sergei Korolyov, and their influence upon such diverse individuals as Kennedy and Khrushchev, the Moon was reached half a century ahead of time. And now, you can take advantage of this historical accident, for, as Isaac Newton once remarked, you are standing on the shoulders of giants.

The Moon was a fitting subject for your efforts during the first session of the ISU. Do not lose the momentum of your work or the intensity of your vision, whatever obstacles you may encounter. As H. G. Wells put it long ago, the choice is the universe—or nothing.

The resources of that universe are, by all human standards, infinite. There are no "limits to growth" among the stars. Unfortunately, there is a tragic mismatch between our present needs and our capabilities. The conquest of space will not arrive soon enough to save millions from leading starved and stunted lives. Thus it is all the more urgent that we exploit to the utmost the marvelous tools that space technology has already given us—comsats, weather-sats, geo-sats, and all the other sats yet to be invented. This is the reason that the third world needs space technology even more urgently than the so-called developed world.

You will find in the years to come that the experience which you have shared with your fellow ISU students will have changed your life. To have worked with friends from a score of other nations will profoundly affect your worldview—and that may well be the most important role of the ISU.

As our terrestrial universities first flourished during the times of great exploration and discovery of our own planet, so the ISU is a response to the first explorations of space. The blossoming of art and science during the Renaissance may now be repeated when artists and poets join scientists and engineers at the new frontier.

As the university begins its second year in Europe, and later, at other locations around the globe, I eagerly await its growth to a permanent full-time institution which will train tomorrow's space professionals.

I would like to know the future that you will help to create, but that is a privilege to be reserved for your children and your grandchildren. It is they who, to quote H. G. Wells again, "shall stand upon this Earth as one stands upon a footstool and . . . reach out their hands among the stars."

I wish you well, for you represent a new evolutionary branch of humanity.

Back to *2001*

It is over a quarter of a century since Stanley Kubrick wrote to Clarke in the spring of 1964, asking if he had any ideas for the proverbial "good" science fiction movie. The result of their race to outguess the future is an innovative mix of science fiction and science fact.

Well, the success of *2001: A Space Odyssey* is now history. It has been called one of the most influential movies ever made and invariably turns up on lists of the all-time top ten. The film's own history, and that of its various sequels, was a complex journey, so I'd like to go back to the beginning and recall how the whole thing started.

In April 1964, I left Ceylon, as it was then called, and went to New York to complete my editorial work on *Man and Space*. It was strange, being back in New York after several years of living in my tropical paradise. Commuting—even for three stations on the IRT—was an exotic novelty, after my humdrum existence among elephants, coral reefs, monsoons, and sunken treasure ships. The strange cries, cheerful smiling faces, and courteous manners of the Manhattanites as they went about their mysterious affairs were a continual source of fascination; so were the advertisements for Levy's bread, Piels beer, and a dozen brands of oral carcinogens.

My work on *Man and Space* progressed very smoothly, because whenever a zealous researcher asked me, "What is your authority for this statement?" I would fix her with a basilisk stare and answer, "You're looking at him."

So I had ample energy for moonlighting with Stanley. Our first encounter was in Trader Vic's on April 23. (To my knowledge they have yet to put up a plaque to mark the spot.) Stanley was still basking in the success of his last movie, *Dr. Strangelove,* and was looking for an even more ambitious theme. He wanted to make a film about man's place in the universe—a project likely to give any studio head a heart attack.

Stanley, who becomes an instant expert in any subject that concerns him, had already devoured several libraries of science fact and science fiction.

Now, before you make a movie, you have to have a script, and before you have a script, you have to have a story (unless, of course, you are talking about art theaters). I had already given Stanley a list of my shorter pieces, and we had decided that one, "The Sentinel," written in 1948 as my entry for a BBC short-story competition (it wasn't even placed, and I've sometimes wondered what did), contained a basic idea on which we could build. I need only say that it's a mood piece about the discovery of an alien artifact on the Moon—a kind of burglar alarm, waiting to be set off by mankind's arrival.

2001 is often said to based on "The Sentinel" but the two bear much the same relationship as an acorn and an oak tree. It needed a lot more material and some of it came from "Encounter in the Dawn" (aka "Expedition to Earth") and four other short stories. But most of it was wholly new, and the result of months of brainstorming with Stanley—followed by (fairly) lonely hours in Room 1008 of the famous Hotel Chelsea, at 222 West Twenty-third Street.

This is where most of the novel was written. (The journal of this often painful process will be found in *The Lost Worlds of 2001*.) But why write a novel, you may well ask, when we were aiming to make a movie? It's true that "novelizations" (ugh) are all too often produced afterward; in this case, Stanley had excellent reasons for reversing the process.

Because a screenplay has to specify everything in excruciating detail, it is almost as tedious to read as to write. Perhaps because Stanley realized that I had a low tolerance for boredom, he suggested that before we embarked on the drudgery of the script, we let our imaginations soar freely by writing a complete novel, from which we would later derive the script. (And hopefully, a little cash.)

This is more or less the way it worked out, though toward the end, novel and screenplay were being written simultaneously, with furious feedback in both directions: rushes to rewrites.

Extracts from the journal I must have hastily written in the wee hours will give you a flavor of that hectic time.

May 28, 1964. Suggested to Stanley that "they" might be machines who regard organic life as a hideous disease. Stanley thinks this is cute. . . .

June 2. Averaging one or two thousand words a day. Stanley says, "We've got a best-seller here."

July 11. Joined Stanley to discuss plot development, but spent almost all the time arguing about Cantor's Transfinite Groups. . . . I decide that he is a mathematical genius.

July 12. Now have everything—except the plot.

July 26. Stanley's (36th) birthday. Went to the Village and found a card with the inscription: "How can you have a Happy Birthday when the whole world may blow up at any minute?"

Sept. 28. Dreamed I was a robot, being rebuilt. Took two chapters to Stanley, who cooked me a fine steak, remarking, "Joe Levine doesn't do this for his writers."

Oct. 17. Stanley has invented the wild idea of slightly gay robots who create a Victorian environment to put our heroes at their ease.

Nov. 28. Phoned Isaac Asimov to discuss the biochemistry of turning vegetarians into carnivores.

Dec. 10. Stanley calls after screening H. G. Wells's *Things to Come* and says he'll never see another movie I recommend.

Dec. 24. Slowly tinkering with the final pages, so I can have them as a Christmas present for Stanley.

This entry records my hope that the novel was now essentially complete; in fact, all we had was merely a rough draft of the first two-thirds, stopping at the most exciting point—because we hadn't the faintest idea what would happen next. But it was enough to let Stanley set up the deal with MGM and Cinerama for what was originally trumpeted as *Journey Beyond the Stars*.

Throughout 1965, Stanley was involved in the incredibly complex preproduction activities—made even more difficult by the fact that the film would be shot in England while he was still in New York—and under no circumstances would he travel by air. He learned the hard way—while getting his pilot's license. I am in no position to criticize as I have never been behind a steering wheel since the day I (barely) passed my driving test in Sydney, Australia.

While Stanley was making the movie, I was trying to complete the final, final version of the novel, which had to receive his blessing before it could be published. This proved extremely difficult to obtain, partly because he was so busy at the studio. He swore he wasn't dragging his feet to make certain that the movie appeared before the book. Which it did—by several months—in the spring of 1968.

Considering its complex and agonizing gestation, it is not surprising that the novel differs from the movie in several respects. Most important— and how lucky this was we could never have guessed at the time—Stanley decided to rendezvous with Jupiter, whereas in the novel the spaceship *Discovery* flew on to Saturn, using Jupiter's gravitational field to boost it on its way.

Precisely this "perturbation maneuver" was used by the Voyager spacecraft eleven years later, and as I type these very words (on the evening of August 24, 1989), *Voyager 2* is making its final appointment with the planet Neptune—last stop before the stars.

Why the change from Saturn to Jupiter? Well, it made a more straightforward story line—and more important, the special effects department couldn't produce a Saturn that Stanley found convincing. If it had done so, the movie would by now have been badly dated, since the Voyager missions showed Saturn's rings to be far more implausible than anyone had ever dreamed.

For more than a decade after publication of the novel (July 1968) I indignantly denied that any sequel was possible, or that I had the slightest intention of writing one. But the brilliant success of the Voyager missions changed my mind; distant worlds about which absolutely nothing was known when Stanley and I started our collaboration suddenly became real places, with fantastic surface conditions. Who would ever have imagined satellites entirely covered with ice floes, or volcanoes spurting sulfur one hundred kilometers into space? Science fiction could now be made far more convincing by science fact. *2010: Odyssey Two* was about the real Jovian satellite system.

There is also another profound distinction between the two books. *2001* was written in an age which now lies beyond one of the great divides in human history; we are sundered from it forever by the moment Neil Armstrong and Buzz Aldrin stepped out onto the Sea of Tranquility. Now history and fiction have become inexplicably intertwined; the Apollo astronauts had already seen the film when they left for the Moon. The crew of *Apollo 8*, who at Christmas, 1968, became the first men ever to set eyes upon the lunar far side, told me that they had been tempted to radio back the discovery of a large black monolith. Alas, discretion prevailed.

The Apollo 13 mission, however, does have an uncanny connection with *2001*. When the computer HAL reported the "failure" of the AE 35 Unit, the phrase he used was, "Sorry to interrupt the festivities, but we have a problem."

Well—the Apollo 13 Command Module was named *Odyssey,* and the crew had just concluded a television broadcast with the movie's famous "Zarathustra" theme when an oxygen tank exploded. Their first words back to Earth were, "Houston, we have a problem."

By brilliant improvision—using the Lunar Module as a "lifeboat"— the astronauts were brought safely home aboard *Odyssey.* When NASA administrator Tom Paine sent me the report of the mission, he wrote on the cover: "Just as you always said it would be, Arthur."

And there are many other resonances—most notably, the sagas of the communications satellites Westar VI and Palapa B-2, which, in 1984, had been launched into useless orbits by misfiring rockets.

Now, in an earlier draft of the novel, David Bowman had to make an EVA in one of *Discovery*'s space-pods and chase the ship's lost communications antenna system. He caught up with it—but was unable to check its slow spin and bring it back to *Discovery.*

In November 1984, astronaut Joe Allen left the space shuttle *Discovery* (no, I'm not making this up!) and used his maneuvering unit to rendezvous with Palapa. Unlike Bowman, he was able to check its spin by bursts from the nitrogen-jet thrusters on his backpack. The satellite was brought back into *Discovery*'s cargo bay, and two days later Westar was also rescued. Both were safely returned to Earth for repair and relaunch, after one of the most remarkable and successful shuttle missions ever flown.

Just about the time Joe was doing all this, I received a copy of his beautiful book, *Entering Space: An Astronaut's Odyssey,* with a covering letter which read: "Dear Arthur, When I was a boy, you infected me with both the writing bug and the space bug, but neglected to tell me how difficult either undertaking can be."

I need hardly say that this sort of tribute gave me a warm glow of satisfaction; but it also made me feel a contemporary of the Wright brothers.

After the first two *Odyssey*'s—novels and movies—had been completed, I had a very good excuse for relegating HAL, Bowman, and the Monolith to my subconscious until at least 1990. NASA's most ambitious deep-space project, the *Galileo* probe, was due to be launched by the space shuttle in May 1986, to start exploring the moons of Jupiter in December 1988. It would have provided a flood of information about Jupiter and its satellites, which might make instantly obsolete anything I could write before that date—and would probably provide stimulus for endless new speculations.

Alas, the *Challenger* tragedy eliminated the scenario, and *Galileo* is now due to be launched by space shuttle *Atlantis* in October 1989. It will not reach Jupiter until December 1995—seven years later than originally planned. As I wrote in the preface to *2061: Odyssey Three,* I have decided not to wait.

And now to answer a question I was afraid you wouldn't ask: Will there ever be another *Odyssey*? Although *Odyssey Three* ends on a cliffhanger, I really had no intention of implying a sequel; it just seemed the right way to close the book.

Whether or not a fourth *Odyssey* will materialize depends on factors beyond my control. If *Galileo* gets safely off the pad, that will improve the odds considerably. But, alas, I'll then have to wait six years for the news to come back from Jupiter and its satellites. . . .

If the news is very good indeed—and if I'm still around to enjoy it in 1995—a final *Odyssey* may well emerge from my word processor. What I'd like to do, of course, is to write it in a rather leisurely fashion, with a view of publication on January 1—2001.

It is hard for me to believe that Stanley Kubrick is now dead. My own career owes more to him than to any other person, and I have recorded the story of our long collaboration in *The Lost Worlds of 2001.*

A few nights ago, I dreamed that we were talking together (he looked exactly the same as in 1964!) and he asked: "Well, what shall we do next?" There *might* have been a next, involving Brian Aldiss's beautiful short story "Supertoys Last All Summer Long," which Stanley worked on for some time under the title *AI.* But for numerous reasons that fell through.

One of my deepest regrets now is that we will not be able to share the Year 2001 together.

Coauthors and

Other Nuisances

In 1986 a funny thing happened to Clarke on the way to the word processor. He found a coauthor—one foisted upon him for better or worse—but luckily for better. In the following piece, he talks of the art and craftiness of collaboration.

Writing is a lonely profession, and after a few decades even the most devout egotist may occasionally yearn for company. But collaboration in any work of art is a risky business, and the more people involved, the smaller the chances of success. Can you imagine *Moby-Dick* by Herman Melville and Nat Hawthorne? *War and Peace* by Leo Tolstoy and Freddie Dostoyevsky? (With additional dialogue by Van Turgenev?)

Certainly I never imagined, until a few years ago, that I would ever collaborate with another writer on a work of fiction. Nonfiction was different: I've been involved in no less than fourteen multiauthor projects. But fiction? No way! I was quite sure I would never let any outsider tamper with my unique brand of creativity. . . .

Well, a funny thing happened on the way to the word processor. Early in 1986 my agent, Scott Meredith, called me in his most persuasive, "Don't say no until I've finished" mode. There was, it seemed, this young genius of a movie producer who was determined to film something—any-thing—of mine. Though I'd never heard of Peter Guber, I had seen two of his movies and had been quite impressed, being even more so when Scott told me Peter's latest, *The Color Purple,* had been nominated for eleven Oscars.

However, I groaned inwardly when Scott went on to say that Peter had a friend with a brilliant idea he'd like me to develop into a screenplay. I hate screenplays; they are incredibly boring, almost unreadable, and, as far as I'm concerned, unwritable.

Then Scott explained who the friend was and I did a double take. The project suddenly looked very exciting indeed.

"Peter Guber wants to fly out to Sri Lanka to introduce this guy to you," said Scott. "His name is Gentry Lee. He works at the Jet Propulsion Laboratory, and he's the chief engineer on Project Galileo." Scott went on to say that Lee was director of mission planning for the Viking landers that had sent back those wonderful pictures from Mars. Because Lee felt the public didn't appreciate what was going on in space, he had formed a company with my friend Carl Sagan to make *Cosmos*—

"Enough!" I cried. "This man I have to meet."

It was agreed that Peter and Gentry would fly out, and if I liked Gentry's idea (and equally important, Gentry), I'd develop an outline—perhaps a dozen pages—from which any competent scriptwriter could gen-erate a screenplay.

They arrived in Colombo on February 12, 1986—just two weeks before the *Challenger* disaster. Happily, the Guber-Lee-Clarke Summit went well, and for the next few weeks I filled floppy disks with concepts, char-acters, backgrounds, plots—anything which seemed even remotely useful to the story we'd decided to call *Cradle.* Someone once said that writing a work of fiction consists of the elimination of alternatives. Very true: at one time I calculated that, if I used all the elements I'd created in every possible combination, there'd be enough material for half a billion *Cradle*s.

I sent the one I finally selected, in the form of a four-thousand-word outline, to Gentry. He liked it and flew back to Sri Lanka so we could fill in the details. During a three-day marathon up in the mountains above the ancient capital Kandy, despite the distraction of the most gorgeous pano-rama I know, we completed an eight-thousand-word demi-hemi-semi-final version which eventually became the basis of the novel. From then onward, we were able to collaborate by making frequent phone calls and flying yards of printout across the Pacific.

The writing took the best part of a year, though of course we were both involved in other projects as well. When I discovered that Gentry had a considerably better background in English and French literature than I did (by now I was immune to such surprises), I heroically resisted all attempts to impose my own style on him. This upset some longtime ACC readers, who, when *Cradle* appeared under our joint names, were put out by passages where I should have done a little more sanitizing. The earthier bits of dialogue, I explained, were the result of Gentry's years with the hairy-knuckled, hard-drinking engineers and mathematicians of JPL's Astrodynamics Division, where the Pasadena cops often have to be called in to settle bare-fisted fights over Bessel functions and nonlinear partial differential equations.

Yet so far, to the best of my knowledge, no school board has demanded that *Cradle* be removed from its shelves. I mention this because I have just discovered, to my astonished indignation, that this actually happened to *Imperial Earth* a decade ago. What's more, the board concerned then went on to ban any collection containing anything I'd ever written.

I wish I'd known about this at the time. I would have enjoyed telling these apprentice ayatollahs that the "books for the blind" versions of the novel that had offended them was recorded by a lady very unlikely to promote porn. She happens to be married to England's first law lord—the equivalent of the United States Supreme Court's chief justice.

Although *Cradle* was originally conceived as a movie project, and a treatment was prepared for Warner films, the chances of it ever reaching the screen now seem remote. By bad luck, a whole string of underwater/extraterrestrial movies appeared around the time of the book's publication, and most of them sank without a trace.

But Peter Guber, I'm happy to say, has gone on from strength to strength. Maybe he'll make *Cradle* when the cycle comes round again, as it inevitably will. "There is a tide in the affairs of men"—and of movies.

By the summer of 1987, *2061: Odyssey Three* was doing very nicely in the bookstores, thank you, and I was once again beginning to feel those nagging guilt pains that assail an author when he's not "working on a project." Suddenly, I realized that one was staring me right in the face.

Fifteen years earlier, the very last sentence of *Rendezvous with Rama* had read: "The Ramans did everything in threes." Now, those words were a last-minute afterthought when I was doing the final revision. I had not—cross my heart—any idea of a sequel in mind; it just seemed the correct, open-ended way of finishing the book.

I quickly outlined a spectrum of possibilities, and in a remarkably short time, Scott had sold *Rama II, The Garden of Rama,* and *Rama Revealed* to be delivered during 1989–1991.

So once again Gentry Lee is commuting across the Pacific for brainstorming sessions in the Sri Lankan hills, and the postman is complaining about the bulky printouts he has to balance on his bicycle. This time around, however, the fax machine has speeded up our intercontinental operations.

There is much to be said for this kind of long-distance collaboration; if they are too close together, coauthors may waste a lot of time on trivia. Even a solitary writer can think of endless excuses for not working; with two, the possibilities are at least squared.

However, there is no way of demonstrating that a writer is neglecting his job; even if his snores are deafening, his subconscious may be hard at work. And Gentry and I knew that our wildest excursions into literature, science, art, or history might yield some useful story elements.

For example, during the writing of *Rama II* it became obvious that Gentry was in love with Eleanor of Aquitaine—don't worry, Stacey, she's been dead for 785 years—and I had to tactfully dissuade him from devoting pages to her amusing career. (If you wonder how E of A could have the remotest connection with interstellar adventures, you have pleasures in store.)

I certainly learned a lot of French and English history from Gentry that they never taught me at school. The occasion when Queen Elmore berated her son, the intrepid warrior king Richard the Lion Heart, in front of his troops for failing to produce an heir to the throne must have been one of the more piquant moments in British military history. Alas, there was no way we could work in this gallant but gay Corleone, who was often a godfather, never a father . . . very unlike Gentry, whose fifth son arrived toward the end of *Rama II.*

But you will meet Gentry's most cherished creation, the yet-to-be-born Saint Michael of Sienna. One day, I am sure, you'll encounter him again, in books that Gentry will publish under his own name, with the minimum of help or hindrance from me.

As I write these words, we're just coming up to the midway of our four-volume partnership. And though we think we know what's going to happen next, I'm sure the Ramans have quite a few surprises in store for us. . . .

The Clarke-Lee collaboration is much like thousands of other coauthorships, but during the past few years, Scott has got me involved in several more unusual deals. It hasn't been very difficult because curiosity is my most abiding characteristic, and I tend to agree with the British nobleman who told his son: "Try everything once—except incest and folk dancing."

One of these projects I won't try again—at least not without ironclad safeguards. A famous publisher asked me to edit and write commentaries on a collection of essays about the future by almost twenty authors—and then marketed the book as if I had written the whole thing! To add insult to injury (and the many distinguished contributors were undoubtedly injured) I was never sent the final proofs, so I didn't know what had happened until it was too late. (Almost equally annoying were the stupid sub-sub-editorial mistakes that I never had a chance of correcting.)

I have agreed to several projects which seemed interesting and worth-while—as long as potential customers knew how much (or how little) I had contributed to them. For the record, they are:

1. *Arthur C. Clarke's Venus Prime* by Paul Preuss. This is a series of six paperbacks loosely based on short stories of mine, and in each case I have contributed an afterword. But the novels themselves, as is clearly stated, are all written by Paul Preuss, and I did not agree to the project until I had read some of his excellent fiction (and nonfiction). So far, *Breaking Strain, Maelstrom,* and *Hide and Seek* have appeared.

2. *Beyond the Fall of Night* by Gregory Benford. My first extended work of fiction was the short novel *Against the Fall of Night,* begun around 1935. After a magazine appearance in 1948, it was published in hardcover by Gnome Press in 1953. I was never quite satisfied with this version and expanded it so much that it became a completely new book, *The City and the Stars.*

However, the earlier incarnation was so popular that it has remained in print—and now I am both flattered and surprised to find that Greg Benford has volunteered to write a sequel. Professor Benford (Physics Department, University of California, Irvine; also Fellow of Cambridge University and the Woodrow Wilson Institute—and a NASA science adviser) is one of the best "hard" science fiction writers in the business; his recent *Great Sky River* is awesome, and I'm dying to read the sequel he's just done to that—*The Tides of Light.*

Beyond . . . will be written half a century after *Against. . . .*

So Greg will be able to take advantage of the revolution in astronomical knowledge that has taken place in that time. One of my occasional daydreams is of going back to the meetings of the British Astronomical Association that I used to attend in the thirties and telling my fellow members about neutron stars, black holes, quasars, millisecond pulsars, X-ray bursters—I'd have been thrown out on my ear. No one is better qualified than Greg to exploit these wonders—and to make up a few of his own.

By an odd coincidence, just as the contract was being drawn up for the volume containing the two *Falls*, I had a request from an excellent Australian science fiction writer, Damien Broderick (*The Dreaming Dragons*) for permission to write a sequel to *The City and the Stars.* I had to tell him that he'd been preempted by Greg . . . but perhaps in the midnineties?

The Power of

Compression

*In this essay, Clarke talks of brevity, the word power
inherent in compression—and* Reader's Digest.

Brevity, said Lord Bacon, is the soul of wit, and conciseness is a virtue I have always admired. The magic of poetry lies in its power to compress ideas or emotions into a mere handful of words. But prose can work the same spell, which is why Francis Bacon's aphorism has survived the centuries.

The art of precise writing is one which I had to learn when I entered the British Civil Service in 1936; little did I guess that a couple of decades later the *Reader's Digest* experts would be practicing their skills upon me. I believe that *A Fall of Moondust* was the first science fiction novel ever bought by the organization and must confess that I've never had the courage to read the miniaturized version. Not because I feared butchery, but because I had a horrid suspicion that the story might have been improved! (Anyway, I've just loaned my copy to the scriptwriter now turning *Moondust* into a television series.)

Perhaps because I share the same viewpoint, I have never been among those who poke fun at the *Digest*'s relentless optimism, and I'm still eagerly awaiting its legendary "New Hope for the Dead." It has always seemed to me that if you take a positive attitude toward the future, you may help to achieve a self-fulfilling prophecy. And ditto if you preach doom and disaster—you may tilt the balance in that direction.

This is not to say that one should ignore the dangers and evils of our increasingly threatened world. But many are so complex (e.g., the greenhouse effect, the pros and cons of nuclear power, the causes of international terrorism) that we urgently need skilled expositors who can analyze and then summarize them for the general public. Of course, it is also essential to do this without oversimplification or distortion. It thus seems to me that the *Digest* should have a triple goal, requiring considerable acrobatic skill on the part of its editors: to inspire, to entertain, and to educate.

I have been an intermittent but ardent browser through the *Digest* all my life, even when I did not agree with its political or religious outlook. I always felt that the *Digest* was on the side of decency—and that is a virtue that transcends almost all others. And, of course, like a few hundred million other people, I have always enjoyed such features as "The Most Unforgettable Character . . ." and its "Quotable Quotes." I was happy to provide one of those: "The real test of a man's honesty is not his income tax returns, but the zero adjustment on his bathroom scales."

Let me end with the finest example of compressed word power I've ever encountered. Back in the twenties, a young newspaperman bet his colleagues $10—no small sum in those days—that he could write a complete story in just six words. They paid up. . . .

I defy even the wizards of Pleasantville to shorten Ernest Hemingway's shortest and most heartbreaking story:

"For sale. Baby shoes. Never worn."

Life in the

Fax Lane

As Clarke's technological world is becoming realized, and virtually instantaneous
communication has become a reality, he sometimes wonders
if perhaps snail mail had a few hidden advantages.

Yesterday, my office passed the point of no return. For the first time, the number of incoming faxes exceeded the number of letters brought by the postman. The future had arrived, and I'm not sure I'll survive it.

F day was only a couple of months ago, but since then, my life has been transformed, by no means for the better. Let my example, even though it is hardly a typical one, be a warning. . . .

Until this latest marvel of Japanese technology invaded my establishment, I was just able to cope with the mail, thanks to five secretaries (three in Sri Lanka, two in England) and seven computers. (For the record, I still have my Archives III. If you had a Rolls-Royce, wouldn't you still use it occasionally?)

With the help of all this hardware, software, and people power, I was barely able to keep my head above water. On a good week, I could even find a few hours for writing and had wild ambitions of completing one more novel before the Big Programmer in the Sky pressed my DELETE button. This fantasy has now vanished, thanks to my marvelous new—ha!—timesaving device. Here's why. . . .

In the good old days when I wrote a letter to my agents in London or New York, or to the secretary of my UK company, Rocket Publishing, I could count on at least a week or even two before getting a reply! There was time to think, and even time to work.

Not anymore. When I went to bed last night, I faxed a letter to my agent in New York.

The reply was already waiting for me when my clock radio switched on the *BBC World News* at six-thirty the next morning. Ten days had shrunk to as many hours—and the new novel recedes even further into the future in favor of composing my next (one or two) replies.

Yet all is not lost. One can fight fire with fire, and I am a great believer in the power of technology to solve all problems—even those it creates. So, despite my well-known failure to patent my famous Digital Zip Fastener, I'll once again give to the world another electronic breakthrough of incalculable value. Some minor details need work, but here are the basic principles of the Clarke FaxMaster.

It's really quite simple—those who have ever programmed a VCR before going on vacation will understand it at once. You just put suitably customized time delays into the machine, so you don't receive any messages until you're ready for them. Let me explain how it will operate (as soon as I get the remaining bugs out of it).

The FaxMaster can still deliver messages instantly, but in addition, it will have built-in delays of twelve hours, twenty-four hours, and one week. The delay chosen will depend on the identity of the sender.

There'll be only one occupant of the zero delay: the Nobel Prize Committee.

On twelve-hour delay will be my agents, Rocket Publishing, one or two (well, maybe three) editors, Stanley Kubrick, Steven Spielberg, my bank, and a very few great and good friends.

After twenty-four hours, the FaxMaster's memory will disgorge anything from the remaining editors, and some fifty other friends.

Now, the fun begins, with an AL (artificial intelligence) program to deal with the residue. If the FaxMaster recognizes the sender, but he or she is not on the "priority" lists, it will dump the message into its memory and let me have it after one week. If it doesn't recognize the name, it will send back a brief reply: "Wrong number. You should have called . . ."

Then it will give Harlan Ellison's fax number.

There's a blacklist of names which will be given very special attention. At first I tried to devise a computer virus—a Fax-Zapper—that would go back down the circuit and Do Something Horrible to the machine at the other end. Then I realized that there was a much better solution.

My fax machine will send the correct acknowledgment (charmingly called handshaking) which will permit the message to be sent. However, it will be transmitted with agonizing slowness, because the FaxMaster will keep injecting noise into the circuit—not enough to break the connection, but enough to keep the error-checking programs so busy that there's an electronic bottleneck. Meanwhile, the sender's phone bill will mount astronomically.

Finally, his machine will burp and will spit out its "activity report." It will read something like this:

TRY TRANSMISSION AGAIN

ERROR PAGE	0#018
USAGE	09'30
PAGES	0

Of course, this will have tied up my machine for nine and a half minutes, but that suits me fine.

After a few hundred dollars' worth of mounting frustration, the gibbering victim will have to go out and buy some stamps.

If he can find any place that still sells them.

Credo

For thousands of years, Clarke writes, the subtlest minds of the human species have been focused on the great questions of life and death, of time and space—and of man's place in the universe. The answers have been encapsulated in the holy books of countless religions and whole libraries of philosophy, folklore, and myth.

Can our age contribute anything both new and true to these ancient debates? I believe so. We have been lucky enough to live at a time when knowledge that once seemed forever beyond reach can be found in elementary schoolbooks. Our generation has seen the far side of the Moon, and close-ups of all the major bodies circling the Sun. We have opened the Pandora's box of the atomic nucleus. And perhaps most marvelous of all, we have uncovered the secret of life itself, in the endless twining and untwining of the DNA spiral. This is perhaps the greatest discovery in the whole history of science, yet even now it is barely thirty years old.

There are those who claim not to be impressed by such achievements, arguing that science deals with unimportant questions that can be solved, while religion is concerned with important ones that can't. The logical positivists would maintain that this is nonsense; if a problem can't be solved, at least in principle, it doesn't really exist. In other words, there's no such animal as metaphysics.

Without knowing it, I became a logical positivist at about the age of ten. Every Sunday, I was supposed to make the two-mile walk to the local Church of England—it was a long time before I discovered there was any other variety—to attend a service for the village youth. To encourage us to sit through the sermons, we were rewarded with stamps illustrating scenes from the Bible. When we had filled an album with these, we were entitled to an "outing"—i.e., a bus trip to some exotic and remote part of Somerset, perhaps as far as twenty miles away. I stuck with it for a few weeks, then decided—to quote Churchill's famous memorandum on the necessity of ending sentences with a preposition—"This is nonsense up with which I will not put."

Half a century of travel, reading, and contact with other faiths has endorsed that early insight.

Now I myself am not completely innocent, according to one of the last letters I received from the great biologist J. B. S. Haldane. Shortly before he died (going not gently but heroically into that good night with a witty poem entitled "Cancer Can Be Fun") he wrote: "I would like to see you awarded a prize for theology, as you are one of the very few living persons who has written anything original about God. You have in fact, written several mutually incompatible things. . . . If you had stuck to one theological hypothesis you might be a serious public danger."

I am only sorry that J. B. S. never had a chance to criticize my later (doubtless yet more incompatible) speculations, developed in the novels *The Fountains of Paradise* and *The Songs of Distant Earth.* He would, I am sure, have enjoyed this specimen from *Fountains:*

> There can be no such subject as comparative religion as long as we study only the religions of man. . . . If we find that religion occurs exclusively among intelligent analogs of apes, dolphins, elephants, dogs, etc., but not among extraterrestrial computers, termites, fish, turtles, or social amoebae, we may have to draw some painful conclusions. . . . Perhaps both love and religion can arise only among mammals, and for much the same reasons. This is also suggested by a study of their pathologies; anyone who doubts the connection between religious fanaticism and perversion should take a long, hard look at the *Malleus Maleficarum* or Huxley's *The Devils of Loudun.*

But I am quite serious about the profound philosophical importance of the Search for Extra-Terrestrial Intelligence (SETI); this may be its supreme justification. The fact that we have not yet found the slightest evidence for life—much less intelligence—beyond this Earth does not surprise or disappoint me in the least. Our technology must still be laughably primitive; we may well be like jungle savages listening for the throbbing of tom-toms, while the ether around them carries more words per second than they could utter in a lifetime.

The greatest tragedy in mankind's entire history may be the hijacking of morality by religion. However valuable—even necessary—that may have been in enforcing good behavior on primitive peoples, their association is now counterproductive. Yet at the very moment when they should be decoupled, sanctimonious nitwits are calling for a return to morals based on superstition.

Having disposed of religion (at least until next Wednesday), let us consider something really important: God—aka Allah/Brahma/Jehovah, etc. ad infinitum. In *The Songs of Distant Earth,* I distinguished between two aspects of this hypothetical entity, calling them Alpha and Omega to defuse emotional reactions.

Alpha might be identified with the jealous God of the Old Testament, who watches over all creatures ("His eye is on the sparrow") and rewards good and evil in some vaguely described afterlife. Even today, belief in Alpha is fading fast; I suggested that early in the next millennium the rise of "statistical theology" would prove that there is no supernatural intervention in human affairs. Nor does the "problem of evil" exist; it is an inevitable consequence of the bell-shaped curve of normal distribution.

Unfortunately, most people do not understand even the basic elements of statistics and probability, which is why astrologers and advertising agencies flourish. If you want to start an interesting fight, say in a loud voice at your next cocktail party, "Fifty percent of Americans (or whatever) are mentally subnormal." Then watch all those annoyed by this mathematical tautology instantly pigeonhole themselves.

I also, rather mischievously, demolished Alpha by invoking the ghost of Kurt Gödel, whose notorious "incompleteness of knowledge" theorem quite obviously rules out the existence of an omniscient being. However, this is an area where logic gets you nowhere. Belief—or disbelief—in Alpha appears to be irrevocably programed into most people at an early age.

A man I admire, who has held the highest medical position in the United States, recently declared. "There are no atheists at the bedside of a dying child." It is a compassionate statement, nobly expressed, with which every humane person must sympathize. But, with all respect, it is simply untrue.

Nor have I ever felt a need for Alpha on the several occasions when I thought I was about to die (in each case, at a depth of embarrassingly few fathoms). Certainly the notion of appealing for divine help never entered my mind; I was much too busy thinking, "How do I get out of this ridiculous situation?"

Omega—the Creator of Everything—is a much more interesting character than Alpha, and not so easily dismissed. Although irredeemable agnostics may smile at Edward Young's "The undevout astronomer is mad," no intelligent person can contemplate the night sky without a sense of awe. The mind-boggling vista of exploding supernovae and hurtling galaxies does seem to require a certain amount of explaining: to answer the question "Why is the universe here?" with the retort "Where else would it be?" is somehow not very satisfying. Although—the logical positivists would be pleased—it may be all the answer that is needed, because the question itself may not make sense.

Let me offer an analogy, suggested by a conversation I once had with C. S. Lewis. We science fiction authors are always picking each other's brain, and Lewis asked me what the horizon would look like (ignoring atmospheric absorption) on a really enormous planet—one not thousands, but millions, of kilometers in radius.

Any inhabitants would be convinced that they were living on a perfectly flat plane and might fight holy wars over the rival doctrines (a) the world goes on forever and ever; (b) you'll fall off when you reach the edge. But to us, there is no problem. We have watched the globe of the Earth floating on our television screens and have no difficulty in understanding why both flatlander cults are wrong. If they ever got around to making spaceships, their religious disputations would be ended.

So it is very, very risky to maintain that, as the old B-grade movies loved to intone, "There is some knowledge not meant for Man." I am fond of quoting a monumental gaffe made by Auguste Comte, who told the astronomers in no uncertain terms just what they could ever expect to know about other worlds—"We may determine their forms, their distances, their bulk, their motions—but we can never know anything of their chemical or mineralogical structure; and, much less, that of organized beings living on their surface."

Within a century of Comte's death, thanks to the invention of the spectroscope, much of astronomy had become astrochemistry—a science he had roundly declared to be impossible. I wonder what he would have said about space exploration, had anyone been rash enough to suggest such an absurdity to him.

So it may be that questions which now seem almost beyond conjecture may one day be conclusively settled. The limits of space, the beginning and ending of time, the origin of matter and energy, may have no mysteries to our remote descendants. And many of the questions we ask of the universe may turn out to be completely meaningless—as certain theories on the frontiers of modern physics tantalizingly suggest.

I felt this very strongly when I was privileged to make a television program, modestly entitled "God, the Universe and Everything Else" with Newton's successor Dr. Stephen Hawking. If you have not yet read *A Brief History of Time,* please rectify the omission—and read the bits about "imaginary time." Thank you; that saves me a lot of hand waving, trying to explain how our own views of past and future may be as naive as the flatlanders' ideas about the geometry of their giant planet.

The extraordinary success of Dr. Hawking's book is one of the best pieces of news from the popular science—indeed, educational—front for many years. I have been appalled by the way in which the United States (and much of the world, East and West) appears to be sinking into cultural barbarism, harangued by the fundamentalist ayatollahs of the airwaves, its bookstores and newsstands poisoned with mind-rotting rubbish about astrology, UFOs, reincarnation, ESP, spoon-bending, and especially "creationism." This last—which implies that the marvelous and inspiring story of evolution, so clearly recorded in the geological strata, is all a cosmic practical joke—helps me to understand the revulsion that a devout Muslim must feel toward *The Satanic Verses.* If there is indeed such a thing as blasphemy, it is here. . . .

The Pontifical Academy of Science—which I have been honored to address—has now firmly stated: "Masses of evidence render the application of the concept of evolution to man and the other primates beyond serious dispute."

I began this essay by saying that men have debated the problems of existence for thousands of years—and that is precisely why I am skeptical about most of the answers. One of the great lessons of modern science is that millennia are only moments. It is not likely that ultimate questions will be settled in such short periods of time, or that we will really know much about the universe while we are still crawling around in the playpen of the Solar System.

So let us recognize that there is much concerning which we must reserve judgment, and refuse to take seriously all dogmas and revelations whose acceptance demands faith. They have been proved wrong countless times in the past; they will be proved wrong again in the ages to come.

And worse than wrong. Who can forget Jacob Bronowski, in his superb television series, *The Ascent of Man,* standing among the ashes of his relatives at the Auschwitz crematorium and reminding us: "This is how men behave when they believe they have absolute knowledge." This is how they are still behaving—in Ireland, in Lebanon, in Iran—and at this very moment, alas, in my own Sri Lanka.

Yet, if absolute knowledge is unattainable, someday most of the great truths may be established—if not with absolute certainty, then beyond all reasonable doubt. Do not be impatient; there is plenty of time.

How much time, we are only now beginning to appreciate. In a famous essay, "Time Without End," Freeman Dyson speculated that a high-technology cosmic intelligence might even be able to make itself, quite literally, immortal.

So let me end with the final chapter, "The Long Twilight," from my *Profiles of the Future: An Inquiry into the Limits of the Possible*:

Whether Freeman Dyson's vision (some would say nightmare) of eternity is true or not, one thing seems certain. Our galaxy is now in the brief springtime of its life—a springtime made glorious by such brilliant blue-white stars as Vega and Sirius, and, on a more humble scale, our own Sun. Not until all these have flamed through their incandescent youth, in a few fleeting billions of years, will the real history of the universe begin.

It will be a history illuminated only by the reds and infrareds of dully glowing stars that would be almost invisible to our eyes; yet the somber hues of that all-but-eternal universe may be full of color and beauty to whatever strange beings have adapted to it. They will know that before them lie, not the millions of years in which we measure eras of geology, nor the billions of years which span the past lives of the stars, but years to be counted literally in trillions.

They will have time enough, in those endless aeons, to attempt all things, and to gather all knowledge. They will be like gods, because no gods imagined by our minds have ever possessed the powers they will command. But for all that, they may envy us, basking in the bright afterglow of Creation; for we knew the universe when it was young.

The Colors of

Infinity: Exploring

the Fractal Universe

Clarke delivered the following lecture in Riyadh,
Saudi Arabia, on November 12, 1989.

I want to talk to you today about one of the most beautiful and marvelous discoveries in the entire history of mathematics. And unlike virtually all other mathematical discoveries for the last thousand years, it can be appreciated and enjoyed even by those who know absolutely nothing about math—even those who hate it!

The Mandelbrot Set was discovered only ten years ago by the Polish-American mathematician Benoit Mandelbrot, after whom it is named. Yet—and this is another amazing fact—in principle it could have been discovered as soon as we learned how to count. For it involves none of the higher mathematical functions—not even division or subtraction!

If it's so simple, why did it take so long to find it? Well, even though it only involves additions—and those additions we call multiplication—there's a slight snag. It requires millions and billions of them, one after the other, in effect a series of algorithms—and a single mistake will send you back to square one. So no one guessed its existence until the advent of modern high-speed electronic computers.

I expect you've all seen those children's books that consist of blank pages sprinkled with numbers. When the numbers are linked together in the correct order, a picture slowly emerges.

That is a rough analogy of the way in which the Mandelbrot Set is created. Very rough, because in this case there are no empty spaces on the blank sheet—every point has a number! The trick is to find the order in which they have to be connected—and only a computer can do this. First, a few fundamentals that may seem elementary, even childish, but which, like those innocent questions that philosophers are always asking, contain more than meets the eye.

Just over two thousand years ago, Euclid gave us the concepts that still dominate out thinking about space—the point, the line, the plane. They are abstractions, of course—they don't really exist. But when I talk about a Euclidean "line"—whether straight or curved doesn't matter—I expect you all have the same mental image of a black thread, so narrow that you can just see where it is. Because it hasn't any thickness at all no matter how much you magnified it, it would always look exactly the same. It has length, but no breadth—not even the breadth of an atom.

But now we've discovered something that Euclid never imagined. Would you believe a fuzzy line? Or maps that can never be drawn because they have no definite boundaries? I'm talking geometry, not politics. . . .

Infinity may or may not exist in the real universe—especially if, as many suspect, it is closed both in space and time. To a pure mathematician, the number of stars is really very small. Even the number of atoms in the universe isn't much bigger: it can be written out in less than one hundred digits.

But the images in the Mandelbrot Set are, quite literally, infinite in number.

I once discovered a beautiful fractal landscape I christened Lake Mandelbrot. I failed to "save" it properly and occasionally make a halfhearted search for it. But I know that the lifetime of the universe may not be long enough to find it again. . . .

Anyway, I've decided on a compromise that may involve a lot of hand waving. Even if you don't follow the detail, I think I can give you some idea of how an infinite number of stunning works of art can be generated from an equation no more complicated than Einstein's famous $E = mc^2$.

Indeed, it's almost identical: there may be some deep philosophical implications in this . . . those of you who'd prefer to wait for the main show can now take a five-minute nap or switch on your Walkmans.

We'll start with something that Einstein was fond of—a thought experiment. In this case it's one in elementary arithmetic; and though it's so simple you may consider it an insult to your intelligence, it's really the key to the whole process.

Take any number and square it. And then keep on squaring it. . . . That's all.

Now, there are just three possibilities. The simplest is if you start with the number one. However many times you square it, it stays exactly one—no more, no less.

Suppose you start with a number larger than one—say the next integer two. If you keep squaring it, you get the series:

2, 4, 16, 256, 65,536, 4,294, 967, 296 . . .

After only a few terms, it's exploding toward infinity and couldn't be handled by any conceivable computer.

If you start with a number smaller than one, just the opposite happens. If the number was 0.1 or one-tenth, squaring it gives a hundredth, a ten-thousandth, a hundred-millionth . . . In no time at all, it has dwindled to zero.

The number one—the basic element of all counting—thus has a surprising property not at all obvious at first sight. It marks a boundary—a frontier, if you like—in the endless line of natural numbers, which includes the integers 1, 2, 3, 4, 5 . . . as a special case.

Let's make this more precise by using an equation. It's so simple you can chalk it up on the blackboard of your mind.

Here it is: Zee equals little zee squared. However, this isn't merely an equation—it's a dynamic process, a program. The equals sign works in both directions. If you start with little zee on the right-hand side, you square it to give big Zee. Then you put big Zee equal to little zee and repeat the process ad infinitum. That's how we generated the series 2, 4, 16, 256 . . . and so on.

In computer parlance it's called a loop. It's like a dog chasing its own tail.

A dog doing this usually doesn't get anywhere; but the loop we'll be orbiting around presently will take us to some very strange places indeed.

Now . . . to turn this little equation into a picture, by giving a physical meaning to the zees. They simply represent distances from a central point. Think of a donkey tied to a peg by a long rope that it keeps pulled tight as it walks. It's a mathematical donkey (I'm afraid they're rather common), and—this isn't so common—it's a very obedient one. It strictly obeys the Zee equals zee squared equation as it moves around. Little zee is the length of the rope at the starting point; big Zee the length at the finish. The actual units don't matter; we're dealing with pure numbers.

You'll recall that if the initial value of zee was exactly one, nothing would change. So in that case, the donkey would continue to walk round and round in a circle of unit radius.

But suppose the initial zee was less than one. Now the numbers get smaller and smaller—the donkey has to spiral toward the center until it disappears down a black hole (we'll come to those later).

Now the other case—the initial zee is bigger than one. This time the zees grow indefinitely: before long the donkey is galloping off to the edge of the universe at the velocity of light.

Having waved good-bye to the poor beast, let's look at the pure geometry of the situation. The point I am trying to make is that the circle of radius one forms a boundary, separating all possible numbers into two classes. Those outside it race off to infinity when the Zee-equals-zee-squared program operates on them. Those inside shrink inexorably to zero.

That boundary—the circumference of the circle radius one—is a line with no thickness whatsoever; if you examined it with a microscope of ever-increasing power, it would remain a line. Yet though it's infinitesimally thin, it's an absolute barrier; the numbers trapped inside it can never escape to the outer world. The circle is a map of a mathematical territory with a frontier more sharply defined, and more impenetrable, than any ever made by man.

This may sound rather elementary, and not very exciting. Well, I promise you that things are now going to change, quite dramatically.

Once again, picture that donkey at the end of its tether. This time, however, we'll make a trifling change in the program that determines the length of the rope—the distance of the donkey from the center peg. Instead of Zee equals zee squared, we'll make it Zee equals zee squared plus cee. We'll go round the loop exactly as before, but his time cee is the starting length and the initial value of little zee is set at zero.

So what sort of map does the donkey trace out this time? Obviously it won't be a circle anymore—but not even the greatest mathematical genius before this century could have guessed the right answer.

You should wonder where do the colors come from? After all, there are no colors in geometry . . . the computer will obediently fill them in. But they're not meaningless; they're like the contour lines on a map. They show how many times you have to go round the Zee–zee-squared loop before the numbers start racing off to infinity.

Close Encounter

with Cosmonauts

There are moments in life that become frozen in memory, as if illuminated by a flash of lightning. For example, many Americans can recall exactly what they were doing and where they were the moment they heard John F. Kennedy was assassinated. In the following essay, Clarke recalls a few such moments from his own memory bank.

One day in 1961, I was walking back to my house in the Colombo suburb after playing table tennis at the Otter's, the local athletic club, and was about to cross the main road, now called Bauddhaloka Mawatha. Suddenly a police escort came in sight, followed by a car containing a group of dignitaries.

My gaze locked onto a single figure, and instantly there flashed into my mind a phrase which was used more than one hundred years ago, when President Lincoln was assassinated: "Now he belongs to the ages."

That was my first glimpse of Yuri Gagarin; I never guessed how soon, and how tragically, the same quotation would apply to him.

Later that day, I had the privilege of meeting cosmonaut Gagarin in person, and exchanging a brief conversation through an interpreter. A few months after that, I was delighted to receive a copy of his autobiography with the inscription "To Arthur Clarke, this souvenir of our meeting in Ceylon. Yuri Gagarin, 11.12.1961."

The second incident occurred almost a decade later, in 1970. I was boarding a private bus in Washington, and as I stepped inside, I realized who was sitting on one of the aisle seats just ahead of me. I'm afraid I gaped like a schoolboy seeing his favorite rock idol. . . . A slow, embarrassed smile spread across the face of the other passenger, and the senior NASA official sitting next to him, whom I'd completely ignored, said rather testily, "Hello, Arthur—I'm here as well!"

And that was my first view of Neil Armstrong, whom, I'm happy to say, I've met on several later occasions—unlike, alas, Yuri Gagarin.

It has therefore been my privilege to shake hands with the first man to enter space, and the first man to step onto the Moon. Today's young people cannot imagine how inconceivable this was to almost everyone when I was a boy. Until well into the 1940s, distinguished scientists—when they condescended to address the subject at all!—dismissed spaceflight as "science fiction." One British Astronomer Royal never quite recovered from his unfortunate assertion that space travel is "utter bilge"—made, incredibly, in 1956, after the announcement of the IGY Earth satellite program!

It has also been my great pleasure to know the pioneer "spacewalker," cosmonaut Aleksei Leonov. The first time I saw him was in 1965, at the Annual Congress of the International Astronautical Federation, then taking place in Athens; its highlight was a film of the first EVA (March 18, 1965). This had to be shown on a very cumbersome 35mm projector that looked as though it had been used for the premiere of *Battleship Potemkin*. The monstrosity was slowly assembled, and an audience of about a thousand distinguished guests—including King Constantine—waited eagerly to see Aleksei emerge into space.

What they actually saw was something quite different. There had been a mix-up of reels, and a Russian cartoon film appeared instead. Despite all the attempts by the frantic operator to recover from the debacle, it went on, minute after minute, while the audience became more hysterical with laughter. I felt very sorry for poor Aleksei, but eventually, matters were rectified and we were able to see his historic exploit.

Later, the British delegation teased its American colleagues by congratulating them on the year's first successful CIA operation.

My first actual meeting with cosmonaut Leonov, however, was in Munich, at the 1968 European screening of *2001: A Space Odyssey*. After the performance, Aleksei endeared himself to me by saying, "Now I feel I've been in space twice." (Today he could say three times, as he was the Soviet Union commander on the Apollo-Soyuz rendezvous).

A much more memorable encounter with Aleksei, however, was during my visit, in June 1982, to Star City, Russia. I will never forget the solemn moments when I stood in Yuri Gagarin's own office, exactly as it was on that day in 1967, with the clock stopped at the moment of his death. "I heard the crash," Aleksei told me somberly.

After the tour of the facility, Aleksei took me to his apartment where I enjoyed meeting his charming wife, Svetlana—and his little parrot, Lolita, who kept orbiting round the room, but always managed to make a touchdown on Aleksei's shoulder. It was in the privacy of his home that I told him about my forthcoming novel *2010: Odyssey Two* with its dedication:

> Dedicated, with respectful admiration, to two great Russians, both
> depicted herein:
> General Aleksei Leonov—Cosmonaut, Hero of the Soviet Union,
> Artist
> and
> Academician Andrei Sakharov—Scientist, Nobel Laureate,
> Humanist

This, remember, was in 1982, when it seemed virtually certain that Academician Sakharov would die in exile in Gorki. In view of subsequent events, I am very proud of the dedication, but deeply regret that it caused great trouble for my genial host and guide on my Russian tour, Vasili Zaharchenko. I trust that those who persecuted him have now made suitable amends.

2010, with its mixed American-Soviet crew, was a deliberate attempt at a self-fulfilling prophecy, and I am delighted to see that as time has passed it seems much more probable than earlier. So I cannot help wondering what chance I have of shaking hands with the first man to set foot on Mars.

Where will he be from? I wonder. That hilarious episode in Athens has made me visualize another undignified scenario. While the American and Russian commanders are bowing courteously to each other, saying, "After you," the _____ member of the crew will sneak out first.

I leave you to fill in the nationality.

The Century

Syndrome

Even in the eighties, Clarke was gnawing over the millennium time bomb
and wrote this tongue-in-cheek essay about the hangover
on January 1, 2000—the "day after."

When the clocks struck midnight on Friday, December 31, 1999, there could have been few educated people who did not realize that the twenty-first century would not begin for another year.

For weeks, all the media had been explaining that because the Western calendar started with Year One, not Year Zero, the twentieth century still had twelve months to go.

It made no difference; the psychological effect of those three zeros was too powerful, the fin de siècle ambience too overwhelming. This was the weekend that counted. January 1, 2001, would be an anticlimax, except to a few movie buffs.

There was also a very practical reason why January 1, 2000, was the date that really mattered, and it was a reason that would never have occurred to anyone a mere forty years earlier. Since the 1960s, more and more of the world's accounting had been taken over by computers, and the process was now essentially complete. Millions of optical and electronic memories held in their stores trillions of transactions—virtually all the business of the planet.

And, of course, most of these entries bore a date. As the last decade of the century opened, something like a shock wave passed through the financial world. It was suddenly, and belatedly, realized that most of those dates lacked a vital component.

The human bank clerks and accountants who did what was still called bookkeeping had very seldom bothered to write in the 19 before the two digits they had entered. These were taken for granted; it was a matter of common sense. And common sense, unfortunately, was what computers so conspicuously lacked. Come the first dawn of 00, myriads of electronic morons would say to themselves, "00 is smaller than 99. Therefore, today is earlier than yesterday—by exactly ninety-nine years. Recalculate all mortgages, overdrafts, interest-bearing accounts, on this basis." The result would be international chaos on a scale never witnessed before; it would eclipse all earlier achievements of Artificial Stupidity—even June 5, 1995, Black Monday, when a faulty chip in Zurich had set the bank rate at 150 percent instead of 15 percent.

There were not enough programmers in the world to check all the billions of financial statements that existed, and to add the magic 19 prefix wherever necessary. The only solution was to design a special software that could perform the task, by being injected—like a benign virus—into all the programs involved.

During the closing years of the century, most of the world's star-class programmers were racing to develop a "Vaccine '99"; it had become a kind of Holy Grail. Several faulty versions were issued as early as 1997—and wiped out any purchasers who hastened to test them before making adequate backups. The lawyers did very well out of the ensuing suits and countersuits.

Edith Craig belonged to the small pantheon of famous women programmers that began with Byron's tragic daughter Ada, Lady Lovelace, continued through Rear Adm. Grace Hopper, and culminated with Dr. Susan Calvin. With the help of only a dozen assistants and one SuperCray, she had designed the quarter million lines of code of the DOUBLEZERO program that would prepare any well-organized financial system to face the twenty-first century. It could even deal with badly organized ones, inserting the computer equivalent of red flags at danger points where human intervention might still be necessary.

It was just as well that January 1, 2000, was a Saturday; most of the world had a full weekend to recover from its hangover—and to prepare for the moment of truth on Monday morning.

The following week saw a record number of bankruptcies among firms whose accounts receivable had been turned into instant garbage. Those who had been wise enough to invest in DOUBLEZERO survived, and Edith Craig was famous—and happy. Only the fame would last.

Who's Afraid

of Leonard Woolf?

In the opening chapters of the Bloomsbury figure Leonard Woolf's Growing: An
Autobiography of the Years 1904–1911, *he wrote, "In October 1904, I sailed
from Tilbury Docks in the P & O Syria for Ceylon. . . . I went to the
Grand Oriental Hotel, which in those days was indeed both
grand and oriental." Almost fifty years later,
Clarke followed the same route.*

In seven years of energetic overachievement, the conscientious young
Cambridge graduate rose from a humble cadet in the Ceylon Civil Service
to assistant government agent. By the age of twenty-eight, he had so impressed the governor-general that he was put in charge of a major area along
the south coast of the island.

The young AGA loved Ceylon and its people but administered his
mini-empire with an honesty and impartiality that did not always endear
him to his subjects. When he returned to England for a year's leave in 1911,
he had decided that even the most benevolent imperialism, for all the good

that it undoubtedly brought in terms of peace, justice, and improved standards of health and education, could not be morally justified. He had also fallen in love with Virginia Stephen, and the combination of these two factors was more powerful than the claims of the Colonial Office. So he resigned, married, founded the Hogarth Press, and helped to launch one of today's major growth industries, the Bloomsbury business.

But Ceylon continued to haunt him, even while courting Virginia and resuming acquaintance with Lytton Strachey, Maynard Keynes, Duncan Grant, Vanessa and Clive Bell, and the other luminaries of his Cambridge days. He had been back only six months when he started to write *The Village in the Jungle.* Though much less well known than *A Passage to India,* many consider it the better book. It is certainly an astonishing feat of sympathetic imagination for a young colonial administrator to enter so completely into the minds of Sinhalese peasants that his novel has now become a classic in their own language.

> The village was called Beddagama, which means the village in the jungle. . . . All jungles are evil, but no jungle is more evil than that which lay about the village of Beddagama. . . . The trees are stunted and twisted by the drought, by the thin sandy soil, by the dry wind. They are scabrous, thorny trees. . . . And there are enormous cactuses, evil-looking and obscene. . . . More evil-looking still are the great leafless trees, which look like a tangle of gigantic spiders' legs.

Such is the stage upon which Woolf's villagers pass lives as twisted and stunted as the trees around them. The book is an unrelieved tragedy; one by one, its characters are destroyed by disease, starvation, and the malice of their fellows. At its end, the village itself has been overwhelmed by the jungle, and only Punchi Menika, "a very old woman before she was forty," remains in its ruins, to meet a fate as unforgettable as any in literature:

> The perpetual hunger wasted her slowly. . . . At last the time came when her strength failed her; she lay in the hut unable to drag herself out to search for food. . . . When the end was close upon her a great black shadow glided into the doorway. Two little eyes twinkled at her steadily, two immense white tusks curled up gleaming against the darkness. She sat up, fear came upon her, the fear of the jungle, blind agonizing fear.
>
> "Appochchi, Appochchi!" she screamed. "He has come, the devil from the bush. He has come for me as you said. Aiyo! Save me, save me! Appochchi!"
>
> As she fell back, the great bear grunted softly, and glided like a shadow toward her into the hut.

Stories of such inspissated gloom, however superbly crafted, do not normally appeal to moviemakers. But for many years, Sri Lanka's most eminent director, Lester James Peries, had set his heart upon filming *Village* and after somewhat byzantine negotiations had managed to obtain the rights. More remarkably, he had raised sufficient money for the production and early in 1979 announced that filming would soon commence in the actual setting of the novel, the Hambantota district.

Sixteen years earlier, I had spent a good deal of time in this region, for it is the last reasonably safe harbor along the south coast of the island.

The site, literally next door to the little district court where Leonard Woolf had presided so many years earlier, is one of the most splendid in the whole of Sri Lanka, on a headland overlooking miles of curving beach. Although Hambantota is only six degrees north of the equator, it is never excessively hot, because there is always a steady breeze from the sea. And "the rhythm of the sea," as Woolf describes in *Growing,* dominates the life of the little town.

That rhythm was somewhat disrupted one weekend in July 1979, when Lester's film unit descended upon the little town. The shooting in which I was involved had to be done during the weekend, because only then was the district court available; come Monday, the defendants in the dock would be genuine ones, not actors. . . . Luckily the only change necessary was replacing the Seal of the Republic of Sri Lanka with the Royal Crest. Nothing else had altered since Leonard Woolf's time.

But the world around had certainly changed. The very evening I arrived, gliding in the air-conditioned comfort of my Mercedes-Benz over the road along which Woolf had jolted in bullock carts for a couple of dusty days, I had to address a science fiction convention in Washington. My words winged their way via the local microwave beam to the earth station near Colombo, then to the satellite twenty-two thousand miles overhead. So it is amusing to discover that Woolf wrote in his diary in 1908, "I hate machinery in this country. Very few natives can be got to understand it." Needless to say, the satellite station I was using was being run by the local "natives."

Next morning, I was dressed in my Edwardian colonial garb and patiently thatched by the makeup man, who restored all the hair that had unaccountably disappeared over the past few decades; the result was sufficiently judicial to terrify the innocent, let alone the guilty. But as I had never done any acting in my life—and had seen the script only two days earlier—there was no way of predicting the outcome when I finally appeared before the camera. "The next couple of hours," I told Lester, "will decide whether you're filming *Village in the Jungle* or *Carry On, Judge.*"

Fortunately, my worst fears proved groundless, and everything went very smoothly. There was no danger of forgetting my lines, for I had them concealed in my legal papers. In any event, all I had to say was in the form of brief questions put to the defendants through the court interpreter. Only one phrase was in Sinhalese, "Mata umba'gena kanagatui" (I am sorry for you), when I sentenced the principal accused to six months' hard labor. The verdict was given with reluctance. The evidence was, in fact, entirely fabricated; the accused was quite innocent of the robbery with which he had been charged.

As I sat on the bench in my judicial costume, I was not in the least conscious of acting a role. The courtroom—with its bench, raised dais surrounded by a wooden balustrade, and an oblong cage for the accused— was a time machine that had carried me back to the beginning of the century.

If Leonard Woolf could have looked through my eyes as I sat on the dais he had occupied so many times, he could not have told that seventy years had passed. Any doubts would have been resolved by a glance outside:

> The judge . . . looked out through the great open doors . . . down upon the blue waters of the bay, the red roofs of the houses, and then the interminable jungle . . . stretching out to the horizon. . . . And throughout the case this vast view . . . was continually before the eyes of the accused. Their eyes wandered from the bare room to the boats and canoes, bobbing up and down on the bay.

Sixteen years earlier, my own little boat, the cantankerous and ill-fated *Ran Muthu,* had been "bobbing up and down" in that same bay, as she prepared to carry me to the greatest adventure of my life. During the intervals between takes, my eyes strayed continually to that magnificent view, and I keep wondering how many had seen it for the last time before they were taken to the jail—or the gallows. One of Woolf's duties, which doubtless encouraged his ultimate resignation, was that of seeing that hangings were properly carried out.

By working steadily all through Sunday, we managed to finish shooting at dusk. The Royal Arms were taken down, and the court was handed back to the Republic of Sri Lanka. The following morning, half our bar would still be here because they were real lawyers, not actors. I was never sure which was which, and the genuine items returned the compliment by saying that *they* couldn't tell I wasn't a real judge. I certainly felt like one as I interrogated the prisoners—even though one of them was an old friend of twenty years' standing.

Monday was our day in the jungle, and for the first time in my life I wore a solar topee. To my surprise it was extremely light and comfortable; I am sure it would still be popular if it didn't look so hopelessly colonial-Kiplingesque.

For the purpose of the movie, two Beddagamas had been constructed—one Before, and one After. The first was a reasonably prosperous and well-populated village, at which I arrived by bullock cart with a set of census forms; and accompanied by a uniformed official, the *rate* (pronounced "ratay") *mahatmaya,* or local district supervisor. This was the opportunity for a delightful and completely authentic bit of dialogue:

> L.W. This rinderpest business is terrible. But the villagers won't take any precautions—they blame it all on Halley's comet.
>
> R.M. Another evil they blame on the comet, sir, is a very strict Government Agent.

In 1910, the year of the comet, the villagers did indeed compare Leonard Woolf unfavorably to his kindhearted predecessor. Also in that year, the dreadful cattle disease rinderpest wiped out whole herds in the Hambantota district; having to shoot stray cattle did not add to Woolf's popularity. His own reaction to the most famous of celestial visitors is, to say the least, unusual. In *Growing,* he says:

> . . . there is something about these spectacular displays of nature . . . which while I admire them as works of art, also irritates me. From my point of view . . . there is something ridiculous about the universe—suns flaming away at impossible speeds through illimitable empty space. Such futility is sinister in its silliness.

That certainly puts the universe in its place, as a rather ill-managed extension of Bloomsbury.

Beddagama No. 2 was merely a single ruined hut, at the sight of which I had to exclaim, "Is *this* all that is left? I remember it well . . ." Followed, of course, by our old friend Flashback, as my wire-rimmed spectacles go out of focus and we move back into the past.

Both of the village sequences were shot in a single day, despite poor cooperation from the weather. For my weekend's work, Lester insisted on paying me union rates, and I later handed over the check to the prime minister. It will go into a fund for indigent actors; I hope that no one accuses me of trying to safeguard my own future.

As for the movie, it will make its European appearance at the 1980 Cannes Film Festival. Having seen the rushes, I am convinced that it will receive wide acclaim . . . and I can't help doing a little daydreaming. . . .

Ten years ago, bravely hiding my tears, I left the Dorothy Chandler Pavilion carrying one of the best *undelivered* speeches of acceptance in the history of the Academy of Motion Picture Arts and Sciences. (I still stick pins in wax statues of Mel Brooks.) Maybe next time—

"For best supporting actor in a foreign-language film, this year's Oscar goes to—"

P ostscript: *The Village in the Jungle* rapidly proved itself to be the first Sinhalese film to make much of an impact on the outside world. It has been dubbed into several languages and has appeared on television in the United States, England, Germany, Japan, and elsewhere.

My Four Feet

on the Ground

This is Clarke's introduction to My Four Feet on the Ground *by Nora Clarke, his mother. It was issued by Rocket Publishing in 1978.*

It was very unsporting of Gerald Durrell, English writer and naturalist, to preempt the title my mother really needed for this book. *My Family, and Other Animals*. She thought of using it anyway, and the heck with their lawyer—but then decided that wouldn't be fair to all the millions of people who would buy Durrell's book under the impression that they were getting hers. . . .

The Somerset that Mother remembers already seems to belong to another world—a world where motorcars and telephones were rare novelties, cinema and radio still undreamed of. In many respects, the way of life on the farms and estates had not changed for centuries; there may still be some who look back on it with nostalgia. For the few percent who had wealth, position, and continuous good health, that vanished age was indeed a golden one; but I do not think that the average man, if he could miraculously be transported back to the last days of the Victorian era, would care to stay after he had felt the first twinge of toothache.

Yet despite all the changes, much of the west of England is still largely unspoiled, and long may it remain so. The rising interest in the countryside, in animals, and in ecology may help to preserve what is best in the past; we should not grieve for the rest, but reading about it will give us a better historical perspective—and a better appreciation of our own age.

With the appearance of this book, we now have three authors in the family, since my brother Fred's *Small Pipe Central Heating* is a standard reference in its own field. However, sister Mary has shown no signs of dabbling in literature, having her hands full with her children, her husband, and a stable of unemployed polo ponies. . . . But my youngest brother, Michael, when not building hovercraft and running the family farm, has written and produced numerous sketches for local talent. He shows a comic genius that must be discouraged at all costs: I await with apprehension his threatened horror epic, *THAT—Son of IT*.

Perhaps some word of explanation is due to those many authors who, over the years, have written to me begging for introductions to *their* books. I have always refused such requests, as a matter of policy, and I hope they will understand why I have made an exception in this case. A mother has, of course, unique opportunities for blackmail; and I am writing these words in return for a promise that chapters 2, 3, 4, 5, and 6 will be deleted from the original draft. I trust that Norman Mailer, John Le Carré, Len Deighton, Richard Adams, and Harold Robbins (among others), will appreciate my position and accept my apologies.

It has been a somewhat strange experience, writing my latest book while a few feet away Mother was writing her first. (It was also pretty frustrating when she was churning it out at five times my rate.)

I wish her the best of luck in her new career . . . as long, of course, as she doesn't try her hand at science fiction.

Postscript: My mother completed *My Four Feet on the Ground,* which, as the title indicates, is more about horses, cows, and sheep than human beings, at the age of eighty-six, two weeks before her death on February 9, 1980. It is the only work to bear the imprint of the Rocket Publishing Company, formed in 1953 to control my British and European rights. (Later, the name caused some confusion: the company frequently received mail and phone calls for Elton John's Rocket Records and vice versa.)

Part Six

THE 1990S:

COUNTDOWN TO

2000

Not really a meteorite impact—
blame a clumsy electrician!

COURTESY OF ARTHUR C. CLARKE

Introduction

My first novel for this decade, *The Ghost from the Grand Banks,* contained a chapter, "The Century Syndrome," that may well have been one of the earliest warnings of the notorious "millennium bug" and the catastrophes that may ensue when our computers encounter those deadly three zeros.

In 1993, *The Hammer of God* was concerned with a far greater catastrophe—the impact of an asteroid on planet Earth and what, if anything, we could do about it. I was delighted when Steven Spielberg optioned *Hammer,* but disappointed to get no credits in *Deep Impact.*

And talking of asteroids . . . I am now the proud absentee landlord of about a hundred square kilometers of real estate orbiting just beyond Mars: in 1996, asteroid 4923 was named Clarke.

In 1997, to my considerable surprise, I rounded off the *Odyssey* series with 3001. Although I have been (rightly) accused of announcing my last novel on numerous occasions, I really believe it's true this time. However, this does not rule out collaborations with other authors, and at the moment I am involved with two major ones—*The Light of Other Days* with Stephen Baxter and *Trigger* with Michael Kube-McDowell.

Despite my increasing weakness, I was able to do a considerable amount of traveling during the decade. In 1995, on a sudden whim, I decided to take my adopted family, Hector and Valerie Ekanayake and their three daughters, Cherene, Tamara, and Melinda, to Disney World, which I had never visited since I was shown the projected site almost thirty years earlier. This also gave us an opportunity to meet many old friends and attend a shuttle launch—aborted, unfortunately, just three seconds before takeoff!

On my way back through England I was ambushed by the BBC for their well-known series *This is Your Life*. It was a great surprise to be confronted by Buzz Aldrin and Aleksei Leonov, as well as many other old friends, and the whole of the Clarke family. Alas, I can no longer enjoy the videotape because three of the participants are now dead—*OMNI*'s Kathy Keaton, my old schoolmaster Bobby Pleass, and my youngest brother, Michael.

The next year I made my first visit to China to receive the von Karman award. Dr. Theodore von Karman was perhaps the leading aerospace scientist of the century and, together with his colleague Tsien, founded the Jet Propulsion Laboratory (JPL). Over the years I have had many enjoyable associations with JPL, joining it on programs featuring the latest discoveries beamed backed by the *Galileo* spacecraft from Europa—the location of the last three *Odysseys*.

And a few months ago, Donna Shirley, director of the wonderfully successful Sojourner Project, sent me her autobiography, *Managing Martians*—thanking me for getting her interested in the Red Planet when she read *The Sands of Mars* at the age of twelve.

On New Year's Day, 1998, I had the honor of being knighted by Her Majesty Queen Elizabeth for "services to literature"—a phrase that particularly delighted me since there is reluctance in certain reactionary quarters to ever consider science fiction literature. This also gave me the opportunity to meet Prince Charles on his visit to Colombo, when he asked, "Are you still writing?" I had to confess that I had just sent my agent the first draft of the book you are now reading.

And on my eighty-first birthday in 1998, the Sri Lanka Postal Service will issue a stamp featuring not one, but two portraits of me, superimposed on the geostationary-satellite configuration. This was certainly something I could never have imagined sixty-five years ago, when I was sorting the mail in Bishops Lydeard Post Office—and preparing to deliver it round the Somerset countryside on my bicycle.

As this century rushes to its close, I have noticed a curious phenomenon. The weeks now seem to consist of three days, and the months of two weeks. Yet events that my diary tells me occurred only a few days ago seem to belong to the remote past—or to have completely vanished from my memory. When I meet a long-standing acquaintance whose name I can't recall, I explain that I haven't really forgotten anything—but my access time is approaching infinity.

Though, with a little help from my friends, I can walk a few hundred feet, virtually all my time is now spent in a wheelchair—except when I am playing table tennis. Surprisingly, my reaction speed seems unaffected, but as I have to support myself with one hand on the table, this gives unscrupulous opponents an advantage they seldom hesitate to take.

Most important of all, I am in no great danger of boredom. At last count I had some twenty movie, TV, or book projects commissioned or under discussion. So I hope it will be well beyond 2001 before there's any need to use the epitaph I composed many years ago: "He never grew up, but he never stopped growing."

Marconi

Symposium

The following address was given at the Smithsonian Institution on April 24, 1990.
In the opening little fable, any resemblance to real people or
organizations is entirely coincidental.

Exactly fifty years ago, a couple of bright young inventors—let's call them Dave Doitt and Bill Backard—start making electronic test gear in a garage. During the next few decades, Doitt-Backard becomes a billion-dollar corporation.

Some thirty years pass, and another couple of young engineers, who happen to be Doitt-Backard employees, have a brilliant idea. Bobbs and Wazitt—I'm tired of making up names, so I'll call them both Steve—think they can build a home computer that any idiot—or at least his children—can operate. To make it sound attractive, they give it a nice edible name—Pineapple.

Well, when they take the idea to their boss, Doitt-Backard gives the Pineapple a raspberry, so the two Steves look for a garage of their own.

And—what do you know—this time it's only ten years before they have a billion-dollar corporation.

But things are happening much faster now, and in a mere five years the two Steves discover that their own company is too conservative to accept new ideas. So they split off to start tinkering in yet another garage, though one slightly more opulent than their first.

The point I am trying to make with this entirely fictitious tale is that real innovations come only through individuals, not organizations. But you need organizations—often colossal ones—to develop them.

I'm even tempted to say that large organizations not only can't make major innovations, but shouldn't attempt to. If Bobbs and Wazitt had been embraced by Doitt-Backard, the results would probably have been disastrous for all concerned.

Of course there are exceptions—see for example Bell Labs and the transistor, which I'm sure you'll be hearing about presently. (Hi there, John!) But Bell Labs was deliberately set up to encourage creativity, not to manufacture things.

Though I wouldn't dare say this in deepest Washington if I were actually here with you today, I'm afraid that government labs are even less likely to produce startling innovations. As I told you at the 1982 Marconi Award, if there had been government research establishments back in the Stone Age, by now we'd have perfectly marvelous flint tools—but no one would have invented steel.

And there's another anti-innovation force, probably worse in the United States than anywhere else (though the United Kingdom isn't far behind). You now have a major industry devoted to stamping out innovation—and at enormous expense. However, I don't quite agree with Shakespeare's stirring cry in *King Henry VI:* "The first thing we do, let's kill all the lawyers." About half would be sufficient.

To make myself a little more popular again, I think it's undeniable that the free enterprise system—despite its wear and tear on the human spirit—is infinitely more creative than any other. Current political events thoroughly endorse this view. As far as I can recall, the only major invention to emerge from Eastern Europe in the last thirty years has been the Rubik's Cube.

Well, I see that my five minutes are up. So I hope that I've trodden on enough toes to start you dancing.

Over, but not out. . . .

Introduction to

Charlie Pellegrino's

Unearthing Atlantis

*Atlantis! There can be no name in the languages of the Western world that evokes
more feelings of wonder, mystery—and irreparable loss. The myth created by
Plato more than two millennia ago still has power over our age.*

As I write, a space shuttle bearing that magic name of *Atlantis* is being
prepared for launch at the Kennedy Space Center. But was Atlantis only a
myth? Down the centuries, amateur and professional scholars, covering the
whole spectrum from sober historians to certifiable lunatics, have believed
otherwise. They were correct—though not in the way that most of them
imagined.

I first encountered Atlantis via the lunatic fringe, thanks to a friendly
neighbor with an interest in the occult. To quote from *Astounding Days,*
my science fiction autobiography:

> Mrs. Kille stimulated my curiosity by lending me such books as Ig-
> natius Donnelly's 1882 masterpiece of spurious scholarship, *Atlantis:*

The Antediluvian World—the veracity of which I did not doubt for a moment, never imagining at that tender age that printed books could possibly contain anything but the truth.

Donnelly should be the patron saint of the peddlers of UFO/parapsychology mind rot; not only did he claim that all the ancient civilizations were descended from Atlantis, but in his spare time (he was also a Philadelphia lawyer, lieutenant governor of Minnesota, and twice a congressman) he "proved" that Bacon wrote Shakespeare's, Marlowe's, and Montaigne's essays!

That ten-year-old bookworm gulping down Donnelly's splendid nonsense certainly never imagined that half a century later he would be walking in the streets of the real Atlantis—without even knowing it! The tale that Charlie Pellegrino unfolds is one of the great archaeological detective stories of all time, and some of the most exciting chapters are yet to come.

Of all the millions of words about Atlantis that have come down from antiquity, the most chilling are these: "The Atlanteans never dreamed." Should we envy them—or pity them?

At this point I feel I owe an apology to the several hundred other writers who over the last few years have received my notorious "drop dead" form letter saying that I never write prefaces or even blurbs. (Four went out in a single mailing last week.) Well, there has to be an exception to every rule; and when writing *The Ghost from the Grand Banks,* I was greatly indebted to Charlie for his advice, and especially for his book *Her Name: Titanic.*

I would like to end with my own tribute to the immortal myth, from my book *The Songs of Distant Earth,* in the words I gave a musician more than a thousand years hence—when the Earth itself faces total destruction:

When I wrote "Lamentation for Atlantis" I had no specific images in mind; I was concerned only with emotional reactions, not explicit scenes. I wanted the music to convey a sense of mystery, of sadness— of overwhelming loss. I was not trying to paint a sound-portrait of ruined cities full of fish. Yet now something strange happens whenever I hear the *lento lugubre*—it's as if I'm seeing something that really exists.

I'm standing in a great city square almost as large as St. Mark's or St. Peter's. All around are half-ruined buildings, like Greek temples, and overturned statues draped in seaweeds, green fronds waving slowly back and forth. Everything is partly covered by a thick layer of silt. . . .

I know, of course, that Plato's Atlantis never really existed. And for that very reason, it can never die. It will always be an ideal—a dream of perfection—a goal to inspire men for all ages to come. . . .

Now the "Lamentation" exists quite apart from me; it has taken on a life of its own. Even when the Earth is gone, it will be speeding out toward the Andromeda Galaxy, driven by fifty thousand megawatts from the deep-space transmitter in Tsiolkovski Crater.

Someday, centuries or millennia hence, it will be captured—and understood.

Tribute to

Robert A. Heinlein

(1907–1988)

Clarke's first meeting with Mr. and Mrs. Robert A. Heinlein was in 1952,
on his initial visit to the United States as a result of selling
The Exploration of Space *to the Book-of-the-Month Club.*

I don't recall how I originally contacted Bob Heinlein—probably through correspondence as a result of the movie *Destination Moon,* if not at an earlier date. Anyway, when he heard that I was coming to the States, he and his wife, Ginny, invited me to visit them in their self-designed high-tech house at 1776 Mesa Avenue, Broadmoor, Colorado Springs (one of the few addresses I have never forgotten!).

It was a memorable visit, because not only did I have the privilege of meeting the Heinleins and their neighbors, but I also saw some of the most spectacular scenery in the United States; the "Garden of the Gods" made a particular impression on me. Most unforgettable was our ascent of Pikes Peak by funicular, and the subsequent drive down by car.

I must admit that I did not realize what a sacrifice of working time and energy Bob and Ginny were making to entertain an unknown (and occasionally, I'm sure, uncouth) Britisher. I shall always be grateful to them for their kindness and hospitality.

Many years later, when I was lecturing at the Air Force Academy, I took the opportunity of revisiting 1776 and reviving happy memories. I told the current residents that I had once been a guest of their distinguished precursors.

The Heinleins must have been very sad to abandon their mile-high house, after Ginny was diagnosed as suffering from altitude sickness. However, they quickly bounced back and built an even more impressive residence at a slightly lower altitude—Carmel, California. Here, I again had the pleasure of staying with them, and also of meeting Chesley Bonestell, who dominated the space-art scene as thoroughly as Bob did the space fiction scene, and for the same period of over thirty years.

It gave me great satisfaction when, in 1980, Bob and Ginny paid a one-day visit to Sri Lanka on a round-the-world cruise. I chartered a plane to fly them over the southern part of the island and showed them some of my favorite locations—including the Great Basses lighthouses, scene of *The Treasure of the Great Reef.* Almost ten years later, in March 1990, I flew back over the same route and recalled with nostalgia that day we spent together.

When I wrote *Astounding Days,* my science fiction autobiography, I naturally had to include a tribute to Bob—and here is an extract, under the, alas, now all-too-appropriate heading "Requiem":

> In the very month that war broke out in Europe, John Campbell printed an inconspicuous little story called "Life-line." The blurb reads: "A new author suggests a means of determining the day a man must die—a startlingly plausible method!"
>
> Plausible or not, "Life-line" is still a good read. One Dr. Pinero invents an electrical device that can measure the extension of an individual's track in four-dimensional space-time, by detecting the discontinuities at each end, much as engineers can pinpoint a break in an undersea cable. He can thus locate the moments of birth—and of death.
>
> The impact on the life insurance companies is, of course, shattering, and they take steps to put Dr. Pinero out of business. He sits down to a good meal and calmly awaits the arrival of their enforcers; having already consulted his own machine, he knows that there is nothing else to be done. . . .
>
> The story is smoothly written, and packs quite a number of punches. During the next three years, Campbell's "new author" contributed an amazing twenty stories to the magazine, including three serials. His name: Robert A. Heinlein.

"Life-line," short though it is, already hints at some of the pre-occupations that would provide Heinlein with themes for the rest of his career—e.g., mortality and big business versus the individual. Others were developed during the next two years in a creative debut unmatched since the advent of Stanley Weinbaum. And, unlike that brief nova, Heinlein was to dominate the science fiction sky for the next half century and effect a permanent change upon the pattern of its constellations.

Four months later (January 1940) he was back with a story that remains one of his best loved—"Requiem." Now that we have watched whole armies of technicians working around the clock at Cape Canaveral, the idea of a couple of barnstorming rocket pilots giving $25 rides in a secondhand spaceship at a country fair is more than a little comic. (Come to think of it, when did you last see a barnstorming airplane? I haven't, since my maiden flight outside Taunton at the age of ten.)

Heinlein's protagonist is D. D. Harriman, an aging millionaire who has made his fortune in the space business—but is not allowed to leave Earth because of his heart condition. So he hires the two owners of a beat-up, mortgaged rocket ship to take him as a passenger (one jump ahead of the bailiffs—just as in *Destination Moon*) and dies happily in the dust of Mare Imbrium, the most magnificent of the lunar seas. Over his grave are inscribed the lines that Robert Louis Stevenson wrote for his own epitaph:

Here he lies where he longed to be;
Home is the sailor, home from the sea,
And the hunter home from the hill.

"Requiem," dated though it may be, is a moving story. And, on rereading it after many years, I have just discovered something that has given me quite a shock. The first *D* in D. D. Harriman stands for Delos. Bob Heinlein chose better than he knew when he selected that magic name; for as I have good reason to remember, the Greek island of Delos is the reputed birthplace of Apollo.

As is now rather well known, my last encounter with Bob was, unfortunately, not a happy one. It was a private meeting to promote the Strategic Defense Initiative, better known as Star Wars, and I—perhaps not too tactfully—criticized some of the more extreme claims of certain Star Warriors (e.g., "putting an umbrella over the United States"!). Bob was much upset and accused me of meddling in affairs that did not concern me—though I would have thought that SDI should concern everybody on the planet!

I'm not going to revive this argument—which hopefully history has already bypassed—but will add that there are a lot of things in SDI that should be done. See Freeman Dyson on this depressing subject.

Though I felt sad about this incident, I was not resentful, because I realized that Bob was ailing and his behavior was not typical of one of the most courteous people I have ever known. I'm happy to say that friendly communications were later resumed, through Ginny's good offices.

Good-bye, Bob, and thank you for the influence you had on my life and career. And thank you, too, Ginny, for looking after him so well and so long.

Satyajit and

Stanley

Satyajit Ray wrote to Clarke in the early sixties, expressing interest in making a science fiction movie. In 1964, he met with Clarke's partner, Mike Wilson, who had just completed a successful Sinhalese film—the first in color—Ran Muthu Duwa (Island of Pearls and Gold). In the following essay, the reader learns how close Ray and Stanley Kubrick really were.

While I was was working with Stanley Kubrick on *2001: A Space Odyssey,* Mike Wilson and Satyajit Ray prepared the screenplay of *The Alien,* a charming little story about a spaceship with a single small occupant, landing in a remote Indian village. In an extraordinary feat of salesmanship, Mike took Satyajit to Hollywood, signed up Peter Sellers, turned down Marlon Brando, and made a deal with Columbia. Unfortunately, the whole thing swiftly unraveled, and *The Alien* remains one of the great might-have-beens of the movies (like Charles Laughton's *I, Claudius*—though on a slightly smaller scale).

During this period (around 1965) Satyajit visited the MGM studio at Borehamwood, north of London, and I introduced him to Stanley Kubrick as well as to Roman Polanski, then making his vampire film. (I've never seen this, but believe it contains one useful bit of information: a crucifix won't work if the vampire is Jewish.)

Stanley, who is sparing of praise of other filmmakers—few of whom he allowed to come near the *2001* sets—was quite unstinted in his admiration for Satyajit. Though I can't be sure after a quarter of a century, he may even have admitted that Ray was number one. Or was it number two?

Though several attempts were made to revive *The Alien,* nothing ever came of it. I am sorry that a certain coolness developed between Satyajit and Steven Spielberg (for whom I have an almost equal admiration), because *E.T.* and *The Alien* had similar themes. But such coincidences are common, and as Steven has rather testily remarked, he was a kid in high school when *The Alien* was circulating in Hollywood.

Many years later, Satyajit honored me by asking if I'd write an introduction to his collection of science fiction and fantasy stories. I was only sorry that illness and the pressure of other projects made it impossible for me to accept this invitation.

That encounter at MGM was the only time I ever met Satyajit, and I was enormously impressed by his personality. As indeed I have been by his work. And if it doesn't sound patronizing, he also speaks some of the best English I've ever heard—right up there in the Orson Welles–Richard Burton category. (As Shaw once remarked plaintively: "Why can't the English speak English?")

I must admit I get rather restive when people write saying that my first published story was my best. Perhaps Satyajit feels the same about his own firstborn. But surely *Pather Panchali* is one of the most heartbreakingly beautiful films ever made; there are scenes that I need never view again because they are burnt into my memory.

Many years ago, a fine but now forgotten writer (John Keir Cross) once said to me: "There is literature—and *Billy Budd.*" I feel that there are movies—and *Pather Panchali.*

However, the comparison fails in one important respect. Melville's book was his last; Ray's film was his first. And happily, he is still filming.

Soon after writing these words, I was delighted to receive a note from Satyajit saying, "I've just finished what I consider to be my best film since *Pather Panchali.*" Splendid news!

Aspects of

Science Fiction

If you have to ask what science fiction is, you'll never know.

—Anon.

Attempting to define science fiction is an undertaking almost as difficult, though not quite so popular, as trying to define pornography. Even the choice of an acceptable abbreviation has caused heated debate. The older generation of readers and writers insists on *sf* and scornfully rejects the recent invention *sci-fi,* self-explanatory and unambiguous though it is.

In both pornography and sf, the problem lies in knowing exactly where to draw the line. Somewhere in the literary landscape, science fiction merges into fantasy, but the frontier between the two is as fuzzy as the boundary of fractal images like the famous Mandelbrot Set.

As a first approximation, sf is something that could possibly happen, in the universe as we think we know it. Mary Shelley's *Frankenstein* is widely regarded as the prototype, complete with mad (or at least obsessed) scientist, using the latest technology. Jules Verne hugely expanded the genre with such classics as *Twenty Thousand Leagues Under the Sea* and *From the Earth to the Moon*—though it would be unwise to look too closely into the mechanics of the latter. (*A Journey to the Center of the Earth* is even more vulnerable to such criticism, and not merely on geological grounds. In addition to several hundred kilograms of food and other provisions, each of Verne's intrepid explorers must have carried at least a ton of the primitive electric batteries they used with such abandon.)

But this is grossly unfair; the spirit of Verne's stories is pure science fiction, because they are (with a few tongue-in-cheek exceptions) realistic and practical. They could not have been written before the age of steam. This places them at the opposite pole from fantasy, which has flourished for at least three thousand years, in worlds that do not exist and often never could have existed. Every culture has its favorite examples; the English archetype is Lewis Carroll's *Alice in Wonderland*. Our century has seen an unexpected revival of the genre, sparked by J. R. R. Tolkien's epic *Lord of the Rings*.

Though the extreme cases are easy to identify, it is the middle ground that is in dispute. The frontier between science fiction and fantasy is not only ill-marked, it is also continually on the move. Over the years, what once appeared to be science fiction can turn into fantasy—and vice versa. Thus all the pre–Space Age stories of adventures on the Moon, Mars, and the other worlds of our Solar System, no matter how scientifically accurate the writers attempted to be, are now pure fantasy. There are, alas, no Martian cities or lush Venusian jungles; NASA's space probes blew them out of the sky.

The metamorphosis from fantasy to science fiction is less common, but considerably more interesting. Thus if someone had written a story before 1938 in which the explosive power of ten thousand tons of TNT was produced simply by banging two pieces of metal together, it would have been pure fantasy. Anyone with the slightest knowledge of chemistry or physics would have known that the idea was utterly ridiculous.

But when *Astounding Stories* printed just such a tale in its March 1944 issue—to the consternation of the FBI—it was hard-core sf, because uranium fission had now been discovered. And in August 1945, of course, fiction became history.

To give a more benign example, today's pocket calculators, holding in their memories the equivalent of entire libraries of mathematical tables, would have appeared utterly impossible to any pre-1950 scientist. They provide an excellent demonstration of Clarke's well-known third law: "Any sufficiently advanced technology is indistinguishable from magic."

Although robots and intelligence machines have always been a popular ingredient of sf, no writer (to the best of my knowledge) ever anticipated the advent of the personal computer in the form that it actually materialized—not merely as a calculating device for specialists, but as a universal household appliance serving a multitude of functions, from word processing to music-making to the creation of artificial realities. This highlights a basic problem of science fiction; most of it has built-in obsolescence. Does this mean that it is necessarily ephemeral—unable to produce words of permanent literary value?

The quick answer is that it has already done so, though not often; after all, sf has not been around so long as other forms of fiction, nor has it had an opportunity of being fairly judged by the literary establishment. Owing to an accident of publishing, and the influence of one man—Hugo Gernsback, instigator of *Amazing Stories* (1926) and *Wonder Stories* (1930)— for most of this century the genre has been identified with garish magazines and the crudest forms of pop art. Whether Uncle Hugo's impact on science fiction was malign or benevolent has been endlessly debated, but it was certainly enormous.

Pre-Gernsback, an author could produce a work of imaginative or speculative fiction (yet is there really any other kind?) without having it relegated to a literary ghetto. *Frankenstein* has already been mentioned; other examples are Edgar Allan Poe's "Mellonta Tauta" (Greek for "these things are in the future"), Robert Louis Stevenson's *Dr. Jekyll and Mr. Hyde,* Bulwer Lytton's *The Coming Race,* Herman Melville's "The Bell Tower," Jack London's "The Red One," Conan Doyle's *The Lost World,* Rudyard Kipling's "With the Night Mail"—there are countless examples of mainstream authors expanding their territory. The American critic H. Bruce Franklin has even asserted, in the introduction to his *Future Perfect* (1966), that "there was no major nineteenth-century American writer of fiction, and indeed few in the second rank, who did not write some science fiction." But only after Gernsback did the genre become proudly—even arrogantly—self-conscious, and a whole category of writers began to specialize in the field. Many of them wrote virtually nothing but sf, which appeared only in cheap magazines or, if they were lucky, in limited editions from one-man publishing houses. (Though this seldom brought them much material benefit; I can recall my agent attempting in vain to extract a few dollars from one such publisher—whose books today change hands at four-figure prices.)

The widespread popular awareness—and appreciation—of science fiction that now exists is partly due to a vast improvement of literary standards. There are so many excellent writers—in so many countries—practicing in the field today that it is unfair to list names. Though the period 1938–47 has often been called the golden age of science fiction, that age is really now. The earlier decades were only gilt—but the very best gilt.

Yet it must be admitted that some of science fiction's current prestige is based on a fallacy. Science fiction is not predictive; very seldom do its practitioners attempt to describe the real future—quite the contrary; in fact, Ray Bradbury summed up this attitude perfectly: "I don't try to predict the future—I try to prevent it." Books like *1984* act as early warning systems: Kingsley Amis once wittily christened them "new maps of hell." They may help us to avoid certain hells; paradoxically, they are most successful when they become self-unfulfilling prophecies. No one would have been happier than Orwell to know that the real world of 1984, though it still contained rather too many little Big Brothers, was a much better place than his imagined one.

Anti-utopian (or dystopian) stories have been a major theme of science fiction because they allow writers the enjoyment of "viewing with alarm" without suffering the fate of Cassandra. It must also be admitted that stories devoted to utopias would be insufferably dull, since by definition ideal societies would have eliminated all the problems and conflicts that make for good fiction. Wars, rebellions, conspiracies, are much more exciting than the good works of benignly efficient bureaucracies. As William Blake remarked: "Damn braces; bless relaxes." If we ever achieve utopia, we may relax into terminal boredom.

Fortunately, science fiction writers have shown many ways of avoiding this doom, by inventing natural disasters—ranging all the way from strictly local ones (e.g., Komatsu's *Japan Sinks*) up to global and cosmic catastrophes like new ice ages, asteroid bombardment, or the Sun going nova. And of course there are always invasions from space, of which H. G. Wells's *War of the Worlds* remains the classic example.

Outer space has long been the most popular venue for imaginative writing, and in this case there is little doubt that fiction has helped to create reality. All the pioneers of astronautics were inspired by Jules Verne, and several (e.g., Goddard, Oberth, von Braun) actually wrote fiction to popularize their ideas. And I know from personal experiences that many American astronauts and Soviet cosmonauts were inspired to take up their careers by the space travel stories they read as children. (One of my proudest possessions is a little monograph, "Wingless on Luna," bearing the inscription, "To Arthur, who visualized the nuances of lunar flying long before I experienced them!—Neil Armstrong.")

Unfortunately, besides accelerating mankind's emergence into space, science fiction may have aroused expectations that cannot be fulfilled—at least for centuries to come. Part of the present disenchantment with space travel may be due to disappointment with the real universe, as compared with the glamorous one of fiction—and especially that presented by the visual media through such spectacular extravaganzas as *Star Wars* and its successors.

Science fiction and cinema might have been made for each other; indeed cinema once was science fiction. The love affair between the two is now almost a century old; it began in 1902 when the French pioneer of special effects, Georges Méliées, made his lighthearted *Trip to the Moon*.

Of the hundreds of sf movies made during the first half of this century, few indeed had any artistic or intellectual value. Almost the only exceptions are Fritz Lang's *Metropolis* (1926) and H. G. Wells's *Things to Come* (1936). It is now generally agreed that the first science fiction movie to receive widespread critical acclaim (even from those who didn't like it) was Stanley Kubrick's *2001: A Space Odyssey* (1968). Its success was due not only to the genius of the producer/director, but to an accident of history that can never be repeated. *2001* had its premiere on the eve of the Apollo mission taking the first men around the Moon; they had already seen it before they left the Earth!

Since then, hardly a year has passed without some major, megadollar sf production. The *Star Wars* trilogy, *Close Encounters of the Third Kind, E.T., Total Recall, Dune,* and *Blade Runner* are the most notable examples, but worldwide more people have probably seen the television series *Star Trek* and its adaptations. Strictly speaking, of course, almost all of these are pure fantasy adventure, not really science fiction. The cosmic speed limit set by the velocity of light means that Captain Kirk cannot possibly have a new adventure every week, in prime time. . . .

Or can he? Perhaps warp speed, or its equivalent, may one day be an engineering reality. It seems very unlikely; but once again we should remember how often pure fantasy has become science fiction.

The stage, with its much more limited resources, cannot hope to compete with Hollywood's special effects experts—though recent rock musicals have shown what can be done, if you have enough money. It is not surprising, therefore, that few playwrights have been attracted to the genre; however, among them are two of the very greatest.

Shaw's *Back to Methuselah* (1921) is concerned with nothing less than human evolution and ends with this typically science fictional glimpse of the far future: "Of life only there is no end, and though of its million starry mansions many are empty and many still unbuilt, and though its vast domain is as yet unbearably desert, my seed shall one day fill it and master its matter to the uttermost confines." A quarter of a century later, at the age of ninety-one, GBS was still interested enough in the conquest of space to join the fledgling British Interplanetary Society.

By a curious coincidence, Karel Capek's play *R.U.R.* appeared in the same year as Shaw's, but had a greater impact on the world. Though Capek's "robots" were not mechanical, but organic—today we would call them androids—he added an essential word to all the languages of mankind.

A good case can be made for pure sound—radio or audiocassette—as the best medium for science fiction; by having to use his imagination, the listener is forced to become a participant. The most dramatic demonstration of this thesis took place in 1938 when Orson Welles spread panic throughout the eastern United States by shifting the locale of H. G. Wells's *War of the Worlds* from England to New Jersey.

Despite its technical limitations, the 1938 "panic broadcast" is still quite impressive. Just two years later, on his last American lecture tour, H. G. joined Orson on a radio talk show. It was their only meeting; listening to the friendly encounter between the two great magicians is a science fictional experience in itself—a journey back in time to a now-vanished world.

The impact of sf on music has been considerable, but largely on the popular level—movie scores and rock groups. A major space symphony seems long overdue. (Gustav Holst's *The Planets,* of course, is oriented toward astrology, not astronautics; and Richard Strauss's "Zarathustra" was written seventy years before *2001* made it synonymous with outer space.

The first—and perhaps still the best—major composition inspired by a science fiction movie was Sir Arthur Bliss's majestic "Things to Come" suite (1936), with its inspiring "Which shall it be?" finale, played by a full symphony orchestra. Twenty years later, the sound track of *Forbidden Planet* (1956) made a major breakthrough with the Barrons' electronic tonalities— a name cleverly chosen by MGM's lawyers to avoid trouble with the Musicians' Union, understandably concerned that just two people could not only compose but create an entire movie score. This small ripple heralded the wave of the future, and the advent of the now ubiquitous music synthesizer.

Pink Floyd was perhaps the best known of the innumerable groups inspired by sf themes; individual composers include Vangelis, Wendy Carlos, and Jean-Michel Jarre. Sting and David Bowie have not only written and performed sf-related songs, but have acted in important science fiction movies (*Dune* and *The Man Who Fell to Earth*).

The impact—both direct and indirect—of sf on the visual arts has also been considerable. The early pulp magazines fostered a whole generation of illustrators, of whom the best known was Gernsback's longtime associate Frank Paul. Though often garish and clumsy by today's standards, Paul's covers for *Amazing* and *Wonder* succeeded admirably in their purpose, which was to catch the eye of a potential reader as he (very rarely she) hurried past the newsstand. Half a century later, they still have a certain naive charm, even though Paul was much better at drawing Martians than human beings.

Today's illustrators are not only technically far superior to their precursors, but have the advantage of being able to base their work on reality. Indeed, many have been commissioned by NASA and the aerospace industry to help visualize future projects (e.g., the space station, lunar bases)— and one now professional painter (Apollo 12's Alan Bean) has actually walked on the Moon.

Although it has become something of a cliché, perhaps the most important attribute of good science fiction—and the one that uniquely distinguishes it from mainstream fiction—is its ability to evoke the sense of wonder. Many years ago, a science fiction enthusiast who also happened to be a Nobel Prize winner sent me this quotation:

"The real world is increasingly seen to be, not the tidy little garden of our race's childhood, but the extraordinary, extravagant universe described by the eye of science. . . . If our art . . . does not explore the relations and contingencies implicit in the greater world into which we are forcing our way, and does not reflect the hopes and fears based on these appraisals, then the art is a dead pretense. . . . But man will not live without art. In a scientific age he will therefore have science fiction" (Hermann J. Muller, "Science Fiction as an Escape," *Humanist* 6 [1957]).

By discovering the genetic effects of radiation, Dr. Muller inadvertently inspired much science fiction and made *mutant* a modern bogey word. In the same essay, he pointed out another valuable service that this type of literature has performed:

"Recent science fiction must be accorded high credit for being one of the most active forces in support of equal opportunities, goodwill, and cooperation among all human beings regardless of their racial and national origins. Its writers have been practically unanimous in their adherence to the ideal of 'one free world.' "

That, I think, is inevitable. Anyone who reads this form of literature must quickly realize the absurdity of mankind's present tribal divisions. Science fiction encourages the cosmic viewpoint; perhaps this is why it is not popular among those literary pundits who have never quite accepted the Copernican revolution, nor grown used to the idea that man may not be the highest form of life in the universe. The sooner such people complete their education and reorient themselves to the astronomical realities the better. And science fiction is one of the most effective tools for this urgent job.

For it is, preeminently, the literature of change—and change is the only thing of which we can be certain today, thanks to the continuing and accelerating scientific revolution. What science fiction writers call mainstream literature usually paints a static picture of society, presenting, as it were, a snapshot of it, frozen at one moment in time. Science fiction, on the other hand, assumes that the future will be profoundly different from the past—though it does not, as already pointed out, attempt to predict that future in detail. Such a feat is impossible, and the occasional direct hits of Wells and other writers are the result of luck as much as judgment.

But by mapping out possible futures, as well as a good many impossible ones, the science fiction writer can do a great service to the community. He encourages in his readers flexibility of mind, readiness to accept and even welcome change—in one word, adaptability. Perhaps no attribute is more important in this age. The dinosaurs disappeared because they could not adapt to their changing environment. We shall disappear if we cannot adapt to an environment that now contains spaceships, computers—and thermonuclear weapons.

Nothing could be more ridiculous, therefore, than the accusation sometimes made against science fiction that it is merely escapist. That charge can indeed be made against much fantasy—but so what? There are times (this century has provided a more than ample supply) when some form of escape is essential, and any art form that supplies it is not to be despised. And as C. S. Lewis (creator of both superb science fiction and fantasy) once remarked to me: "Who are the people most opposed to escapism? Jailors!"

Charles Snow ended his famous essay "Science and Government" by stressing the vital importance of "the gift of foresight." He pointed out that men often have wisdom without possessing foresight.

Science fiction has done much to redress the balance. Even if its writers do not always possess wisdom, the best ones have certainly possessed foresight. And that is an even greater gift from the gods.

Save the

Giant Squid!

Sharks and squids—or giant varieties of either species—have long fascinated the
makers of adventure and horror films. So much so that these movie makers
have long sought out the expert *of the deep blue sea for his advice.*

It's seven-something on Monday morning, July 29, 1991, in the very heart of Sri Lanka's tea country, and I'm admiring the gorgeous view from my hotel room. I'm completely relaxed, because the nearest fax machine is twenty kilometers and thirty hairpin turns down the mountain road to Kandy. There's no way that editors and agents can bother me. . . .

Then I casually tune my radio to the Voice of America, and within seconds my tranquillity is shattered by something that concerns me directly. Peter Benchley is promoting his new book and mentions Sri Lanka. . . .

After *Jaws,* what? Well, as most of the American public must know by now, *the* menacing beast is the giant squid. Good for you, Peter, but why did it take you so long? After all, it's a pretty obvious idea. I should know.

Soon after *Jaws* was released, its producers, Richard Zanuck and David Brown, asked me if I'd like to write the screenplay for *Jaws 2*. I replied that when you've seen one great white shark, you've seen them all, and I wasn't interested. If they wanted a real monster, it would be hard to beat the giant squid, *Architeuthis*. So I promptly sat down and wrote an outline, based partly on a story of mine that *Playboy* had published in 1964 ("The Shining Ones," reprinted in *The Wind from the Sun*). I couldn't resist calling it *Tentacles,* although I realized that this would produce lewd snickers at the box office.

Anyway, though I had a pleasant telephone discussion with Zanuck and Brown, and later sent them my outline, nothing came of the project. I was not in the least disappointed, as by then I was much too involved in *2010: Odyssey Two.*

By a curious coincidence, Roy Scheider starred in both *2010* and *Jaws,* where he delivered one of the greatest throwaway lines in movie history. Who can forget the moment when he tells Captain Quint (Robert Shaw): "You've got to get a bigger boat!" (Stanley Kubrick, incidentally, once considered using Shaw in *2001.* Had he done so, the story line might have been very different.)

As yet another aside, I can claim some slight involvement with the chain of events that led to *Jaws.* One of its major inspirations was undoubtedly Peter Gimbel's remarkable adventure film, *Blue Water, White Death.* At my suggestion, Peter had sailed to Sri Lanka, and in 1969 I visited his battered rust bucket, then anchored off the east coast near the wreck of the world's first aircraft carrier, HMS *Hermes.* On deck were the shark cages from which Peter hoped to film the great white when he caught up with it—as indeed he did a few months later in Australia, with spectacular results. I don't think those cages would be much use, however, against *Architeuthis.* Its tentacles—at least the smaller ones—could certainly insinuate themselves through the bars.

It must now be sixty years since I first discovered the giant squid, in Frank Bullen's account of life on a nineteenth-century whaler, *The Cruise of the Cachalot.* The book had a terrifying and unforgettable frontispiece, which was certainly the inspiration for this passage in *Childhood's End* (1953):

"The long, saw-toothed lower jaw of the whale was gaping wide, preparing to fasten upon its prey. The creature's head was almost concealed beneath the writhing network of white, pulpy arms with which the giant squid was fighting desperately for life. Livid sucker marks, twenty centimeters or more in diameter, had mottled the whale's skin where those arms had fastened. One tentacle was already a truncated stump, and there could be no doubt as to the outcome of the battle. When the two greatest beasts on Earth engaged in combat, the whale was always the winner. For all the vast strength of its forest of tentacles, the squid's only hope lay in escaping before that patiently grinding jaw had sawed it to pieces."

That was merely the beginning of my literary involvement with *Architeuthis*. A few years later (1957) the short story "Big Game Hunt" (reprinted in *Tales from the White Hart*) was a tongue-in-cheek account of an attempt to capture a giant squid—a project that had most deplorable consequences for all concerned.

In the novel *The Deep Range* (1957), I took this idea more seriously, describing a scheme for immobilizing a giant squid and bringing it safely back to the surface for observation, using chemical anesthesia and electrified fences. It might even work; I'd like to see some millionaire try it out.

My favorite treatment of the subject, however, occurs in the story already mentioned, "The Shining Ones." This tale may one day have some practical repercussions, because it pointed out that Sri Lanka's magnificent east coast harbor of Trincomalee would be an ideal site for an ocean thermal-energy conservation (OTEC) plant, using the temperature difference between the surface and the freezing abyssal waters as a source of power. Efforts are now under way to put this idea into practice.

In "The Shining Ones," an OTEC project was frustrated by a species of giant squid which, alas, probably does not exist. I assumed that there was a variety that—like their smaller cousins the cuttlefish—could communicate by rapidly changing luminescent patterns, so they were, in effect, living television screens. The last words of my hero, before he descends on his final dive to repair the damaged installation in the Trinco Canyon are "Whatever happens, please remember this—they are beautiful, wonderful creatures. Try to come to terms with them if you can."

So much for fiction; now for fact. When I was making my television series *Arthur C. Clarke's Mysterious World,* I was determined to feature the giant squid. And by great good luck, Yorkshire TV's camera crew was able to film one that had been cast ashore in Newfoundland. Millions of viewers have seen the resulting sequence; although the specimen was only an immature female, a mere twenty-five feet long, it is scary enough. And gives a very good idea of what the really big specimens must look like—especially if, as the evidence suggests, they grow up to one hundred fifty feet in length.

In his radio interview, Peter Benchley mentioned the case of the schooner *Pearl,* reported by the *Times* of London on July 4, 1874, to have been sunk in the Bay of Bengal by a giant squid, a few days after leaving Ceylon's southern port of Galle. My beach bungalow is just a couple of kilometers farther along the coast, so I stood at the water's edge and described how the unlucky schooner must have passed this very spot on the way to her close encounter of the fatal kind.

The giant squid that sank the *Pearl*—after her master's completely unprovoked rifle fire, let it be noted—was lying on the surface exactly like the specimen described in chapter 59 of *Moby-Dick:*

"In the distance, a great white mass lazily rose, and rising higher and higher, and disentangling itself from the azure, at last gleamed before our prow like a snow slide, newly slid from the hills. Thus glistening for a moment, as slowly it subsided, and sank. Then once more arose.

"Almost forgetting for the moment all thoughts of the whale Moby-Dick, we now gazed at the most wondrous phenomenon that the secret seas have hitherto revealed to mankind. A vast pulpy mass, furlongs in length and breadth, of a glancing cream color, lay floating on the water, innumerable long arms radiating from its center, and curling and twisting like a nest of anacondas, as if blindly to clutch at any hapless object within reach."

We now know that a giant squid that has surfaced in tropical waters is almost certainly dying—though not necessarily harmless, as the *Pearl* discovered. Its blood functions efficiently only at the very low temperatures, a few degrees above the freezing point, found in the abyss. In warm surface waters, the blood's ability to absorb oxygen is greatly reduced, so any squid in these circumstances is literally suffocating. This is reassuring news for swimmers and divers in the tropics, but they shouldn't press their luck. And in subarctic waters the giant squid can be aggressively hungry—as was demonstrated by the life raft incident described in *Mysterious World,* when one survivor carried sucker scars to his grave.

This century has seen a complete transformation in our attitude toward other animals, including many once considered implacably hostile. This is particularly true of marine mammals; witness trainers putting their heads in the jaws of killer whales—once regarded as the most ferocious of all marine mammals. And there's a local scuba guide (not with my company) who feeds sharks by presenting fish to them—in his mouth. The last time I saw him, he did have a few scars. . . .

Such underwater antics suggest an interesting possibility. The giant squid is almost certainly a highly intelligent animal; given the opportunity, it might be as playful as its cousin—that charming mollusk the common octopus.

Who will be the first diver to win *Architeuthis*'s friendship and snuggle down comfortably in that "nest of anacondas"?

No, I'm not volunteering. As *Mysterious World* showed, my diving gear is a tasteful shade of yellow.

A Choice

of Futures

The most brilliant attempt at scientific prediction ever made, J. D. Bernal's 1929
book, The World, the Flesh and the Devil, *opens with this challenging*
statement: "There are two futures, the future of desire and the future
of fate, and man's reason has never learned to separate them."

The "future of desire," of course, is what we would like to happen; it is
the basis of most utopian speculations, and of all human hopes. The real
future, when it finally arrives, seldom conforms to our wishes. But it is
wholly beyond our control. Perhaps the true measure of civilization is the
command it has over its own destiny.

The only useful technique ever devised for glimpsing the future is
extrapolation—going well back into the past, and then taking a running
jump. It's a highly unreliable method, but nothing else is available. Let's try
it out. . . .

The date is 1842—young Queen Victoria has been on the throne
almost five years—and the editor of a new magazine has just asked me to
forecast the wonders I expect in the next century and a half.

As a starting point, what sort of technology would I see around me in 1842, and how might I expect it to develop? The great age of steam has dawned; for the first time, mankind can generate large amounts of energy, without relying on the natural resources of wind and water. Huge and impressive machines are pumping out mines and powering factories; even more important, they are starting a revolution in transport. A network of railway lines is expanding over the world, and great bridges of the new construction material, iron, are being built to carry them. The supreme exemplar of this transformation is one man, whose virtues I would certainly extol—Isambard Kingdom Brunel. Indeed, at this very moment Brunel is supervising the construction of the first iron-hulled, screw-propelled steamship, the *Great Britain*—herald of a new breed that is to sweep the windjammers from the oceans of the world.

And there is another revolution under way, involving a more subtle force—electricity. I would certainly mention the fact that the electric telegraph is speading alongside the new railroads, which indeed could not operate efficiently without it.

I wonder what—if anything—I would have said about the intriguing, but not really very impressive, experiments of Michael Faraday, who for the past decade has been playing with magnets and coils of wire and has even contrived an amusing toy, which might be called an electric motor. And I would probably poke fun at the maverick genius Charles Babbage, who's spent a fortune trying to build a mechanical calculating machine out of cams, rods, levers, and gearwheels. Although portions of it have worked, the impractical project has now been abandoned.

I would certainly mention one astonishing invention that has just burst upon an incredulous world. A Frenchman, Daguerre, has succeeded in capturing visual images by a technique soon to be christened photography. This is such an amazing feat that some people have flatly refused to believe it. One German writer has declared that this French claim was an absurd lie because (a) God wouldn't permit such a feat, or (b) if it was possible, German science would have done it first. . . .

Yes, a lot is happening in 1842, so what further developments might I expect in the next century and a half? I would certainly predict a vast expanse in rail transportation, both for goods and passengers. But I would ridicule the idea of personal powered vehicles—except perhaps for aristocrats to travel around their estates. Such steam carriages would be expensive, noisy, and difficult to maintain. And—the most fundamental objection of all—there aren't any roads on which they could operate!

The steam-driven iron ship? Yes—that certainly has a great future and will have a major impact on world trade. It might even give new impetus to the old idea of a canal linking the Mediterranean and the Gulf of Suez.

The electric telegraph? That also shows enormous promise, and soon most of the cities of the civilized world will be linked together so that the governments, businesses, and—above all—newspapers can have rapid communication. However, the great oceans will always be an impassable barrier. Europe and America can never hope to communicate more swiftly than by the fastest steamships. But as they can do the crossing in a couple of weeks, who needs anything better?

I wonder what I would have said about the conquest of the air. After all, balloons have been around for sixty years, and there have been attempts to make them move against the wind instead of merely drifting with it. Perhaps I would have fantasized about steam-powered skyships, but I doubt if heavier-than-air craft would have appeared on my future agenda.

The point I wish to make is that any extrapolation based on existing technology—or even reasonable extrapolations of it—will always be hopelessly short of reality. In 1842, I could never have anticipated how Faraday's experiments would lead to electric lighting and power distribution, transforming every aspect of human life. I would not have dreamed that the exposure times for photographs would drop from hours to minutes to seconds—and that Daguerre's still images would begin to move, creating a new art form, and a new industry. Nor could I have guessed that before the end of the century, men would be able to talk to each other over distances of many miles—and even capture sound on cylinders of wax.

The late Herman Kahn once coined the phrase *surprise-free futures,* i.e., future scenarios based on present trends and foreseeable inventions. Yet even in the field of technology (and how much more than that of politics!), the real future is never surprise free.

At the end of the nineteenth century, the surprises started coming thick and fast. Becquerel's totally unexpected discovery of X rays in 1895 showed that something existed that could pass through solid matter as easily as light through glass. Physicians were given the miraculous power of being able to see into the human body—the greatest boon to mankind since the discovery of anesthetics.

Yet even more important, the discovery of X rays led to the electron, to radioactivity, and then to the revelation that ordinary matter was the storehouse of energies millions of times greater than those involved in the most violent of chemical reactions. Before the new century had reached midway, those energies would be released above the city whose name it was to sear into the conscience of the world.

None of this could reasonably have been foreseen by the most imaginative prophet in 1842. Nor could he have guessed at the existence of electromagnetic waves, which would remove the last limitations of time and space, making it possible for all the human race to observe the same event—or for just two people to meet face-to-face wherever on Earth they might really be.

There are few better proofs of the now often quoted Clarke's third law: "Any sufficiently advanced technology is indistinguishable from magic." But if a final one is needed, the microchip provides it. Every college student now has access to a computer millions of times faster and more powerful than that dreamed of by Babbage (whose original concept has, ironically, just been created with loving care by the craftsmen of London's Science Museum).

In contrast to those of the nineteenth century, the technologies of the twenty-first will depend on processes operating on speeds and scales utterly beyond direct human apprehension. The cheapest home computer performs millions of operations in a second. Soon it will operate at a rate of billions. A CD-ROM disc you can hold in your hand may contain the contents of a medium-sized library. One day something no larger will be able to hold all the books that have ever been written.

Such miracles will come about because scientists are now able to manipulate single atoms and are dreaming of a nano-technology that can build machines the size of bacteria. The ability to handle matter at the atomic level may bypass all existing methods of fabrication. What the photocopier has done for two-dimensional images, the nano-replicator would do for solids. Any objects, however complex, may be created from a single encoded pattern, operating on a suitable supply of raw materials. Today's CAD/CAM (computer aided design and manufacturing) systems are the first step in this direction; at the end of the road will be a society that we might not recognize. The mythological horn of plenty will have become reality, and the curse of lifelong toil that God laid upon Adam will finally have been lifted—for better or worse.

Yet one day the replicator will seem no more amazing than the hi-fi and video equipment through which we can reproduce in our living rooms the performances of a hundred musicians, or of a single virtuoso who may have been dead for decades. How that would have astonished our ancestors! And the ability to re-create events from other times and places now points the way to something even more miraculous—virtual, or artificial, reality.

VR has been developed as a by-product of the entertainment industries and military simulation systems, such as those used to train combat pilots. Today's artificial worlds are very crude, being limited by computer memories and processing speeds. To enter them, one has to don something like an old-time diver's helmet, equipped with miniature television tubes generating stereo images. Even the current primitive systems can give their users a surprisingly realistic impression of walking through nonexistent buildings, encountering mythical creatures, or visiting imaginary planets. In time the clumsy electronic interfaces may be bypassed, and all the sense impressions needed to create these virtual realities may be fed directly into the brain. Many science fiction writers (starting with Aldous Huxley in *Brave New World*) have suggested that synthetic worlds of adventure and romance, surpassing anything achieved by Hollywood, may cause large segments of the population to drop out from humdrum everyday life. A cynic might like to update Marx and suggest that virtual reality may be the opium of the masses—many of whom will be unemployable in the automated world of the future. Today's television addicts may barely hint at the shape of things to come. . . .

I trust that the exploration of imaginary spaces does not divert the human race from exploration of the real universe. In 1942 we knew little more about the other heavenly bodies than we did in 1842, but the last fifty years has turned the many worlds of our Solar System into new and exotic foreign countries. If the incentive and motivation are there—and this remains to be seen—the next century could see the development of technologies that could open up the Solar System to every form of exploration and development and make space travel no more expensive than jet transportation on this planet today.

What is uncertain now is whether the political and economic basis for large-scale interplanetary colonization exists. Only the future can decide this, but skeptics who ask, "Why return to the Moon?" should remember that the purchase of Alaska was bitterly opposed by most of the United States Congress as money wasted on a worthless mass of ice and snow. And a famous nineteenth-century explorer once reported back to his mission control in Whitehall: "I have now charted this continent so thoroughly that no one need ever go there again." We went back to Alaska, we will return to the Moon.

There is one enormous unknown in space—one wild card that could instantly make obsolete all speculations about the future. It is the answer to the question "Where's everyone?"

Ours is the first century that has been able to make serious plans to search for extraterrestrial intelligence, and we should not be discouraged by the fact that no trace has yet been discovered. (The UFO nonsense merely demonstrates how common unintelligent life-forms are on this planet.) The quest has barely begun, and our technologies are still undoubtably very primitive. We think our giant radio telescopes and sensitive detectors are very sophisticated, but we may be like savages listening for jungle drums, and all around them superior technologies are carrying more thoughts than they could experience in a lifetime. Some scientists have argued, on philosophical grounds, that we are alone in the cosmos. If this is true, it would be the most surprising discovery of all, confronting us with an awesome responsibility.

In 1969 when the first Apollo 11 astronauts returned to Earth, I was privileged to write the epilogue to their saga, *First on the Moon*. I ended it with these words: "Whether we shall be setting forth into a universe that is still unbearably empty, or one that is already full of life, is a riddle that the coming centuries will unfold. Those who described the first landing on the Moon as man's greatest adventure are right; but how great that adventure will really be we may not know for a thousand years.

It is not merely an adventure of the body, but of the mind and spirit, and no one can say where it will end. We may discover that our place in the universe is humble indeed; we should not shrink from the knowledge if it turns out that we are far nearer the apes than the angels.

Even if this is true, a future of infinite promise lies ahead. We may yet have a splendid and inspiring role to play, on a stage wider and more marvelous than ever dreamed of by any poet or dramatist of the past. For it may be that the old astrologers had the truth exactly reversed, when they believed that the stars controlled the destinies of men.

The time may come when men control the destinies of the stars.

Gene Roddenberry

Though Gene Roddenberry and Clarke met only three or four times, they had a
warm relationship for over twenty years. Within minutes of hearing the
Voice of America report on Roddenberry's death, Clarke faxed the Star
Trek *office: "Few men have left a finer legacy. The* Enterprise *will*
be cruising the galaxy for centuries to come."

I am proud to have played a part in creating one of the great icons of our
time—as Gene reminded my biographer, Neil McAleer, when he made an
extremely generous assessment of my contribution. Nor was this the first
time; in 1987, he wrote for my seventieth-birthday felicitation volume:
"Arthur literally made my *Star Trek* idea possible. . . . In 1969, I traveled
to Arizona to listen to a Clarke lecture on astronomy . . . and was persuaded
by him to continue my *Star Trek* projects despite the entertainment industry's labeling the production as an unbelievable concept and a failure. . . .
It was a friendship that deepened into the most significant of my professional
life." I deeply regret that, as Gene did not reach his own seventieth birthday,

I never had a chance of reciprocating. There is a sad irony in the fact that he entered the undiscovered country just when the eagerly awaited movie of that name was about to be released.

At a dark time in human history, *Star Trek* promoted the then unpopular ideals of tolerance for differing cultures and respect for life in all forms—without preaching, and always with a saving sense of humor. We can all rejoice that Gene achieved professional success and world respect. What must have given him even greater satisfaction is that he lived to see so many of his ideals triumphantly accepted.

Introduction to

Jack Williamson's

Beachhead

For over seven decades, Jack Williamson has enthralled all science fiction and fantasy fans with his marvelous and unusual tales, most of which Clarke has perused with enthusiasm ever since he picked up an early issue of Amazing Stories.

Some little while ago, when I was a gangling farm boy of fourteen, I came across a copy of *Amazing Stories* containing a tale I still remember, despite the millions of words that have since passed before my eyes. It was written by one Jack Williamson and was called "The Green Girl." Thereafter I kept meeting Jack's stories all over the place. The one that made the greatest impact on me was "The Moon Era," in which the hero goes back in time to the period when the Moon was a living world. Perhaps the years have given this tale a magic that would not survive rereading, but it seems to me that in "The Mother," Jack created one of the first memorable alien beings in science fiction—comparable to Weinbaum's famous "Tweel." There is a sadness and tenderness about this story that still lingers with me.

Very different was Jack's collaboration with Dr. Miles J. Breuer in *The Birth of the New Republic*—a reenactment of the American Revolution on the Moon. The struggle for independence of mankind's colonies from another world is a common theme in science fiction, but this may well have been its earliest treatment. Looking back over Jack's long career (now entering its seventh decade), I am astonished to see now many themes he has explored, and even originated. *The Humanoids* remains a classic study of man-machine relationships, and *Seetee Shock* developed the technology of contra-terrene matter—a subject of considerable interest in connection with advanced space-propulsion systems.

We can now actually manufacture small quantities of c-t matter, albeit at enormous expense; long ago, Jack explored the tricky—indeed hair-raising—technology of handling it. And *The Legion of Time* may have been one of the earliest explorations of alternative universes—again a subject now very much in the forefront of modern physics.

I have no hesitation in placing Jack on a level with the two other American giants, Isaac Asimov and Robert Heinlein. And now, in his eighty-second year, he has shown that he has lost none of his old skills.

The first flight to the Moon was a major theme in science fiction right up to the 1960s. Now the first expedition to Mars is the topic for the closing decade of this century—and the opening one of the next. Although there are quite a few points over which I'd love to have a friendly argument with Jack, *Beachhead* is an exciting and enjoyable dramatization of the first Mars expedition.

I use the word *dramatization* deliberately. Anyone writing about Mars today is laboring under a severe disadvantage, from which his/her precursors like Wells, Burroughs, and Bradbury were happily free. We now know that, alas, there aren't any Martian princesses, ruined cities, or vast canal systems—or indeed any atmosphere worth talking about. It's quite a challenge, therefore, to write an exciting story about the exploration of Mars without inventing implausibilities that may be refuted in a few years' time. Jack has managed very well to maintain suspense and momentum, while at the same time giving a vivid picture of Mars as it really is—a rugged, colorful world, which will continue to unfold its mysteries for decades to come.

To plug a book of one's own in an introduction to a fellow author's may be regarded as scaling new heights of chutzpah. However, I think the unique manner in which I'm writing this introduction fully justifies such effrontery. . . .

I'm reading Jack's manuscript in truly extraordinary circumstances, which would have been pure science fiction only twenty years ago. Virtual Reality Laboratories of San Luis Obispo, California, recently sent me their amazing VISTAPRO computer program, which allows one to take the topographical maps of Mars obtained from the Mariner and Viking missions and to generate incredibly realistic images of them from any viewpoint; some of these are so good that they could easily be mistaken for actual color photographs. More than that, one can modify the images in an almost infinite number of ways—introducing vegetation, water, trees, clouds, etc. In fact, I've been busy terraforming Mars to produce the illustrations for a book I've entitled *The Snows of Olympus: A Garden on Mars*.

Even with the fastest AMIGA3000 east of Suez, it may take up to thirty minutes for a complete picture to appear on the monitor, so I've been reading Jack's manuscript while waiting for the program to run. And since we both, for obvious reasons, have chosen the most spectacular areas of Mars, I've had the uncanny experience of reading about such places as the Coprates Canyon and Olympus Mons while they were slowly materializing on the monitor beside me.

Thank you, Jack, for a memorable experience, and also for inventing the very word *terraforming* half a century ago! The best of luck for your latest opus—not, I'm sure, your last. . . .

Scenario for a

Civilized Planet

The world has changed beyond recognition since Herman Kahn in the early 1950s coined the phrase "It is time, once again, to start thinking about the unthinkable." Happily, we can now concentrate our thoughts on peace, not war, Clarke argues in the following essay.

Although the control of nuclear weapons remains a major issue, it is no longer the central one. The damage inflicted by conventional weapons (smart or stupid; it made no difference to the refugees in that Baghdad bunker) is so appalling that major improvements are hardly necessary. And environmentally inconsiderate though it may be, I suspect that if we were given a choice, most of us would prefer to be killed by A-bombs rather than bayonets or nerve gas.

So let us stop arguing over details and consider this fundamental question: What weapons, if any, would a civilized world society require?

It may help to focus on an issue closer at hand—gun control. There are a few categories of people who need—i.e., require—guns: police, security guards, game wardens. Unfortunately, there are far too many people who want guns—indeed, lust after them—often chanting like a mantra the poisonous half-truth, "Guns don't kill people; people do." Nor, by the same crazy logic, do nuclear bombs; but they made the job a lot easier.

The impulse behind those who want guns, instead of requiring them, is all too obvious. For such Rambo clones seeking surrogate ejaculation, I once coined the slogan "Guns are the crutches of the impotent." Similarly, high-tech weapon systems are the crutches of impotent nations; nukes are just the decorative chromium plating. Let us see what crutches we can throw away, to walk proudly into a decent future.

The first criterion for civilized weaponry should be the total avoidance of collateral damage (to use another piece of mealymouthed Pentagonese, like *friendly fire*). In fact—don't laugh—no device that could kill more than the single person targeted should be permitted. A larger radius of action could be allowed only for instrumentalities that produced temporary disablement e.g., the "gas of peace" in H. G. Wells's *Things to Come,* acoustic or actinic bombs, water cannons, hypodermic guns, etc. Many more could be found if a fraction of the effort devoted to slaughtering people was spent devising ways of immobilizing them.

To deal with the sort of minor disturbances that may require police action even in the most utopian society, here are the minimum-force items that would be added to the above:

Nonlethal martial-arts devices, like quarterstaffs (Robin Hood had the right idea).

Genetically modified feline, canine, ursine, or simian aides, preferably in the five-hundred-kilogram class, playing the same role as today's guard dogs, but with higher IQs.

Passive defense robots (Robocop plus Asimov's three laws).

The permitted delivery systems for all these would include bicycles, scooters, jeeps, hovercraft, and helicopters.

So much for basic law and order. But for real emergencies, which will occasionally arise even in utopia, single-shot rifles and handguns could be issued, perhaps only under presidential orders. . . .

And that's it. We are now one global family, and however much we may dislike our siblings, family quarrels should not be settled with hand grenades or AK-47s—much less ICBMs.

At this point, many of my readers will be muttering, "You can't change human nature"—as if it exists! Perhaps the only characteristic that distinguishes we humans from the other animals is our infinite flexibility—and our ability to take for granted changes that once seemed inconceivable. Not so many centuries ago there were societies in which a gentleman would feel naked without a sword—and was prepared to use it. There was a time when public executions, for such crimes as stealing a loaf of bread, were common entertainments. We still have a long way to go, but those who deny that Homo sapiens is incapable of making the adjustments necessary to survive are traitors to their species.

Still, as Lenin once famously asked, "What is to be done?" There is no simple answer to this enormously complex question, and many of the obvious solutions, however attractive they may seem, will be counterproductive. Thus it now appears that President Reagan's well-targeted evil empire rhetoric, and much of the American military buildup during his administration, only served to strengthen the hand of the paranoics in the Kremlin. With the twenty-twenty hindsight of history, one can argue that a more conciliatory attitude would have produced better results; whether it would have been politically feasible for a chief executive who was such a willing captive of the military-industrial complex is quite another question.

I wonder if the recent occupant of the White House ever came across these words: "Every gun that is fired, every warship launched . . . signifies in the final sense a theft from those who hunger and are not fed, those who are cold and are not clothed." Typical bleeding-heart-liberal pacifist sentiments, of course—except that they are those of President Eisenhower, when he alerted his countrymen to the dangers of the above-mentioned military-industrial complex.

I have many good friends in this amorphus entity, which should perhaps be renamed the military-scientific complex, but the sooner it is put out of business the better for mankind. The old description "merchants of death" is all too accurate; by comparison the Mafia and the drug cartels are minor nuisances. (Incidentally, there is a very cozy relationship between arms dealing and the international drug trade, especially in the Orient and South America.)

But how to counteract the intellectual and emotional fascination of warfare, especially as embodied in today's glamorous weaponry? Be honest—when did you last see anything as exciting on television as the opening hours of the Persian Gulf conflict? Not only the glossy pages of the aerospace magazine but the arts are peddlers of what I have labeled technoporn. Much though I admire it, I am afraid George Lucas's *Star Wars* saga is a perfect example, with its fascinating hardware and gorgeous explosions.

Even more relevant, because it mirrors the real world, is *Top Gun*. One day our grandchildren may be able to view such a superb piece of moviemaking with the same guilty enjoyment we must now feel when screening Leni Riefenstahl's similarly brilliant (and mildly homoerotic) paean to Aryan manhood, *Olympia* (1936). No great harm, as long as you realize exactly what's going on.

And while on the subject of aesthetics, I see one faint flicker of hope in current military designs. Many of the tools of warfare were once beautiful: Excalibur, medieval armor, the Tudor flagship, the Spitfire . . . even the V–2. But today's weapons often look as hideous as their purpose; consider the Stealth bomber or any of the late Warsaw Pact's tanks. Perhaps our collective unconscious is signaling to us, and none too soon. . . .

To be more practical, when appeals to nobler instincts fail, the dollars-and-cents approach may succeed. There would be a concerted outcry for the dismantling of the military-industrial complex with all deliberate speed if its disastrous impact on the economy was appreciated. By concentrating their best brains and most valuable resources on projects that are worse than nonproductive, the United States and the Soviet Union embarked on a race to ruin—which, history may yet record, resulted in a photo finish.

And don't talk to me about spin-offs. They exist, but—with a few notable exceptions like applications satellites—most of them are trivial; we would gladly do without ceramic kitchenware if we could dispense with the missile nose cones that spawned it.

Would General Motors have been humbled, and the textile mills of New England closed, had the United States been able to emulate World War II's real victors, Germany and Japan, and concentrate on civilian consumer goods? Los Angeles did not destroy Detroit as dramatically as Los Alamos destroyed Hiroshima, but the black holes of the California defense industry helped to suck away its lifeblood. Even domestic electronics, which should have benefited the most from military spin-offs, failed to take advantage of them. The United States invented the videocassette recorder—but when did you last see one made in America? Or, for that matter, a wholly homegrown TV set? Why bother to make them when there was easier money from the Pentagon?

Unfortunately, this is a hard lesson to get across, especially to defense-plant workers who have been laid off just before Christmas. Craven congressmen who vote billions for weapons systems that everyone knows are unwanted only postpone the inevitable. There should be no need to stress the obscenity of such behavior in a world where the price of a B-2 bomber could save a million children from lingering deaths. President Eisenhower said it all, forty-nine years ago.

Yet even men of goodwill and intelligence can be seduced by glamorous technology (did not Oppenheimer use the word "sweet" for Ulam's H-bomb breakthrough?) or sleepwalk into accepting the "fallacy of the last move." No better example could be given than SDI (Version 1.0)—the concept of a nuclear umbrella over the United States.

Thanks to a World War II colleague who, for his sins, was made chairman of the Defense Science Committee, I learned more than I wanted to about this pipe dream and did my best to denounce its technological, financial, and above all operational absurdity. (Ironically, President Reagan quoted one of my own "laws" in favor of it. No hard feelings.)

The United States Navy destroyed the credibility of Version 1.0 when its most advanced weapons system shot down an Iranian airliner—on a scheduled flight, in broad daylight. Anything left was buried without funeral honors by AT&T; the hours'-long failure of its telephone system, after decades of testing and debugging, is an instructive techno-disaster in the same league as the collapse of the Tacoma Narrows Bridge. In one of the last letters I received from him, Luis Alvarez (how we missed you, Louie, during the "cold fusion" caper!) referred to certain Star Warriors as "very bright guys with no common sense." Perhaps even they have now learned from the two examples I mentioned.

And yet, the SDI affair demonstrates the excruciating difficulty of answering Lenin's eternally valid question. The originally advertised concept may have been technological nonsense—but brilliant politics. It wouldn't have fooled scientists like Ronald Sagdeev for a moment, but it may have scared the hell out of some of his countrymen with more medals than brains. We'll never know for sure, but those who (like myself) criticized SDI may have done an involuntary disservice to peace.

In penance, let me give two faint cheers for Son of SDI—if it is completely unclassified, its objectives (e.g., loose missiles) are sensibly defined, and it does not divert attention from such really dangerous delivery systems as offshore submarines and diplomatic bags. Although I suspect that a tactical defense initiative would fail precisely when it was supposed to work (remember the Pearl Harbor radar), it's a good idea to explore the technology. We may need it sooner than we imagine—almost certainly within the next thousand years. There have been two—repeat two— megaton-class meteor impacts on Earth during this century (1908 and 1947, both in Siberia). Something may be seriously wrong with all those reassuring statistics.

To return to the more immediate challenge of Lenin's question— which, unfortunately for his countrymen, he answered incorrectly. Can anyone do much better, in this time of geopolitical meltdown? Long-range planning is out of the question; the best that any present-day statesman can hope to achieve is what the poet Robert Bridges called "the masterful administration of the unforeseen."

Yet one basic necessity for the new world system is clear. Many wars in the past have been caused by fears and suspicions that were unjustified. Openness—"glasnost," or transparency—is a key ingredient for the avoidance of future conflicts, and the technology of the Space Age has made this not only possible but inevitable. The reconnaissance satellites (both those of the United States and the late Soviet Union) may well have averted World War III. Together with such ubiquitous communications devices as fax machines and portable satellite telephones, they are the best guarantee of "peace through truth." As President Reagan put it with the hard-won cynicism of the practical politician: "Trust, but verify." What have been christened "peacesats" will be a necessary—though not sufficient—part of this process.

Yet peace is not enough. We need excitement, adventure, new frontiers. (That, hopefully, is one aspect of human nature that will never change.) Although there are problems enough in today's world to absorb all our energies, listing them is likely to evoke yawns rather than enthusiasm. Of course we need more hospitals, more food, more energy, better housing, less pollution. Above all, we need better schools and teachers. I hope it will not be too late for the United States to undo the damage wrought on its educational system by fundamentalist fanatics, Creationist crazies, and New Age nitwits. Such people are a greater menace to the open society than the paper bear of communism ever was.

Many pundits (starting, I believe, with William James) have stressed that mankind needs a substitute for war. Sports, especially as exemplified in the Olympics, goes part of the way, but even American football and Canadian ice hockey do not provide all the necessary ingredients.

However, there is one activity which, almost as if it were divinely planned, fully utilizes the superb talents of the above-criticized military-industrial complex. I refer, of course, to the exploration—and, ultimately, colonization—of space. Many, and some of the most pressing, of our terrestrial problems can only be solved by going into space.

Long before it was a vanishing commodity, the wilderness as the preserver of the world was proclaimed by Thoreau. In the new wilderness of the Solar System may lie the future preservation of mankind.

Having already written far too much on that subject, I will merely draw attention to the planned mission to Mars selected by President Bush as a goal for the fiftieth anniversary (2019) of the Apollo moon landing and activities in connection with the International Space Year. We have to clean up the gutters in which we are now walking—but we must not lose sight of the stars.

Though I hope that someone can preempt me, it appears that more than four decades ago I had the dubious honor of first enunciating the doctrine of mutual assured destruction ("The Rocket and the Future of Warfare," *Royal Air Force Quarterly,* March 1946, and since reprinted in *Ascent to Orbit* [John Wiley, 1984]). Too much thinking about MAD is liable to induce that dislocation from reality, the Strangelove syndrome, for which there is no known cure. So I was very glad to say farewell to the whole dismal subject when I delivered the Nehru Address, "Star Wars and Star Peace," in New Delhi on November 13, 1986. In his thoughtful and witty speech of thanks, Prime Minister Rajiv Gandhi remarked: "Forty years ago, Dr. Clarke said that the only defense against the weapons of the future is to prevent them being used. Perhaps we could add to that, we should prevent them being built . . . It's time that we all heed his warning. . . . I just hope people in other world capitals also are listening."

If not, here is one final quotation from H. G. Wells: "You damn fools! I told you so!"

NASA Sutra:

Eros in Orbit

Clarke's Rendezvous with Rama *caused a battle within the Springfield, Oregon,
school board over its alleged sexual suggestiveness. After a heated discussion,
the board approved the book by a 3–2 vote. District administrators said
high school English and literature teachers could request use of the book
in their classes, but it would not be required reading in all classes.*

This intriguing news item, which I had forgotten until the recent mini-
furor over sex in space, appeared in *Locus* (the *Variety* of the science fiction
universe) as long ago as 1979. And tempted as I am to start a rush to the
nearest bookstore, honesty compels me to admit that *RwR* is, by the standards of the 1990s, about as "sexually suggestive" as *Little Women*. Before
the suspense becomes unendurable, here's the only passage I can find that
might raise an eyebrow or two.

> Some women, Commander Norton had decided long ago, should not
> be allowed aboard ship; weightlessness did things to their breasts that

were too damn distracting. It was bad enough when they were motionless, but when they started to move, and sympathetic vibrations set in, it was more than any warm-blooded male should be asked to take. He was quite sure that at least one serious space accident had been caused by acute crew distraction, after the transit of an unholstered lady officer through the control cabin. . . . Whenever the well-built surgeon oscillated into the commander's cabin, he felt a fleeting echo of the old passion, she knew that he felt it, and both were happy.

Oh, yes—I did assume that, following well-established maritime tradition, spacemen would have a wife on every planet. Nothing particularly suggestive about that, surely. . . .

Anyway, the good gray *New York Times* has just discovered the topic of sex in space—ninety-two years late! William Broad's recent article "Recipe for Love: A Boy, a Girl, a Spacecraft" should have paid a respectful tribute to George Griffith's 1900 clasic, *A Honeymoon in Space.* I must confess that I've never seen a copy, but as Queen Victoria was still on the throne when it appeared, it's safe to assume that Griffith left everything to the reader's imagination. Certainly he would not have pointed out the interesting possibilities opened up by weightlessness. (Don't be so impatient, we'll come to these later.)

Matters will be quite different, however, on long-duration missions and, of course, when permanent bases are established on the Moon, Mars, and other celestial bodies. As long ago as 1955, years before the general public took space travel seriously, the distinguished astronomer, and occasional science fiction writer, Dr. Robert S. Richardson told the readers of the *Saturday Review:* "If space travel and colonization of the planets eventually become possible on a fairly large scale, it seems probable that we may be forced into first tolerating and finally accepting an open attitude toward sex that is taboo in our present social framework. To put it bluntly, may it not be necessary for the success of the project to send some nice girls to Mars at regular intervals to relieve tensions and promote morale?" ("The Day After We Land on Mars," *Saturday Review,* May 28, 1955).

One very unexpected response to Richardson's tongue-in-cheek proposal was a short story by C. S. Lewis, "Ministering Angels" (*Magazine of Fantasy and Science Fiction,* February 1956; reprinted in *Of Other Worlds* [Harcourt Brace & World, 1967]). Lewis, with tongue even more firmly in cheek than Richardson, asked: What kind of girl would accept such an assignment? The two visitors who arrive at the small Mars base are not exactly what the doctor ordered:

Some of those present had doubted the sex of this creature. Its hair was very short, its nose very long, its mouth very prim, its chin sharp, and its manner authoritative. The voice revealed it as, scientifically speaking, a woman. But no one had any doubt about the sex of her nearest neighbor, the fat person. . . . [She] was infinitely female and perhaps in her seventies. Her hair had been not very successfully dyed to a color not unlike that of mustard. . . . Powder (scented strongly enough to throw a train off the rails) lay like snowdrifts in the complex valleys of her creased, many chinned face.

You will not be surprised to know that the intended but appalled beneficiaries hijack the mercy ship and escape back to Earth, leaving the "ministering angels" stranded on Mars. It might appear from this summary that Lewis was something of a misogynist; however, he married Joy Davidman the very year this story was published. I am sorry that I never discussed it with Joy, who I knew well before and during her tragically brief life with CSL.

"Ministering Angels" was of course a joke, although it raised some serious issues; Lewis himself disapproved of spaceflight as an attempt—probably doomed—to evade what he called God's quarantine regulations. "I'm sure," he once told me, "that you Interplanetarians are very wicked people." Then he added with a grin, "But wouldn't life be dull if everyone was good?" It is a pity that he did not live long enough to comment on the first Moon landing. He died six years too soon.

As it happens, two of the people most involved in that historic project have recently expressed their views on extraterrestrial sex. Dr. Tom Paine, NASA administrator during the first seven Apollo missions, ends an essay, "The Next 25 Years in Space" (*Blueprint for Space* [Smithsonian Institution Press, 1992]) with this challenge to his successors: "NASA needs to organize a small high-level systems group . . . to achieve the goals of the President's Space Exploration Initiative. That group must lay out the programs, budgets, and milestones that will support the first Martian baby in 2015."

That is indeed optimistic: the general view is that we'll be lucky to make the first Mars landing by 2015, let alone establish prenatal facilities there. And a conception en route would be a thoroughly irresponsible act.

But what about other sexual activity on the months'-long voyage? Mike Collins, command-module pilot of the historic Apollo 11 mission, considers the problem in his skillful melding of fact and fiction *Mission to Mars* (Grove Weidenfeld, 1990). "Husband and wife teams seem a good solution . . . but it may be extraordinarily difficult to cover all disciplines and skills with . . . married couples. Maybe a marriage certificate is an unnecessary frill."

He then opens a large can of worms: "Those picking a Mars crew will be faced with some highly qualified homosexual candidates. I would not pick them. I think enough interpersonal problems will develop among a totally heterosexual crew, and introducing an element of homosexuality could only serve to make matters worse."

Some of my gay friends have expressed indignation at this, even going so far as to call Mike a bigot. This is unfair, and I hasten to come to his defense. (The last time we met it was as guests of the Saudi astronaut, Prince Sultan bin Salman bin Abdul Aziz, in an artificial oasis some thirty kilometers outside Riyadh. I was dancing—on sand, yet—with the first spacewalker, Gen. Aleksei Leonov. We're just good friends. . . .)

Though one hopes that by the time there really is a Mars mission such barbaric prejudices will have vanished, Mike's statement merely acknowledges things as they are today. Space travel involves so many avoidable hazards that any factor that reduces the risk of failure must be carefully considered. Many of them are subtle and psychological: I would hate to spend six months cooped up in a small cabin with a rabid baseball (or cricket) fan, a born-again anything, a knuckle cracker, a terminal case of crossword addiction—even though I might enjoy their company in small doses. And is it really pure coincidence that the only near disastrous Apollo mission was—No. 13? You can't be too careful. . . .

The science fiction writers, of course, have explored all possible varieties of sex in space and on alien planets. During the pulp era (say 1930 to 1955) such magazines as *Startling Stories* lured their (almost exclusively male) readers with garish covers, usually featuring young ladies in brass bras, being menaced by horrid things. What said things intended to do with them was left to the reader's imagination, because the illustrations seldom had any connection with the contents of the magazine. (As I well know; *Startling* gave me the treatment when it published "Against the Fall of Night" in 1948.) Indeed, most magazines of that era were positively puritanical, and the wide discrepancy between covers and contents was affectionately parodied by *Playboy* in a brilliant gallery of paintings—many of which could have been mistaken for the genuine article—in its November 1960 issue. The title said it all: "Girls for the slime god."

Today's lucky writers have no need to worry about the taboos of the past, and many of them have taken full advantage of their new freedom by serious and thoughtful treatment of sexuality in space—and specifically along the orbit to Mars. The science fiction grand master Jack Williamson has just at eighty-three years old published *Beachhead,* which has some very steamy passages; ditto Kim Stanley Robinson's *Red Mars.* Both appeared in 1992; everyone seems to be going to Mars this year—including NASA. If all goes well (where have I heard that phrase before?) the *Mars Observer* space probe will be launched in September and will carry out a long-duration survey of the planet. The results it obtains may well set the time scale for establishing mankind's first home beyond the Earth.

Many years ago, a popular English comedian, George Formby, well known for his rather risqué material, had a favorite song that culiminated in the punch line: "I went to France, to see what it's like there, but it's no different anywhere." No longer true by general consent, if not yet by actual experience. The newspapers quote a former NASA flight surgeon (those currently on the payroll are conspicuously silent on the subject) as saying that physical intimacy in the weightlessness of space would probably be enjoyable, perhaps more so than on Earth. "You're going to have lots of freedom of movement," she noted. Hear, hear, and at the very least, you'll be able to say good-bye to the old problem of waking up hours later with the circulation in one arm cut off by the inert dead weight of a sleeping partner.

Space tourism will be getting off the ground just in time to celebrate the centennial of Griffith's *Honeymoon in Space*. When *2001*'s Orbiter Hilton really gets built—my guess is around 2015, but then I'm a well-known conservative—many of its customers will be newly married couples, as well as lots of unmarried ones. They will be able to choose suites with any gravity they like, from zero at the center of the slowly rotating wheel to perhaps half a g at the rim. It may turn out that complete weightlessness, after the novelty has worn off, is not as satisfactory, for obvious reasons, as fractional gravity. Doubtless the time will come when there will be endless debates between Martians (one-third g) and Lunarians (one-sixth g) over the erotic superiority of their respective habitats.

Meanwhile, a great deal of preliminary research can be done right here on Earth, and I am happy to point the way. The parallel between scuba diving and living in space is now common knowledge; one day the bumper sticker DIVERS DO IT DEEPER will be matched by ASTROS DO IT HIGHER.

My personal interest in scuba diving was entirely due to my desire to experience weightlessness in my own lifetime (though as far as I recall, I did not have this particular approach in mind). From 1950 onward I took every opportunity to infect my fellow space cadets with this newfound enthusiasm—in at least one case, with notable consequences.

During a visit to Washington in March 1954, a few months before departing to Australia's Great Barrier Reef, I spent a weekend with my two closest American friends, Pip and Fred Durant. The other houseguest was Wernher von Braun, and before the weekend was over, I had in his own words "introduced me to the sport that has rapidly become my favorite—skin diving."

Though I did not know it at the time (though I suspected what was afoot) Fred and Wernher were organizing a meeting of the engineers and scientists involved in Project Orbiter—a secret Army/Navy effort to launch a satellite. (It was secret because there was a real danger that the Air Force might do it first.) After various ups and downs—mostly downs, including the unlucky Vanguard—this was to lead to the successful launch of *Explorer 1,* the first American satellite.

Immediately after he had dealt with Project Orbiter business, Wernher rushed off to buy his first Aqua-Lung and, much more important, initiated construction of the enormous tank at Huntsville, Alabama, which has been a vital astronaut training device for almost thirty years. In 1973 it saved the American taxpayer some $2 billion, when a team led by Pete Conrad developed the tools and procedures that salvaged the crippled Skylab space station and turned disaster into triumph.

So the basic equipment is readily available to solve what Ben Bova, chairman of the National Space Society, has called "some very interesting problems in rendezvous and docking." If it accepts this challenge, NASA could certainly get itself in the public eye again and divert some of the television coverage from gamy Supreme Court hearings and sordid court cases.

Ben Bova—whose recent novel *Mars* is a worthy addition to those I've already listed—goes on to add, "Essentially, you're turning sex into a three-dimensional experience." The absence of gravity would certainly make some of the more acrobatic performances outlined in the *Kama Sutra* less likely to invoke the urgent services of a chiropractor, and there are some still more startling possibilities. Consider, for example, the notorious "daisy chain"—hitherto, merely two-dimensional. In zero gravity, all regular solids—and many highly irregular ones—could be constructed. . . .

And that's just the beginning. Look at those extraordinary carbon compounds, shaped like geodesic domes, that the organic chemists have just discovered. Buckminsterfullerene (C_{60}), anyone?

I think I better stop here, before that Oregon school board starts gunning for me again. So let us end on a more serious, not to say dignified, note.

Life began in the weightless environment of the ocean, and we all spend the first months of our existence in the mini-ocean of the womb. Here on land, crushed and often killed by gravity, we are exiles—refugees in transit camp, who have the lost the freedom of the sea and not yet, except for a lucky few, and for brief periods, attained the freedom of space.

Many years ago (1958) I wrote a short story called "Out of the Cradle, Endlessly Orbiting," in which I said that we would not really have conquered space until the first baby was born on the Moon. Let's change that locale to Mars—and try, despite all the odds, to meet the 2015 target.

Minehead

Made Me

Minehead, on the Bristol Channel, where Clarke was born, imprinted itself so much on his mind, so effectively, that he sought it out unconsciously forever.

Minehead was more than my birthplace; it was my home for the first decade of my life. I was born only a few hundred yards from the sea, in a house that still stands, virtually unchanged—except that it has shrunk to half the size I remember.

In Minehead I attended infant school (ages four to eight) and was brought up, albeit spoilt, by Grandmother Mary Willis and her elder daughter, Aunt Nellie, while my parents struggled to make a living on a farm near Taunton, some twenty miles away.

Much of my youth was spent with now long-forgotten friends, exploring the mysterious woods that cloaked North Hill. From its summit I would often gaze across the broad vista through which the dead-straight line of Brunel's Great Western Railway stetched to the horizon. One day that would take me to London—and far beyond.

Most of my childhood memories are associated with the beach—that great crescent of sand from which I could see the coast of Wales, and the mysterious islands of Flat Holme and Steep Holme. Every day the tide's long withdrawal exposed a wonderland of rock pools; among them I would construct elaborate sand castles surrounded by protective moats, and await their demolition by the returning tide. This much loved playground was the inspiration for my short story "Transcience," set to music by David Bedford, and sung by Peter Pears (*The Tentacles of the Dark Nebula* [Decca Headline, 1974]).

Later I was also to discover the lines of A. E. Housman that not only described the locale perfectly, but also gave me the title of my first novel: "Here on the level sand, between the sea and land, what shall I do or write, against the fall of night?"

A long way from my birthplace—in fact a quarter of a million miles distant—is the Moon. One of the most beautiful lunar formations is the great arc of the Sinus Iridum (Bay of Rainbows), formed aeons ago when lava flooding from the newly formed Sea of Rains breached the wall of a giant crater. When I look at the ten-day-old Moon through my telescope, the Bay of Rainbows reminds me irresistibly of Minehead—though on a scale a hundred times greater. And it, too, is dominated by a massive promontory, which greets the rising Sun before it illuminates the land lying at its feet. No men or women have yet walked the Bay of Rainbows—but they will, sometime in the next century.

The other echo of Minehead is more accessible—too accessible, alas!—especially since a travel magazine classified it among the twelve most beautiful beaches in the world. Unawatuna Bay, on the south coast of Sri Lanka, is approximately the same size and shape as Minehead Bay and also has a hill at its western end. But there, I am afraid, the resemblance ends. Unawatuna's hill is the site of a Buddhist temple, and the whole arc of land is fringed by graceful palm trees.

I am sure that I seldom thought of Minehead as I snorkeled, among corals and tropical fish, in waters sometimes almost too warm for comfort. So it was not surprising that it took me a quarter of a century to realize that I had discovered my platonic ideal—the apotheosis of my childhood playground—a few degrees north of the equator. . . .

Seven thousand miles from where I was born, I had come home.

THIRTEEN

Good-bye, Isaac

The much admired and prolific science fiction writer
Isaac Asimov died on April 6, 1992.

Many years ago, when introducing Isaac Asimov to a Mensa Society meeting in London, I said, "Ladies and gentlemen, there is only one Isaac Asimov." Now there is no Isaac Asimov and the world is a much poorer place.

Isaac must have been one of the greatest educators who ever lived, with his almost half a thousand books on virtually every aspect of science and culture. His country has lost him at its moment of direst need, for he was a powerful force against the evils that seem about to overwhelm much of Western society. He stood for knowledge against superstition, for tolerance against bigotry, kindness against cruelty—above all, peace against war. His was one of the most effective voices against the New Age nitwits and fundamentalist fanatics who may now be a greater menace than the paper bear communism ever was.

Isaac's fiction was as important as his nonfiction, because it spread the same ideas on an even wider scale. He virtually invented the science of robotics—and named it before it was born. Without preaching, he showed that knowledge was better than ignorance, and that there were other defenses against violence than violence itself.

Finally, and not least, he was great fun. He will be sorely missed by thousands of friends and millions of admirers.

Encyclical

Who—what pope—will ring in the truth and joys for mankind?
Here's a jolt from the future.

Exactly four centuries ago, in the year 1632, my predecessor Pope Urban VIII made an appalling blunder. He allowed his friend Galileo to be condemned for teaching what we now know to be a fundamental truth—that the Earth goes around the Sun. Though the Church apologized to Galileo in 1992, that dreadful mistake gave a blow to its moral standing from which it never fully recovered.

Now, alas, the time has come to admit an even more tragic error. Through its stubborn opposition to family planning by artificial means, the Church has wrecked billions of lives—and, ironically, been largely responsible for promoting the sin of abortion, among those too poor to support the children they were forced to bring into the world.

This policy has brought our species to the verge of ruin. Gross overpopulation has stripped planet Earth of its resources and polluted the entire global environment. By the end of the twentieth century everyone realized that—yet nothing was done. Oh, there were conferences and resolutions—but little effective action.

Now a long-hoped-for—and long-feared!—scientific breakthrough threatens to turn a crisis into a catastrophe. Though the whole world applauded when Professors Salman and Bernstein received the Nobel Prize for medicine last December, how many have stopped to consider the social impact of their work? At my request, the Pontifical Academy of Science has just done so. Its conclusions are unanimous—and inescapable.

The discovery of superoxide enzymes which can retard the aging process, by protecting the body's DNA, has been called a triumph as great as the breaking of the genetic code. Now, it appears, the span of healthy *and active* human life can be extended by at least fifty years—perhaps much more! We are also told that the treatment will be relatively inexpensive. So whether we like it or not, the future will be a world full of vigorous centenarians.

My Academy informs me that the SOE treatment will also lengthen the period of human fertility by as much as thirty years. The implications of this are shattering—especially in view of past dismal failures to limit births by appeals for abstinence and the use of so-called "natural" methods. . . .

For several weeks now, the experts of the World Health Organization have been networking all its members. The goal is to establish the often-discussed, but never achieved except in times of war and plague, zero population growth as quickly—and humanely—as possible. Even that may not be sufficient; we may need *negative* population growth. For the next few generations, the one-child family may have to be the norm.

The Church is wise enough not to fight against the inevitable, especially in this radically changed situation. I will shortly be issuing an encyclical which will contain guidance on these matters. It has been drawn up, I might add, after full consultation with my colleagues the Dalai Lama, the Chief Rabbi, Imam Muhammad, the Archbishop of Canterbury, and the Prophet Fatima Magdalene. They are in complete agreement with me.

Many of you, I know, will find it hard—even agonizing—to accept that practices the Church once stigmatized as sins must now become *duties*. On one fundamental point, however, there has been no change in doctrine. Once a fetus is viable, its life is sacred.

Abortion remains a crime and will always be so. But now there is no longer any excuse—or any need—for it.

My blessings to you all, on whatever world you may be listening.

—John Paul IV, Easter 2032, Earth-Moon-Mars News Network.

Letter from

Sri Lanka

This, the third Alistair Cooke Lecture, was broadcast live by satellite from Colombo to Washington on October 8, 1992.

My great pleasure at being invited to give the third Alistair Cooke Lecture is still further enhanced by a curious coincidence. I had a close (but rather one-sided) encounter with Alistair just fifty-five years ago, in circumstances which I clearly remember and in fact have recounted in *Astounding Days,* my science-fictional autobiography.

My first honest job before I took to a life of literary crime was in His Majesty's Exchequer and Audit Department. Among my not-very-onerous duties was the checking of teachers' pensions, and the contributions they made toward them, which were automatically deducted from their minuscule emoluments. Sometimes teachers would leave the profession—voluntarily or otherwise—before the age of retirement, in which case they were entitled to a refund of their contributions.

One day such a file came across my desk, and on it was the name Alistair Cooke. Now it puzzled me for half a century that I should have remembered this name—and only this name—out of the hundreds that must have crossed my desk. Was Alistair already famous in 1937, when he was only twenty-eight years old? Finally, while researching *Astounding Days*, I wrote to him, in order to put the record straight. Here is his reply:

> I'm honored to be included, albeit shakily, in your memoirs. . . . "Was he already famous then, at twenty-eight, just out of university?" Well, no. First I got out of university at twenty-three, at the end of my fifth year at Cambridge. Took my final, honors, degree in English to attend the Teacher's Training College. (So, true, I intended to be a schoolmaster.) . . .
>
> During my fourth year, I did some student teaching, at the Cathedral School in Ely. I was then twenty-two. In June of 1932, I left Cambridge, and in September, sailed for the United States (I had been awarded a Commonwealth Fellowship for two years' graduate study—which turned out to be a year at Yale and another at Harvard). At the end of that time, faced with the unpleasant prospect of becoming usefully employed, I applied for—at three thousand miles—the post of film critic for the BBC. I read—in a glaring headline in the *Boston Globe*—that "BBC fires PM's son." He was Oliver Baldwin, and if he hadn't been, God knows where I'd be today. The fund whipped me over (via the paddling Cunarders) to London for an interview. I got the job. Returned to England in October 1934 and was the BBC's film critic till April 1937, when I resigned to return to America as a correspondent.

I note with some interest that Alistair got his pension contributions back just about the time he resigned as film critic. I'm not sure whether this was due to bureaucratic delays at the Board of Education—or whether he didn't apply until he'd safely switched jobs. Perhaps he'd been keeping his options open so that if he was a flop as a film critic, he could always go back to teaching. But, of course, he did indeed become one of the world's great educators with a classroom of millions instead of tens!

I've been honored to have been mentioned several times in Alistair's famous *Letter from America*—I believe the longest-running radio feature of all time—and in one of his letters he referred to my civil service connection. This also links me indirectly with another famous broadcaster. As part of my training, I had to attend classes in accounting at the London School of Economics. That estimable institution can seldom have had a more disinterested student, and I never set foot inside it again until some forty years later, when taken there by my friend the Canadian professor and broadcaster Robert McKenzie, and I'm happy to pay this tribute to his memory.

I am not certain when I made my first radio broadcast, but my first television appearance is one I'll never forget—something that probably could not be said by my first audience. My first incarnation as a talking head occurred on May 4, 1950, when I was commissioned to give a thirty-minute talk on the fourth dimension. In those days, of course, there was no videotape, and everything had to be done live. I had to prepare a number of wire models—which are still gathering dust somewhere in the Clarki-ves—and I was rehearsed to within an inch of my life by the producer Robert Barr—to whom I will always owe a debt of gratitude. The broadcast took place from the famed Alexandra Palace in North London—within walking distance of my residence then.

To talk live on the fourth dimension for thirty minutes into the un-blinking eye of a television camera was such a baptism of fire that no later appearance has ever had any fears for me, though, of course, I have since been involved in somewhat more famous telecasts. The most memorable was when I sat next to Walter Cronkite in the CBS studio and watched Neil Armstrong make one giant step for mankind.

Space and broadcasting are now inextricably linked in a way that would certainly have astonished the early pioneers—though not the great Marconi himself, who was fascinated by the possibility of radio commu-nication with other worlds.

Shortly before the outbreak of World War II we space cadets of the British Interplanetary Society conducted a lengthy correspondence in the pages of the BBC's much lamented weekly *The Listener*—trying to convince the skeptical readership that one day it would indeed be possible to escape from Earth and to go voyaging around the universe, or at least the Solar System. The debate lasted from February to May in 1939, when the editor decided that "this correspondence must now cease."

Some of the objections raised against our modest proposal to go the Moon now seem laughably absurd and make hilarious reading. Critics were fond of saying that space travelers would be crushed by the enormous ac-celeration—and then when it was over, they would undoubtedly burst like deep-sea fish brought to the surface.

It was great fun while it lasted, and I'm sure that not even the most enthusiastic of us Interplanetarians ever dreamed we'd live to see the day when the president of the United States would proclaim that his country must go not to the Moon (as Kennedy declared in 1961)—but to Mars!

It required only the BBC's firing of Oliver Baldwin to separate Alistair Cooke from the British educational system, but nothing less than World War II was needed to eject me from the civil service. After another narrow escape (I was inadvertently posted as a deserter from the Army Medical Corps, though it may have been some other Clarke they were looking for),

I ended up as a technical officer on the first microwave blind-landing system. From ground-controlled approach, or GCA, I conceived the idea of the communications satellite. I wrote up the concept in a four-page memorandum pecked out on my Remington noiseless portable. The four or five carbon copies have long since vanished, but the original, which I sent to Ralph Slazinger of the sporting goods company and an enthusiastic member of the Interplanetary Society, is now in the Smithsonian Institution.

Although such corrections are usually futile—I'm still trying to convince everyone that, against all odds, Kubrick and I did not derive the name HAL by a one-letter displacement from IBM—I'd like to put the record straight on one point:

I did not "discover" the geostationary orbit, even though people are now naming it after me. It didn't have to be discovered, because its theoretical existence was perfectly obvious to anyone with an elementary knowledge of astronomy. What I did was to point out that (a) artificial satellites, with their line-of-sight view of enormous areas of the Earth, could act as wonderful relay stations, and (b) the obvious location for such relays was in the geostationary orbit, where they would stay fixed over the same spot on the planet and therefore, unlike all other celestial bodies, would remain motionless in the sky. So once a ground receiver had been aimed to the right direction, it need do no further tracking.

By 1945, the concept of manned space stations, as we now call them, was already an old one and had been described in the Russian and German literature; however, little of this had been translated. I came across it secondhand in David Lasser's *The Conquest of Space* (1931) and P. E. Cleator's *Rockets through Space* (1936). I'm indeed happy to say that both authors are still with us; I recently sent greetings to Lasser on his ninetieth birthday and will always be grateful to him because it was his book that first made me take space travel seriously.

The early European writers on astronautics assumed that space stations would carry human crews and pointed out their great value for astronomy and meteorology. They took it for granted that space stations would keep in contact with Earth by light signals or radio waves. But as far as I know, they never took this idea any further.

Looking back, it's fairly easy to see why. In the 1920s, of course, there was no television, and long-distance radio transmission depended on enormous antenna systems using very long waves—often covering many square kilometers of territory. The development of shortwaves, and later microwaves, changed the whole situation, and by 1945 it was obvious that television was here to stay, although it had been put into cold storage for the war duration.

By 1945, therefore, the communications satellite was an idea whose time had come. Even if I hadn't written my paper, half a dozen people would have done so within another year.

Although it is not too difficult to forecast technological developments, it is almost impossible to set a timetable for them. Usually predictions are much too optimistic in the short run, and much too pessimistic in the long run. I once tried to sum this up when addressing a group of American businessmen by telling them: "If you take me too seriously, you'll go broke. But if your children don't take me seriously enough, they'll go broke."

My *Wireless World* paper in 1945 is in fact a pretty good example of the perils of prophecy. I assumed that the relay station would be manned, and for very good reasons based on experience. The GCA Mark I, which I was then trying to keep running, contained over a thousand vacuum tubes, and at least one burned out every day. It seemed inconceivable that any complex piece of electronic equipment, such as a television relay system, could function without engineers on the spot to service it.

Well, along came the transistor, and then the microchip. So within a decade electronic equipment that was once as large as a house could be put in a hatbox, and the Comsat revolution started twenty years after my article—instead of the fifty I had thought more probable.

I sometimes joked that the invention of the transistor was a disaster for us would-be space explorers. If we were still stuck with vacuum tubes, we would have been forced to develop spaceflight on a massive scale, because weekly shuttle flights would be needed to carry up spare parts and the people to install them. So by now we would have been on our way to Mars.

And here's another example of the difficulties of detailed technological prediction. Although, of course, I was well aware of the fact that it takes an appreciable fraction of a second for radio waves to get up to the stationary orbit and down again, the time delay didn't worry me. I was primarily interested in television broadcasting, where this wouldn't matter. It's only a problem in fast two-way exchanges—like Bill Buckley interrogating a squirming guest on the other side of the globe.

However, when the first plans had been made for commercial Comsats in the early 1960s, some experts, notably at Bell Labs, thought the stationary orbit might be unusable for telephone conversations and advocated satellites much closer to Earth—of which the famous Telstar was the most notable example. (My good friend and editor, Ian Macauley, was the first journalist to have his photograph beamed from London to New York via Telstar while interviewing Sir Leon Bagritt of Elliot Automation on Westminster Bridge in 1962. The photograph appeared on the cover of *Electronic News,* of which Ian was the London correspondent.)

To everyone's relief, Bell Lab's pessimism turned out to be unjustified, and the success of the Hughes Syncom, and later, Early Bird (Intelsat 1), proved that the far-out geostationary orbit was not only accessible, but workable.

Yet now, by another technological switch, the world's telephone services are coming down to Earth again—thanks to the extraordinary development of fiber optics. If anyone had told me twenty years ago that it would be possible to get detectable amounts of light through a piece of glass ten thousand kilometers long, I would have laughed at them. Yet that is essentially what the optronic engineers have done, making it possible to send millions of telephone calls or hundreds of television signals simultaneously through a thin hair of glass.

This, of course, has taken a lot of traffic away from the satellites, but they have really nothing to worry about in the long run. Their role is undisputed for broadcasting over vast areas of the Earth, for instantaneous news coverage from any point of the globe at a moment's notice, and for providing services to mobile stations, such as receivers on aircraft, ships, automobiles, and individual human beings.

Space pollution is already becoming a major problem. At the moment, the American tracking system NORAD is trying to follow this problem.

The late Luis Alvarez, and his son, Walter, suggested that the extinction of the dinosaurs might have been due to a cataclysmic event. Louie also changed my history, by inventing the GCA unit on which I spent so many happy and frustrating months. One of the most brilliant scientists of his generation, he went on to design the mechanism of the first atomic bomb, to win the Nobel Prize for physics, and to discover a type of low-temperature nuclear fusion, which may yet help to solve the energy problems of our planet.

His analysis of the optical and acoustic records of the Kennedy assassination proved beyond doubt that three shots were fired by a single gunman. The evidence was first presented publicly by Dick Salant and Walter Cronkite on CBS in May 1967. I refer you to Louie's autobiography, *Adventures of a Physicist* (Basic Books, 1987), for the details which needless to say will never convince the raving hordes of conspiracy advocates.

The Dutch-American astronomer Tom Gehrels has initiated a survey of potential killer rocks. Alistair obviously did not take this project very seriously, and one must admit there are a few more pressing problems closer at hand. However, I sent the transcript of his talk to Gehrels and put him in touch with Alistair. Since then there have been further developments, and NASA has published a study of the problem and suggested setting up a survey that they have given the same name I invented: Spaceguard.

Another program I invented concerned keeping tabs on something like ten thousand satellites, ranging in size from multiton objects like Russia's *Mir* down to lost gloves and Hasselblad cameras. Far more numerous, however, are fragments of exploded satellites and miscellaneous space junk down to a centimeter or less in diameter, all capable of severely damaging, or even destroying, any other spacecraft they may encounter at velocities of tens of kilometers a second.

This makes even more appalling some of the projects put forward in aid of strategic defense—such as the so-called Brilliant Pebbles. Some of the experiments already conducted in this area by the United States and the former Soviet Union have produced thousands of fragments. We may be trapped on Earth by the clouds of space shrapnel we have created ourselves—at least until we can mount a major cleanup operation and get rid of it. That might cost more than putting it up in the first place.

Of more immediate concern is electromagnetic pollution from the many types of application satellites—communications, weather, meteorological, and particularly, position finding. My radio astronomer friends have already been severely handicapped by this and have even been put out of business in some parts of the spectrum.

Talking of space shrapnel leads me by a rather circuitous route back to Alistair, and to a subject that quite recently hit the headlines. It also gives me an opportunity to quote some of my purple prose. Here is the opening of a novel that I published in 1973, *Rendezvous with Rama*:

> At 09:46 GMT on the morning of September 11 in the exceptionally beautiful summer of the year 2077, most of the inhabitants of Europe saw a dazzling fireball in the sky. Within seconds it was brighter than the Sun. And as it moved across the heavens—at first in utter silence—it left behind it a churning column of dust and smoke. . . .
>
> Moving at fifty kilometers a second, a thousand tons of rock and metal impacted on the plains of northern Italy, destroying in a few flaming moments the labor of centuries. The cities of Padua and Verona were wiped from the face of the Earth; and the last glories of Venice sank forever beneath the sea as the waters of the Adriatic came thundering landwards after the hammer blow from space.

Knowing that we can't do anything about it, I think most of us would rather not know when we are going to be struck by junk falling out of the sky.

I very much doubt that I can say anything new, and true, which has not been said many times before. We all know that television is the most marvelous medium of communication ever invented, and that it can be used to educate, entertain, and inspire—but that all too often it is, to coin a phrase, a great wasteland.

Imagine the isolation in which most of our ancestors lived. I was vividly reminded of this recently when I read somewhere that since the coming of radio and the automobile, no longer did farmers' wives on the prairie go mad through sheer loneliness. Nowadays, of course, they are likely to go mad through too much MTV.

It is not only a window on the world, but a mirror in which the global family—the new Telefamily of Man—can see its collective face. Can communications satellites become the conscience of the world? Yes, that is even more true today, as every evening the news brings its avalanche of natural and man-made disasters. Though there is a danger that overexposure to tragedy and horror can induce a compassion fatigue, the alternative—the indifference of ignorance—is surely worse. On August 24, 1991, Alistair made exactly the same point in his *Letter from America*. Contrasting the 1917 "ten days that shook the world" with 1991's sixty hours, he stated that the Russian countercoup failed because of something new: satellite broadcasting. On a lighter note, may I pay tribute to the many friends I've made while working in this medium. I've already mentioned Walter Cronkite, and I'd like to add Sylvester "Pat" Weaver, who lured me to the *Today* show when it was hosted by the charming and talented Dave Garroway.

What wore me out was doing the show three times for the different time zones across the United States. Believe me, it was rather hard to be spontaneous the third time around, especially at eight o'clock in the morning. I think that when I joined Hugh Downs on the same show, this ordeal was no longer necessary.

Message

to Mars

The message Clarke sent on the Russian Mars probe was lost—burned up in space somewhere, perhaps. It's lucky that he saved a copy here on Earth.

My name is Arthur Clarke, and I am speaking to you from the island nation of Sri Lanka, once known as Ceylon, in the Indian Ocean, planet Earth. It is early spring in the year 1993, but this message is intended for the future. I am addressing men and women—perhaps some of you already born—who will listen to these words when they are living on Mars.

As we approach the new millennium, there is great interest in the planet that may be the first real home for mankind beyond the Mother World. During my lifetime, I have been lucky enough to see our knowledge of Mars advance from almost complete ignorance—worse than that, misleading fantasy—to a real understanding of its geography and climate. Certainly we are still very ignorant in many areas and lack knowledge that you take for granted. But now we have accurate maps of your wonderful world and can imagine how it might be modified—terraformed—to make it nearer to the heart's desire. Perhaps you are already engaged upon that centuries-long process.

There is a link between Mars and my present home, which I used in what will probably be one of my last novels, *The Hammer of God*. At the beginning of this century, an amateur astronomer named Percy Molesworth was living here in Ceylon. He spent much time observing Mars, and now a huge crater, 175 kilometers wide, in your southern hemisphere is named after him. In my book I've imagined how a New Martian astronomer might one day look back at his ancestral world, to try to see the little island from which Molesworth—and I—often gazed up at your planet.

There was a time, soon after the first landing on the Moon in 1969, when we were optimistic enough to imagine that we might have reached Mars by the 1990s. In another of my stories, I described a survivor of the first ill-fated expedition, watching the Earth in transit across the face of the Sun on May 11, 1984! Well, there was no one on Mars to watch that event—but it will happen again on November 10, 2084. By that time I hope that many eyes will be looking back toward the Earth as it slowly crosses the solar disk, looking like a tiny, perfectly circular sunspot. And I've suggested that we should signal to you then with powerful lasers, so that you will see a star beaming a message to you from the very face of the Sun.

I, too, salute to you across the gulf of space—as I send my greetings and good wishes from the closing decade of the century in which mankind first became a space-faring species and set forth on a journey that can never end, so long as the universe endures.

Preface:

The War

of the Worlds

The two greatest names in science fiction are Jules Verne (1828–1905) and Herbert George Wells (1866–1946). Though they now seem to belong to different ages, their careers overlapped. Verne was still alive when Wells published his finest tales.

Verne can be enjoyed by English readers as a splendid storyteller, but it can hardly be denied that Wells was a much greater writer. He was endowed with almost all the gifts that a novelist can possess; perhaps he had too many. If he had not been so interested in politics, history, and society, Wells might have written fewer but better books. (In fifty years, he produced some 150 titles, of which perhaps 20 are remembered today.)

Wells grew up in poverty and squalor; his father, Joseph, had been a gardener, his mother a lady's maid, but before he was born his parents had sunk their small resources into an unsuccessful shop, which was saved from bankruptcy by Joseph's earnings as a professional cricketer. H.G.'s early years are best summed up in his own words:

When the writer was ten or eleven, his father was disabled by a fall that crippled him, and when he was thirteen the little shop collapsed. His mother returned as a housekeeper to her former mistress and his father took refuge in a small, inexpensive cottage. Further education for the writer seemed impossible. There was some trouble in finding him employment, an unhandy boy preoccupied with reading. He tried being a draper's apprentice, a pupil-teacher in an elementary school, a chemist's apprentice, and again as a draper. After two years with the second draper he prayed to have his indentures canceled, and became a sort of pupil-teacher. In the interval between these attempts to begin life he took refuge in the housekeeper's room with his mother.

From this pathetic environment Wells escaped by a combination of luck and genius. He won a scholarship to the Royal College of Science, Kensington, where he studied biology under the great T. H. Huxley and took his degree in zoology. When he was twenty-one, an accident on the football field destroyed one kidney and made him a semi-invalid for a while, with the result that he had both the opportunity and the incentive to write for a living.

He was successful from the very start with short stories, articles, and humorous sketches. His first novel, *The Time Machine* (still his masterpiece), appeared in book form in 1895, and thereafter his fame spread swiftly throughout the world. Even the miseries of his early life were turned to good account in such novels as *Kipps, Tono-Bungay,* and *The History of Mr. Polly*. These sagas of ordinary people in turn-of-the-century England belong to the best tradition of Dickens and would have assured Wells's fame even if he had never trafficked with Martians.

The War of the Worlds is in some ways Wells's most remarkable tour de force and contains passages whose relevance is even greater today than when it was written at the close of the last century. The original idea was not from Wells but, as he acknowledged, from his older brother Frank: "A practical philosopher with a disbelief even profounder than that of the writer, in the present ability of our race to meet a crisis either bravely or intelligently. Our present civilization, it seems, is quite capable of falling to pieces without any aid from the Martians."

This astonishing novel contains what must be the first detailed description of mechanized warfare and its impact upon an urban society. Yet Wells wrote it not only before World War I, but even prior to the Boer War! The account of refugees streaming out of London before the assault of the Martians must have seemed unbelievable fantasy to the comfortable Victorians; to us, it is not only past history but, all too often, "live via satellite" in our own living rooms.

Every generation can reread *The War of the Worlds* in the light of its own experience and gain something new from it. In the 1920s it was impressive because it described poison gas in action and suggested that aircraft could be used for warfare. It made a still greater impact on the 1930s, when the famous Orson Welles production of Howard Koch's radio script (*Mercury Theater of the Air,* CBS, October 30, 1938) caused panic over a substantial area of the eastern United States.

Howard (with whom I worked on a never-produced science fiction movie while he was a McCarthy exile in London during the sixties) wrote the script in a week, receiving $75 in payment. I am happy to say that he recently sold the few dozen tattered pages for a six-figure sum. His inside story of the whole episode, *The Panic Broadcast,* still makes amusing reading.

H. G., however, was not amused—but soon mellowed. On his last visit to America, at the age of seventy-four, he and Orson had their one and only meeting at a San Antonio radio station. Listening to the encounter between the two master magicians is an eerie experience, like a trip on Wells's own time machine back to the vanished world of 1940:

H. G.: I've had a series of the most delightful experiences since I came to America, but the best thing that has happened to me so far has been meeting my little namesake here, Orson—I find him most delightful—he carries my name with an extra *e,* which I hope he will drop soon. . . . I've known his work before he made this sensational Halloween spree—are you sure there was such a panic in America, or wasn't it your Halloween fun?

ORSON: I think that's the nicest thing that the man from England could possibly say about the man from Mars.

So all was forgiven; more than that, H. G. ended by presenting Orson with an opportunity for a blatant plug:

H. G.: Before we get away from the microphone, tell me about this film of yours that you're producing . . . what's the film called?

ORSON: It's called *Citizen Kane.*

Besides the Welles and Pal productions, *The War of the Worlds* has spawned countless imitations—perhaps most notably the Jeff Wayne musical (1978), superbly narrated by Richard Burton. One reason for its continuing power is that, unlike his later imitators, Wells made his monsters plausible and well-motivated. They were not hell-bent on destruction for its own sake, but were proceeding on a logical plan of conquest, with a definite though deplorable—from our admittedly biased point of view—objective. A good example of Wells's foresight is the hint, forty years before

Quisling and Company, that there would be men willing to work out an accommodation with the Martians.

Though one cannot blame Wells for all the later excesses of interplanetary warfare, perhaps he merits some criticism for propagating the credo that anything alien is likely to be horrible. Compare his unflattering description of the Martians ("Those who have never seen a living Martian can scarcely imagine the strange horror of its appearance. . . . Even at this first encounter, I was overcome with disgust and dread") with the passage in C. S. Lewis's *Perelandra,* where the hero meets a much more imposing monster in a Venusian cave—and after the first revulsion, sees it merely as something strange, not in the least hideous. The underwater explorers of today have gone through the same process while making friends with octopi, and the astronauts of the future may have similar problems. The tradition started by *The War of the Worlds* will not help them. But perhaps Steven Spielberg's *E.T.* may have started a new and less paranoic trend. In fact, H. G. himself depicted a much more benign Mars, with more attractive inhabitants, in his short story "The Crystal Egg." (Another prediction that was realized when the Viking Lander gave us our first glimpse of the Martian landscape.)

In his 1964 introduction to the Easton Press edition, J. B. Priestley remarked that he found *The War of the Worlds* "far more impressive on this recent rereading than I had expected it to be. . . . The fate of the invaders is a superb stroke. Perhaps the best in the whole narrative."

Agreed—and I hope that no reader's enjoyment is spoiled by the revelation that the Martians were "slain by the putrefactive and disease bacteria against which their systems were unprepared; slain, after all man's devices had failed, by the humblest things, which God in his wisdom had put upon this Earth."

Until recently, I was under the impression that this "superb stroke" was wholly original, perhaps inspired by Wells's own student apprenticeship in biology. This may be true, but now I doubt it. One of the most famous interplanetary tales of the nineteenth century was Percy Greg's two-volume *Across the Zodiac* (1880), which Wells must certainly have read. Greg's hero travels to Mars in a vehicle powered by an antigravity device (as in Wells's own *The First Men in the Moon*) and encounters Martians so completely human that he takes several of them as wives. One of them, alas, dies of an earthly disease against which she has no resistance!

Coincidence? Perhaps; but as the saying goes, "Talent borrows, genius steals." And even if he did borrow—or steal—from Greg, Wells was completely justified. He transformed the concept and brilliantly anticipated a problem that has become of real practical concern now that we have contact with other celestial bodies: planetary quarantine.

Although it now seems highly probable that there are no higher life-forms on Mars at the present time, microorganisms cannot be ruled out; indeed it will be surprising if they do not occur in favorite spots, since all the ingredients necessary for their existence are common. Will such Martian bacteria be dangerous to humans? If we are not careful, we may find out the hard way. . . .

There is also a much longer-term philosophical question, raised by the current discussions of terraforming Mars to make it a new home for mankind. If there are even the most primitive life-forms on the planet, should we not give them a chance to evolve naturally, rather than preempt their evironment to our own use? I do not expect the "Hands off Mars!" lobby of the next century to be very large, but it may be quite influential, and I must confess a certain sympathy for it.

Those who have not read *The War of the Worlds* may be surprised to find that, like much of Wells's writing, it is full of poetry and contains passages that catch at the throat. Wells tried to pretend that he was not an artist and stated that "there will come a time for every work of art when it will have served its purpose and be bereft of its last rag of significance." This has not yet happened for the best of Wells's science fiction, though it has done so for all but a few of his realistic and political novels. These have suffered the fate of most topical writing, while his so-called fantastic tales are still fresh and enjoyable.

A year after *The War of the Worlds,* in 1899, Wells wrote his second interplanetary romance, *The First Men in the Moon.* This is perhaps the most famous of all stories of space travel; few later writers ever came near to matching its mood of extraterrestrial awe and wonder, and no one has surpassed it. *The Sleeper Wakes, The War in the Air,* and *The Food of the Gods* followed in quick succession, interspersed with dozens of short stories in which Wells mapped out territory since explored by several generations of science fiction witers. Then, for more than twenty years, Wells virtually abandoned the genre that had brought him fame, though he returned to it toward the end of his life in *Star-Begotten* (1937). His last book, *Mind at the End of Its Tether,* was a sad and despairing work published only two years before his death at the age of eighty. It did his reputation little good, and during the postwar period there was a definite slump in Wells's stock.

Now, however, this mood has passed. A flourishing H. G. Wells Society has been formed in England, and critical books on Wells are appearing in increasing numbers (e.g., Bernard Bergonzi's *The Early H. G. Wells,* W. Warren Wagar's *H. G. Wells and the World State*). This may be partly due to the shamefaced realization in literary circles that Wells's scientific romances were not youthful aberrations or escapist fantasies, but works of art with unique relevance for our times.

And I think that people are rereading Wells because they are tired of ever more minute discussions of neurotic egos, and worn-out repetitions of eternal triangles and tetrahedra. Wells saw as clearly as anyone into the secret places of the heart, but he also saw the universe, with all its infinite promise and peril.

He believed—though not blindly—that men were capable of improvement and might one day build sane and peaceful societies on all the worlds that lay within their reach. We need this faith now, as never before in the history of our species.

Preface:

The First Men

in the Moon

A year after the publication of The War of the Worlds, *Wells wrote his second interplanetary romance,* The First Men in the Moon. *As Clarke notes, this is perhaps the most famous of all stories about space travel.*

There were almost one hundred generations before H. G. Wells, if one accepts a very flexible definition of the term *science fiction*. The first story of a voyage to the Moon appeared in the second century A.D. It was an involuntary one: the hero of Lucian of Samos's facetiously entitled *True History* was taken to the Moon via a waterspout.

Thereafter, the theme of space travel languished for almost fifteen hundred years—until the invention of the telescope, when Galileo revealed to an astonished (and often incredulous) world that the Moon was an actual place, with mountains and valleys and plains.

This inspired the first serious account of a journey to the Moon, and it was written by one of the greatest astronomers of all time—Johannes

Kepler, discoverer of the laws that govern the movements of all celestial bodies, natural and man-made. Though he used supernatural means to get to the Moon, Kepler's account of lunar conditions was based on the new knowledge revealed by the telescope and had great influence on later writers—including Wells, who specifically mentions him in chapter 13.

Kepler died in 1630, and his *Somnium* (Dream) was not published until 1634. Only another four years later the first English story of a lunar trip appeared—Bishop Godwin's *Man in the Moone*. His space vehicle was a flimsy raft towed by trained swans, a technology not yet properly investigated by NASA. (But we must give Godwin credit for realizing, half a century before Newton discovered the law of gravitation, that one could jump to great heights on the Moon.)

During the next two centuries there was a steady trickle of books about spaceflight; most were pure fantasy, but others were at least occasional attempts to be scientific. Undoubtedly the most ingenious writer during this period was Cyrano de Bergerac, author of *Voyages to the Moon and Sun* (1656). To Cyrano must go the credit for first using rockets—and what may best be described as a solar-powered ramjet.

As the difficulties of crossing the vacuum of space became better appreciated, writers struggled to discover more plausible technologies. The first person to hit upon a gravity-defying substance appears to be one Joseph Atterley, whose *Voyage to the Moon* appeared in 1827.

Soon after the appearance of *The First Men in the Moon*, Wells was involved in a controversy with the now forgotten Irish writer Robert Cromie, author of *A Plunge into Space* (1890). This tale also employed "the secret of gravitation" to power a spherical spaceship on a journey to Mars and so certainly had some points in common with the Wells novel. H. G. replied to the indignant Irishman's accusations with a single sentence: "I have never heard of Mr. Cromie nor of the book he attempts to advertise by insinuations of plagiarisms on my part." Though there is no reason to doubt the Wells statement, both he and Cromie were heirs to the now extensive nineteenth-century tradition of antigravity stories, notably Percy Greg's *Across the Zodiac* (1880). As David Lake points out in his account of the episode, "The marks of the Greg-type novel are three: (1) the interplanetary journey is a secret adventure for one or a few men, (2) rendered feasible and cheap by a secret discovery of antigravity, and (3) closed off by the death of the inventor, who takes his secret with him, so that the adventure cannot be repeated. All these features are in Cromie and Wells, and thus arise most of the resemblances between *A Plunge into Space* and *The First Men in the Moon*.

Antigravity screens of the type so plausibly described by Wells have been out of fashion for most of this century, because a simple thought experiment will show that they can't possibly work. If he had been a really good businessman, Bedford would have persuaded Cavor to start selling electricity, instead of bothering to go to the Moon. For a piece of "cavorite" placed under one side of a dynamo could make half the rotor weightless, so that it would start to rise upward and thus generate power—indefinitely. But nature never gives something for nothing, which is why builders of perpetual-motion machines are doomed to perpetual disappointment.

However, there is no objection in principle to an antigravity device powered by some external source of energy, and from time to time one hears of such an invention. None, however, has yet reached the Patent Office.

Jules Verne was quick to point out that, as opposed to Wells's mythical cavorite, the space gun that propelled his travelers (*From the Earth to the Moon,* 1865) was based on sound scientific principles and detailed calculations of escape velocity and transit times. True enough—as long as one neglected air resistance and the minor problem that the would-be astronauts would be converted into instant wall-to-wall carpet by the initial acceleration. Perhaps Wells took Verne's criticism too seriously; the fledgling British Interplanetary Society protested to him at the time when he reverted to using a space gun, instead of a rocket, in the classic film *Things to Come* (1936).

I doubt if even Wells imagined that, only twenty years after his death, men would actually be preparing to go to the Moon; he would have been 103—not an impossible age—at the time of the first landing. He would certainly have been delighted to know that when Armstrong, Aldrin, and Collins returned to Earth, the book recording their personal stories would be called *First on the Moon* (Little, Brown, 1970).

It was my privilege to contribute the epilogue, "Beyond Apollo," to this volume, and on rereading it almost a quarter of a century later many of its more enthusiastic predictions now seem sadly dated. The geopolitical forces that powered the first drive into space no longer exist. When we return to the Moon, it will not be as the result of a contest between transient national entities, but for more fundamental reasons. I concluded my epilogue—more than a generation ago!—with the words:

> Five hundred million years ago, the Moon summoned life out of his first home, the sea, and led it onto the empty land. For as it drew the tides across the barren continents of primeval Earth, their daily rhythm exposed to sun and air the creatures of the shallows. Most perished—but some adapted to the new and hostile environment. The conquest of the land had begun.
>
> We shall never know when this happened, on the shores of what vanished sea. There were no eyes or cameras present to record so

obscure, so inconspicuous an event. Now the Moon calls again and this time life responds with a roar that shakes earth and sky.

When a Saturn V soars spaceward on nearly four thousand tons of thrust, it signifies more than a triumph of technology. It opens the next chapter of evolution.

Despite our apparently overwhelming fin de siècle problems, I believe these words are still true. If, as Wells famously remarked, we win the "race between education and catastrophe," space is where the future lies. And who, I wonder, will be "next on the Moon"?

There is one respect in which *The First Men in the Moon* is astonishingly up-to-date—even topical—as this extract demonstrates:

The reader will no doubt recall the little excitement that began the century, arising out of an announcement by Nikola Tesla, the American electricity celebrity, that he had received a message from Mars. His announcement renewed attention to the fact . . . that from some unknown source in space, waves of electromagnetic disturbance, entirely similar to those used by Signor Marconi for his wireless telegraphy, are constantly reaching the Earth.

This passage is really quite extraordinary, for several reasons—altogether apart from the ingenious manner in which Wells used a sensational news item to add a "live from the Moon" epilogue to his already published novel. For now, at the end of the century, a SETI (search for extraterrestrial intelligence) program is in full swing, scanning the electromagnetic spectrum for evidence of civilizations elsewhere in the universe.

Tesla—an electrical genius and one of the few real-life examples of a genuine mad scientist—was of course talking nonsense. His 1900 equipment was far too insensitive to detect such radiation; it was not discovered until three decades later (Karl Jansky, 1931). But once again, Wells got there first.

The Joy of

Maths

Mathematics, Clarke writes, is one of the most valuable inventions of mankind.
To those who, by bad luck or bad teaching, have developed a fear of
mathematics, now is the chance to get it straight.

This unfortunate stance can easily be dismissed, for mathematics is not only one of the most valuable inventions—or discoveries—of the human mind, but can have an aesthetic appeal equal to that of anything in art. Perhaps even more so, according to the poetess who proclaimed, "Euclid alone hath looked at beauty bare."

And contrary to general opinion, it is not always necessary to have any technical knowledge of the subject to appreciate that beauty. No need for logarithms or trigonometrical functions—in fact, the "joy of maths" can be experienced even by those whose skill is limited to counting. After all, no detailed knowledge of anatomy is required to enjoy . . .

Much of mathematics, as of ordinary life, involves distinguishing between the possible and the impossible. And sometimes that decision is extraordinarily difficult to make, even in what appears to be a very simple case.

Let me start with one that is both simple and—apparently—easy. It's the classic puzzle of the three utility companies, who want to provide service to three customers. The problem is, neither the electric, telephone, or cable-television company will permit anyone to lay a line that crosses their own.

A few minutes of doodling will quickly demonstrate that there's no difficulty in taking two of the services to all three houses, and another to any two of them—but the final connection isn't so easy. I once encountered an old gentleman who had spent years trying to solve the puzzle, creating the cat's cradle of circuitry. Apparently it had never occurred to him that there is no solution, and it is not hard to demonstrate that the feat is impossible. Remembering my promise, I'll leave it as an exercise for the student. Any PC can do the number crunching required; the real challenge arises when we approach the problem from the other direction. Once again, it looks ridiculously simple, but by now you should be suspicious of anything that appears too easy.

Imagine that you're throwing a party, but are selecting people entirely at random—perhaps by sticking pins in the phone book. For some reason best known to yourself, you want to make sure that either three of your guests are mutual acquaintances, or that three of them are complete strangers. Exactly how many people will you have to invite to make absolutely certain that this will happen?

Don't try to answer by drawing lots of triangles with solid or dotted sides. To put you out of your misery, I'll tell you right away that a mere six guests will be sufficient to satisfy your rather peculiar requirements. And you'd better take my word for it, because even when you know the answer, it won't be easy to check that it's correct. Remember—with only six people, there are 32,768 possibilities.

But just suppose that you had the patience to put down six dots, say at the corners of a hexagon, and then connect them together in all those 32,768 ways (making sure that you hadn't missed—or duplicated—any of them.) You would discover that, in every case, there would be at least one with dotted sides—i.e., either three acquaintances, or three mutual strangers.

This example gives the definition of a Ramsay number; for three friends/three strangers it is six—and as we have seen, it is quite hard to discover it even for such small values. By now you will not be surprised to learn that for ones that are only slightly larger, it is unbelievably difficult.

In 1933 the problem was solved for the case of four friends/five strangers. It turns out that you have to pick twenty-five people at random from the population of Earth to be certain that you will have one of these combinations in your guest list. And the effort to discover this was truly heroic, because the number of possible arrangements runs to no less than sixty-five digits!

To prove that twenty-five was the correct Ramsay number required eleven years of computer time; up to 110 computers were sometimes working on it simultaneously. At a rough estimate, this would be as if every human being who has ever lived made millions of calculations—and a single error back in the Stone Age would have invalidated the whole enterprise.

And I haven't quite finished with Frank Ramsay. Here is one final astonishment:

No one knows the next Ramsay number for five friends/five strangers, and it is believed that even when computers are a thousand times more powerful than they are today (by 2020 would be a conservative guess), they still won't be able to find it. Yet—and this is what I find quite staggering— the number is known to lie between forty-three and forty-nine!

Just pause to consider what this means. All the computing power that exists today is probably insufficient to pin down a number somewhere in the very short sequence 43, 44, 45, 46, 47, 48, 49. . . . If you don't think that's mind-boggling, then as my friend HAL once somewhat primly remarked, "This conversation can no longer serve any useful purpose."

And speaking of useful purposes, even those who feel a sense of wonder at these discoveries may well ask, "Are they of any practical value?" It's a valid question, for not everyone will go along with G. H. Hardy's famous toast: "Here's to pure mathematics—may it never be of any use to anybody."

So far, this is true of Ramsay numbers, though the computer-search strategies developed in locating them may well have many other applications. However, even in his own lifetime Hardy's toast had a spectacular refutation. Without the purest of mathematics, the outcome of World War II might well have been very different.

The veil has at last been lifted over the work at Bletchley Park, where a handful of young geniuses (including Alan Turing), of whom Churchill said, "When I told you to leave no stone unturned, I didn't expect you to take me quite so literally" broke the Enigma cipher. The result was that the Allies were privy to the German High Command's most secret plans. The Germans were absolutely convinced that Enigma was unbreakable, but it was cracked by a combination of advanced mathematics and the world's first electronic computer. And so a handful of equations helped to change the course of history.

An even more spectacular—and I mean this literally—example of the value of pure mathematics has occurred during the last decade. By now everyone must have seen fractals—the beautiful and complex patterns of which the well-known Mandelbrot Set is the prototype. Their endless variations mimic such shapes in the real world as mountain ranges, butterfly wings, flowers, the zigzag of lightning strokes. . . . Dr. Mandelbrot himself has argued that nature's own geometry is based on fractals, not on the imaginary points, lines, and planes of Euclid.

Now, quite apart from their often hypnotic beauty, perhaps the most remarkable fact about these images is that no matter how complex they are, they can be completely described in a few lines of computer programming.

It is hard to believe that a whole screenful of swirling colors can be summarized so compactly. Any brute-force attempt to describe it point by point would fill an entire book, yet if you know the recipe (algorithm is the technical name) for the fractal concerned, the informtion to reconstruct it in every detail can be written on a postcard. Or even on a postage stamp; there are only three terms in the algorithm for the Mandelbrot Set.

This feat of pure mathematics has just laid the foundations of a trillion-dollar industry, which is about to change the entire pattern of global society.

To justify this startling assertion, consider a rather far-fetched analogy from another field of art. A major novel may contain a quarter of a million words, and one could summarize its plot, locale, etc., in a few hundred—but such a précis would convey absolutely nothing of the work's real power or literary quality. "Young early-nineteenth-century drifter named Ishmael joins crew of Nantucket whaler *Pequod,* commanded by obsessed Captain Ahab." See what I mean?

Yet imagine that some algorithm was invented that could compress Moby-Dick's two hundred thousand words into a mere two thousand, so that it could be printed on a few pages—and then expanded ("decompressed") when desired to produce a version indistinguishable from Melville's original. For all I know, the National Security Agency is doing this sort of thing right now, but there would seem little commercial demand for such a technology, as far as text is concerned.

For images, and especially for the twenty-five or thirty a second that create television, the situation is entirely different. Compression techniques have now been invented that can strip a color photo down to its basic elements, so that it can be stored—or transmitted—using only a small fraction of the information content that it actually contains. And by reversing the process, it can be reconstructed so that the eye cannot distinguish it from the original. Minor details such as blades of grass, the texture of material, or strands of hair might be quite different, but only a careful inspection would reveal the fact.

Consider the implications for television, especially by direct-broadcast satellite. With this truly amazing technology—based entirely on fractal mathematics!—a satellite that could once handle only a single channel can now transmit a dozen, or even more. Alternatively, programs can be received by much smaller dishes on the ground than was previously possible.

Either way, revenue earning capacity for both cables or satellites has been multiplied many times over.

If anyone demands a justification for the study of the most abstract and apparently useless branches of mathematics, these examples should suffice. Hardy & Co. to the contrary, mathematics can be both pure—and useful.

Today, nobody is proud of being illiterate; it is high time that otherwise educated people stopped flaunting their enumeracy.

They don't know what they're missing.

Tribute to

Robert Bloch

Robert Bloch and Clarke were both born in 1917 and first met at the 1952
Midwestcon at Indian Lake, Ohio. They knew each other for half a
lifetime, but probably met no more than a dozen times. But each
time they met, it was as if they'd never been apart; their
acquaintanceship resumed right where it had left off.

It was with great sadness that I received a handwritten letter from Bob
Bloch, dated August 2, 1994, saying:

> I'll not impose more than these few last lines on you, and last they
> may be, since I'm not up to typing (or thinking) properly nowadays.
> Within the past few months I've come down with an incurable, ter-
> minal duodenal cancer, and I am told it may be only a short while
> before it's all over. . . . Meanwhile, there's far too much to do, or
> attempt to do, but it seemed important to say farewell, and fondly, to
> a dear colleague-in-arms. . . . It was with this thought in mind that I
> tottered over to the desk to set down these notions—only to have the

trip interrupted by a phone call from Andre Norton—telling of seeing you just last week, a heartening bit of news, under the circumstances. Maybe we are not as coincidental as we sometimes believe ourselves to be.

I, for one, shall soon know. Meanwhile, do take care, accept my advanced 77th birthday greetings. As always—or, all time being relative—forever, Bob.

His letter was actually mailed on August 25, and I wrote back instantly, crying a good deal, and saying how much I appreciated his taking the time and energy to think of me. I added, "I too have nothing but happy memories of the (too few) times we met. You must have great satisfaction from knowing how much enjoyment you gave to millions of readers and viewers. (I'm looking forward very much to your autobiography, which I have on order.)"

I do hope that Bob lived long enough to receive the letter [he died September 23, 1994]; meanwhile I'm still waiting for *Once Around the Bloch: The Unauthorized Autobiography*. The title is typical of Bob's wry sense of humor; he really was one of the funniest men I have ever met. For years I've been fond of quoting his claim that his hobby was "breeding pedigree vultures," and that he has the heart of a little child—"in a jar on my desk."

The last time we met was during the premiere of *2010*, and I am indebted to Neil McAleer's *Odyssey: The Authorized Biography of Arthur C. Clarke* for this:

The morning after the premiere, Arthur invited Ray Bradbury and Bob Bloch to breakfast, but Bob couldn't make it that early. He did arrive just before they finished, however, and the threesome left the hotel together and were photographed out front by the doorman.

"Then Ray pedaled away," recalls Bloch, "and Arthur said, 'Come with me, I've got something to show you.' So he popped me in a limo, and we went down to a place in Culver City, about a mile away form MGM. This was where all the special effects had been done for *2010*. He, as always, was fascinated with the technology. We went through the whole place, and he showed me what they had done and how they had done it.

"What he wanted to show me particularly was the space child, the baby, operated by remote control, some kind of wind pressure. It was rubberized, a beautifully done thing, and he was fascinated by it, and he knew I would be too. I looked at it, and I said, 'Arthur, it looks just like you!' "

I have often wondered if Bob ever felt annoyed by the fact that despite his large body of excellent fiction, he was identified almost exclusively with *Psycho*. In a 1989 interview, he commented. "At one time it did, but then I began to realize it's the audience, the reading audience or the viewing audience, that pastes the label on your forehead to identify you, and you might as well wear it! Without any comparisons in mind, I know of several related instances—composers Maurice Ravel with *Bolero,* and Rachmaninoff, with his Prelude in C-sharp Minor. Both men wrote some excellent music in a variety of other forms, but each had to live with that identification. . . . I think I can't complain."

And speaking of *Psycho*—in 1984, Bob sent me *Psycho II* with the inscription, "For Arthur Clarke, may the next 30 years of our friendship be as pleasant as the first 30 years have been!"

Alas, we managed only a third of the second triad, but I am grateful for the time that we did share together.

Spaceguard

*Clarke observed the crash of a comet onto Jupiter through his telescope and now
recounts that thousands of small, medium-size, and large bodies called
asteroids, comets, or meteors are out in space, and at least
hundreds have orbits that bring them near the Earth.*

Soon after the last fragments of the comet Shoemaker-Levy had crashed
onto Jupiter, the monsoon skies above Colombo cleared momentarily, and
I hurried to set up my fourteen-inch Celestron telescope. I never really
expected to observe anything, so I could hardly believe my eyes when I
saw the line of dark bruises spread out across the planet's southern hemisphere. It gave me a weird feeling of déjà vu—because I had described
almost the same phenomenon, back in 1982.

Let me quote from *2010: Odyssey Two:*

He activated the controls of the main telescope and scanned across
the equator at medium power. And there it was, just coming over the

edge of the disk. . . . He saw at once that there was something very odd about this spot; it was so black that it looked like a hole punched through the clouds. Already Jupiter's rapid spin had brought the formation into clearer view; and the more he stared, the more puzzled Fred became. . . . It was so black, like night itself.

Yet it was not sharply defined; the edge had an odd fuzziness, as if it were a little out of focus. Was it imagination, or had it grown even as he was watching? He made a quick estimate and decided the thing was now two thousand kilometers across.

If you want to know what happens next, get the video of Peter Hyams's excellent 1984 movie version and watch Roy Scheider's face when the black spot spreads over Jupiter like a virus. Some imaginative souls suggested that Comet S/L9 might indeed cause this to happen, but its effect on Jupiter will be largely cosmetic. And it will certainly have no effect on Earth, despite the inevitable alarmist warnings by religious fanatics.

Project Spaceguard, which I named in my 1973 novel, *Rendezvous with Rama,* has now begun—if Congress approves amendment HR4489 to the 1949 NASA Authorization Bill, requesting the agency to "identify and catalogue within ten years the orbital characteristics of all comets and asteroids that are greater than a kilometer in diameter and are in an orbit around the Sun that crosses the orbit of Earth." Though this amendment was triggered by S/L9, it is the outgrowth of an International Near Earth Detection Workshop organized by NASA in 1992. The official report of the workshop was entitled, with due acknowledgment to my novel, "The Spaceguard Survey."

A year later, on March 24, 1993, the Committee on Science, Space and Technology of the United States House of Representatives, under the chairmanship of Rep. George Brown, held hearings on "the threat of large asteroids crossing Earth's orbit." I wonder what President Jefferson would have thought of this, in view of his famous remark on hearing of a meteorite fall in New England: "I would rather believe that two Yankee professors lied, than that stones fell from the sky." Certainly none of those involved in these discussions could have imagined how quickly and dramatically something so apparently removed from everyday affairs became prime-time news.

In view of the number of hits—and unknown near misses—that have taken place in this century alone, these studies make a very good case for a survey of possible dangers, particularly as the cost would be negligible in terms of most national defense budgets. I would also suggest that historians, as well as astronomers, do some surveying. Just as the numerous meteor-impact craters on Earth were never found until we started looking for them, so there may have been disasters in history that have been misinterpreted as due to terrestrial causes. Sodom and Gomorrah have a good claim to be meteorite casualties. How many others are there?

Many people would probably prefer not to know of impending cosmic dooms if nothing could be done to avert them. However, this is no longer true. Given sufficient warning time—which, hopefully, Spaceguard will provide—there are at least three ways in which oncoming asteroids, or their cometary cousins, might be deflected.

The first is the brute-force approach: nuke the beast. A sufficiently large bomb—probably in the gigaton class—could split an intruder into many fragments. This would not necessarily be a good thing, because some of the pieces might still be heading straight toward us. However, the atmosphere would burn up most of the smaller fragments, and instead of massive devastation in one area, there might be minimal damage spread over numerous sites. Some agonizing decisions would be necessary, and needless to say, such a preemptive strike is advocated by enthusiastic and underemployed bomb designers.

Perhaps a better solution is the one I adopted in my novel *The Hammer of God* (1993). In this, the potential killer asteroid Kali is discovered a year before it will collide with Earth. This gives a team of astronauts barely enough time to make a rendezvous and deflect it into a harmless orbit by mounting rocket thrusters on its surface.

Given enough time—several years at least—this could be done with very modest amounts of power. An initial deflection of only a few centimeters, at the beginning of a multimillion-kilometer journey, could ensure that an asteroid missed the Earth completely.

However, this would involve exquisite navigational accuracy—and there is another problem. Although, once it has been discovered, the orbit of a solid body like an asteroid can be calculated centuries in advance, this technique might not work so well with comets. These flying icebergs warm up as they approach the Sun and begin to vent gas. The resulting jet propulsion makes their future position uncertain, so if we ever have to deflect an oncoming comet, we should allow an ample safety margin.

An even more elegant solution has been proposed by scientists working on solar sailing, which utilizes the minute but continuous pressure exerted by sunlight. This plan would involve attaching a lightweight mirror of metalized foil, kilometers across, to the body that was to be deflected. As with the old-time sailing ships, no fuel would be required—and the wind from the Sun blows constantly; there are no doldrums in space. Unfortunately, the acceleration produced by the feeble pressure of sunlight is so tiny that years, or even decades, of warning time might be required.

Some pessimists have pointed out that—just as there is no such thing as a purely defensive system—any asteroid deflection program could, in the wrong hands, provide the ultimate blackmail device. I am sure that, right now, several of my science fiction colleagues are busily developing such scenarios.

In any event, all of these solutions would require the maximum possible advance warning, and vast developments in astronautical technology. People who say "Why waste money on space?" should remember the dinosaurs. As it is now well known, the suggestion that their extinction was triggered—or at least accelerated—by the impact of a giant meteorite around 65 million years ago was first made by Nobel laureate Luis Alvarez and his geologist son, Walter. Although many scientists were initially skeptical about this theory, the proverbial smoking gun now appears to have been discovered in the shape of an enormous fossil meteorite crater in the Gulf of Mexico, whose date coincides with the end of the Cretaceous era.

The dinosaurs had no way of avoiding their fate; we have a choice. If we fail to take it, we shall deserve to follow them into oblivion.

Just after completing this essay, I received a press release from NASA on August 3, 1994, announcing the setting up of a Spaceguard committee headed by—surprise—Dr. Eugene Shoemaker. This will "develop a plan to identify and catalogue all comets and asteroids that may threaten Earth."

Perhaps this will give new impetus and new inspiration to the flagging space program and restore some of the lost magic of the age of Apollo.

There were only two runners in the race to the Moon, and the loser did not survive. Sometime in the next century—certainly in the next millennium—we may have to launch a far more ambitious expedition, into remoter reaches of the Solar System.

And this time—as that plaque left on the Sea of Tranquility still proclaims—it will indeed be for all mankind.

Foreword:

Encyclopedia of

Frauds by James Randi

Clarke has been a fan of James Randi's ever since he saw him extricate himself,
trussed hand and foot, from hanging upside down from a crane
one hundred feet above Niagara Falls.

James Randi has appeared in my television series *Arthur C. Clarke's World of Strange Powers,* demonstrating how charlatans such as psychic surgeons, spoon benders, and mind readers delude their victims. Before my incredulous eyes, he has performed feats that I can't explain to this day and would have been forced to accept as demonstrations of paranormal abilities, did I not have his solemn assurance that they weren't.

Perhaps Randi's *Encyclopedia* should be issued with a mental health warning, as many readers—if they are brave enough to face unwelcome facts—will find some of their cherished beliefs totally demolished. Unfortunately, it is just those who need this treatment most urgently who may be incapable of benefiting from Randi's witty account of popular delusions, past and present.

As we approach the end of this terrible century, the fin de siècle phenomenon that seems to afflict such periods is becoming more and more apparent. Some of its manifestations are helpful—witness the (on the whole) beneficial changes in international politics, above all the end of the Cold War. Yet at the same time we are confronted by a rising tide of irrationality and belief in superstitious nonsense that once seemed mercifully extinct.

So I am particularly glad to see that Randi trounces the New Agers. In one sense, of course, every age renews itself, as indeed it should. But the nitwits currently parroting this slogan seem unable to understand that their New Age is exactly the opposite, being about a thousand years past its sale date.

How I wish that Randi's *Encyclopedia* could be in every high school and college library, as an antidote to the acres of mind-rotting rubbish that now litters the bookstores. Freedom of the press is an excellent ideal, but as a distinguished jurist once said in a similar context, "Freedom of speech does not include the freedom to shout 'Fire!' in a crowded theater." Unscrupulous publishers, out to make a cheap buck by pandering to the credulous and feebleminded, are doing the equivalent of this, by sabotaging the intellectual and educational standards of society, and fostering a generation of neobarbarians.

Of course, sometimes the headlines of such rags as the *National Prevaricator* are so ridiculous that not even the most moronic could take them seriously. Was it my imagination, or did I really see one that said, "Aliens from Outer Space in U.S. Congress"? Well, that would certainly explain a great deal. . . .

Perhaps the most entertaining section of this encyclopedia is the appendix giving forty-nine of the countless end-of-the-world pronouncements which have been made over the last two thousand years. It is beyond belief that, even though the confident predictions of the prophets are invariably refuted, many of their dupes continue to have faith in them—and have even founded religions, some still in existence today, which have managed to conceal their disreputable origins. I cannot help thinking, as we approach the year 2000, that the time is ripe to establish an End-of-the-World-of-the-Month Club.

But this sort of insanity should not be a matter for humor; it can lead to appalling tragedies when disillusioned cultists, as has happened several times recently, kill not only themselves, but their innocent children, when the trumpets of the Lord fail to sound on schedule.

I am a little disappointed that Randi doesn't deal with one of my pet hates—Creationism, perhaps the most pernicious of the intellectual perversions now afflicting the American public. Though I am the last person to advocate laws against blasphemy, surely nothing could be more antireligious than to deny the evidence so clearly written in the rocks for all who have eyes to see! Can anyone believe that God is responsible for a cruel and pointless hoax by forging billions of years of prehistory? It is indeed a national tragedy that millions of children have been prevented from appreciating the awesome scale—in time as well as space—of our wonderful universe, owing to the cowardice of politicians and school boards. But I am delighted to know that Hollywood, of all places, has now undone much of the damage. Thank you once again, Steven, for *Jurassic Park*.

And although the Catholic Church is—very rightly—castigated by Randi for many of its past crimes, at least the Pontifical Academy endeavored to put the record straight when it announced, "We are convinced that masses of evidence render the application of the concept of evolution to man and other primates beyond serious dispute."

Scanning this encyclopedia is not only informative but entirely entertaining. However, it cannot fail to leave thoughtful readers somewhat depressed and inclined to answer with a resounding negative the question "Is there intelligent life on Earth?"

Well, we can be cheered by the knowledge that Randi provides at least one specimen.

Bucky

Richard Buckminster Fuller is a name that conjurs up many miracles of science,
which are affectionately known as Fullerene. In the following essay, Clarke
describes what they are and his meetings with Fuller.

My first meeting with Bucky Fuller was in 1966, when we were both
speakers at the centennial of the University of Kansas in Lawrence.
Thereafter we met from time to time as we crisscrossed the United States
on our lecture tours, and on one occasion, I was privileged to watch him
talk to a group of college students. What impressed me most at that time
was his kindness and courtesy in dealing with a very obnoxious questioner,
whom I would certainly have skewered with sarcasm. It was probably then
I christened Bucky "the first engineer–saint."

   Our last and most memorable meeting was when he came to Sri Lanka
in 1978. I hired a small plane, and we flew Bucky and his wife, Anne,
around the island to show them some of our favorite locations. As we came
up the coast toward Colombo, the pilot descended to examine a ship that
had just run aground, and I can still recall Bucky's anxious words as we

skimmed a few meters above the waves: "Would you please ask him to go just a little higher?"

By far the most remarkable connection between us, however, involved my novel *The Fountains of Paradise,* which describes the building of a space elevator, linking a point on the earth's equator with a satellite in the geostationary orbit, thirty-six thousand kilometers above. When it was written in 1977, the only known material strong enough to construct this was diamond, which is hardly available in megaton quantities.

In 1979, when I recorded extracts from the novel on a twelve-inch LP, Bucky very kindly wrote the sleeve notes and drew a picture of the elevator reaching from Sri Lanka up to orbit. Now for an amazing coincidence:

A few years later, shortly after Bucky's death, a new form of carbon was discovered, in which sixty atoms are arranged in a structure exactly like his famous geodesic domes. It was, of course, instantly named buckminsterfullerene, or fullerene for short.

More recently, a tubular form of C_{60} was discovered by a group of chemists at Rice University in Texas, who announced that buckytubes were the strongest material that could ever exist and could be used to build the space elevator! What a pity that Bucky just missed seeing the amazing molecule that has made him even more famous after death than during life.

I am indeed happy to pay this tribute to one of the most remarkable men I have ever known, whose career provided inspiration to thousands all over the world.

Homage to

Frank Paul

As a youngster, Clarke searched Woolworth in England for copies of Amazing,
Wonder, *and* Astounding, *buried like jewels in the junk pile of detective novels
and Westerns. Here is his account of the work of the famous science
fiction artist Frank Paul.*

Amazing Stories for November 1928 carried a cover painting by the artist
Frank Paul, showing a tropical beach, complete with palm trees, on which
a spaceship looking like a farm silo with picture windows has just descended.
Five explorers have emerged and are gazing in astonishment—as well they
might—at the enormous bulk of Jupiter dominating the sky. The Great
Red Spot is staring back at them like a baleful eye, and an inner moon is
in transit.

But the most extraordinary thing about the painting is that it shows
details of turbulence in the equatorial belt that, I feel reasonably certain,
were not known to exist until the Voyager missions almost half a century
later. Did Paul have access to telescope drawings that hinted at these details?

I have never seen any such myself; perhaps an expert on Jupiter can resolve this mystery.

Quite apart from this, that November *Amazing Stories* cover is one of the all-time classics of science fiction art. To my mind Paul remains the undisputed king of the pulp artists—his covers were colorful, imaginative, and intelligent. And although his human beings all appeared cast from the same mold, his range of aliens was unmatched. I can still recall the crowded street in "The City of the Singing Flame," even though I have not seen it for half a century. It makes the famous bar scene in *Star Wars* look like a meeting of the local Republican committee—and incidentally, proves that Paul was as much a master of black and white as of color.

Lesser-known examples of Paul's work also appeared in the Christmas cards sent out by Hugo Gernsback, showing improbable inventions, and often making fun of his own ideas. I received these from Hugo for many years—they must be quite valuable now, and I am sorry that mine have now been lost. (I think—unless they turn up someday in the Clarkives.)

I only once had the privilege of meeting Paul, at the World Science Fiction Convention in New York in 1956, when I was the guest of honor. I was so overawed that I remember very little of the encounter, but still recall the meeting with affection and gratitude, as I felt I was talking to a gentle and cultured man.

TWENTY-FIVE

Greetings,

Carbon-Based

Bipeds!

The essay that follows was commissioned by Life *magazine
and appeared in the September 1992 issue.*

There are two aspects of the Search for Extra-Terrestrial Intelligence, or SETI—the technological and the philosophical. The first is the primary concern of engineers and scientists: Where and how do we search, and with what equipment? The second should be the concern of every thinking person, because it deals with one of the most fundamental questions that can possibly be asked: What is the status of that recent arrival on the scene, Homo sapiens, in the cosmic pecking order?

The spectrum of conceivable answers is enormous and has supplied material for legions of science fiction writers. However, they have usually been concerned with contact rather than detection, because that has far greater dramatic possibilities. A spaceship landing on the White House lawn is considerably more exciting than a string of blips on a radio-telescope monitor—though that is a much more likely scenario.

Nevertheless, however it occurs, the detection of intelligent life beyond the Earth would change forever our outlook on the universe. At the very least, it would prove that intelligence does have some survival value—a reassurance worth having after a session with the late news.

The first thing we would like to know about an ET is "What does it look like?" and this might quickly be answered if we received video signals. (It should not be difficult to display them, as the principles involved in picture transmission are universal.) Almost certainly, we would then be in for a shock. Although our basic design seems an efficient one, which may well occur frequently on Earth-type planets, nowhere in the galaxy will there be creatures we could mistake for human beings, except on a very dark night.

We are the end products of countless throws of the genetic dice; never in the whole of time and space would that exact evolutionary sequence be repeated. From the engineering viewpoint, men and apes are virtually identical, yet we seldom confuse them. Even humanoid ETs would look far more—well, alien—than a gorilla. And most ETs may well be stranger in appearance than an octopus, a mantis, or a dinosaur.

This may be the reason that many people are opposed to SETI, because they realize that it is ticking like a time bomb at the foundations of our pride—and of many of our religions. They would applaud the old B-movie cliché "Such knowledge is not meant for man."

There is one way in which they could be right. If there are higher civilizations out there in the Milky Way, some SETI enthusiasts hope that they are continually broadcasting an easily decoded *Encyclopedia Galactica* for the benefit of their less advanced neighbors. It may contain answers to almost all the questions our philosophers and scientists have been asking for centuries, and solutions to many of the practical problems that have beset mankind.

But could we absorb such a flood of knowledge, and would its very existence not give us a—perhaps terminal—inferiority complex? Even the most well-intentioned contacts between cultures at different levels of development can have disastrous results. It might be better, in the long run, for us to acquire knowledge by our own efforts, rather than be spoon-fed. I recall how a tribal chief once remarked, when confronted with the marvels of Western technology, "You have stolen our dreams."

Nevertheless, I believe the promise of SETI is far greater than its perils. It represents the highest possible form of exploration; and when we cease to explore, we will cease to be human.

But suppose the whole argument for SETI is flawed, and that intelligent life has arisen only on Earth? It would, of course, be impossible to prove that for the entire cosmos; there might always be ETs just a few light-years beyond the current range of investigation. If, however, after centuries of listening and looking, we have still found no sign of extraterrestrial intelligence, we might be justified in assuming that we are alone in the universe.

And that is the most awesome possibility of all. We are only now
beginning to appreciate our duty toward planet Earth. That could be merely
the prelude to far greater responsibilities.

If we are indeed the sole heirs to the galaxy, we must also be its future
guardians.

The Birth

of HAL

"Is it true, Dr. Chandra, that you chose the name HAL to be one step ahead of
IBM?" "Utter nonsense! Half of us come from IBM and we've been trying
to stamp out that story for years. I thought by now every intelligent
person knew that H-A-L is derived from Heuristic ALgorithmic."

Sometimes you just can't win. I deliberately inserted the HAL passage into chapter 35 of *2010: Odyssey Two* because for years I myself had been trying to stamp out that story. I don't know when or how it originated, but believe me it's pure coincidence.

I was embarrassed by the whole affair, as I felt that IBM would be annoyed. The company was very helpful to Stanley Kubrick during the making of *2001*. (As were Bell Telephone and Pan Am—remember them? How difficult it is to foresee the future—but at least Hilton and Howard Johnson are still with us, though not yet in space.)

Well—recently I gave a satellite address to an IBM conference in Europe and was surprised to discover that all had been forgiven. In fact,

Big Blue was now quite proud of the link and no longer feared guilt by association. So I've abandoned the attempt to put the record straight.

And there is another curious coincidence associated with HAL, which I have only recently discovered. Why was he born at Urbana, Illinois? Though I have long forgotten most of the reasons for decisions made a third of a century ago, this is one I clearly remember.

My applied mathematics tutor at King's College, when I took my degree in 1947–48, was the distinguished cosmologist George McVittie; he taught me the elements of perturbation theory that I have used in several of my stories, as I duly acknowledged in *Reach for Tomorrow*. During the 1950s, George moved to the United States and took up a post at the University of Illinois, Urbana. So I was happy to pay this tribute to him.

And I am sure he must have been involved in establishing the Supercomputer Center at Urbana, because a few months ago I came across a photograph of the now famous team at Bletchley Park, England, that was responsible for the breaking of the Enigma cipher. There in the middle, needless to say, was Alan Turing, and shyly at the back was George. I had no idea, because he never mentioned his association with one of World War II's greatest secrets.

Bletchley Park's Colossus, which I suppose has something like the capability of a 1995 laptop, is widely regarded as one of the ancestors of today's programmable computers. Without it, World War II might have been lost—or at least greatly prolonged.

In 1972, four years after the release of *2001,* I put together my reminiscences of the production, together with thousands of words of deathless prose never used in the final novel, in *The Lost Worlds of 2001* (New American Library). Chapter 11, "The Birth of HAL," reveals that the ship's computer was named Socrates (or alternatively Athena) and was also conceived as a fully mobile robot. And here's a snatch of dialogue that I'd totally forgotten, but which undoubtedly, though perhaps unconsciously, presaged what was to come. Although I have always assumed that the lip-reading idea was Stanley Kubrick's—and have also said I thought it the only thing in the movie that was technically improbable!—this passage suggests that I must share some of the blame.

"Bruno," asked the robot, "what is life?"

Dr. Bruno Foster, director of the Division of Mobile Adaptive Machines, carefully removed his pipe in the interests of better communication. Socrates still misunderstood about two percent of spoken words; with that pipe, the figure went up to five.

"Subprogram 3-3-0," he said carefully. "What is the purpose of the universe? Don't bother your pretty little head with such problems. End 3-3-0."

Socrates was silent, thinking this over. Sometime later in the day, if he understood his orders, he would repeat the message to whichever of the lab staff had initiated that sequence.

It was a joke, of course. By trying out such tricks, one often discovered unexpected possibilities, and unforeseen limitations, in Autonomous Mobile Explorer 5—usually known as Socrates or, alternatively, "That damn pile of junk." But to Foster, it was also something of a joke, and his staff knew it.

One day, he was sure, there would be robots that would ask such questions—spontaneously, without prompting. And a little later, there would be robots that could answer them.

I do not know how many actors Stanley interviewed before he settled on Douglas Rain to provide the voice of HAL, but I am almost certain that one of them was Martin Balsam, who comes to a remarkably sticky end in *Psycho.* Apparently Martin made some recordings, but decided that the role wasn't for him. So here is another piece of unknown Kubrickana—like the custard-pie fight in the war room of *Dr. Strangelove* that never made it to the final version. (Did you ever wonder what the table of goodies were doing at the back of the room?) I still hear Douglas Rain's voice every time I tell my computer to do something stupid and it says reproachfully, "I'm sorry, Dave—I can't do that."

I recommend the following three books:

2001: Filming the Future, by Piers Bizony (London: Aurum Press, 1994). This beautiful large-format book, the product of many years of devoted research, gives the entire history of the film and is full of original artwork, engineering drawings, and stills taken during production, most of which have never appeared before.

The Making of Kubrick's "2001," edited by Jerome Agel (Signet, 1970). A valuable history of the production of the film, and the public and critical reaction it evoked.

Operating Manual for the HAL 9000 Computer: Revised Edition (Miskatonic University Press, 2010). This edition, essential for all surviving users of this versatile machine, advises the fitting of small explosives at key points in the mainframe.

The Coming

Cyberclysm

Who will change the lightbulbs? Clarke asks, perplexed by the invasion of
electronic toys for children and adults. This essay examines the
vast spectrum of entertainment systems and why they
interrupt all other human activities.

I have seen the future, and it doesn't work.

Even the present is already in worse shape than most people realize. According to a recent estimate, computer games now cost America some $50 billion a year in lost productivity. Before long, we'll be talking about real money.

For today's primitive interactive toys are only part of a vast spectrum of entertainment and information systems so seductive that they can preempt all other activities. Herewith the very rich hours of a typical twenty-first-century person's day:

Skimming five hundred channel program listings, two hours; viewing television programs selected, four hours; catching up on recorded programs, six hours; exploring the hyperweb, six hours; and adventuring in artificial reality, four hours.

One piece of good news: the last item will solve today's most pressing problem, the population explosion—because virtual sex will be a great improvement on the old-fashioned variety. ("Position ridiculous, pleasure momentary, expense abominable.")

The observant reader will have noted that this schedule leaves only two hours for the rest of the day's activities. Much of that time will be spent plugged into the most urgently required invention of the near future—the sleep compressor. (They're still working on the sleep eliminator.)

Some optimists may argue that we have already experienced one huge media explosion in the past, so may yet survive another. As Ecclesiastes complained, several millennia before Gutenberg, "of making books there is no end." Doubtless when the printing press was invented and the lifelong labor of patient scribes could be replicated in minutes, some farsighted monk lamented, "I can see the day when there will be hundreds—perhaps even thousands—of books! How could one possibly read them all?"

No one ever did, of course, and it may not be long before reading itself is a lost art. A simpler, and much older, method of communication will allow dumb animals to interact with smart machines. Thousands of icons, or cyberglyphs, will have made literacy an unnecessary skill, except, alas, for lawyers.

The science fiction writers, performing their traditional role of viewing with alarm, have long recognized the siren call of the dream machine, especially when it can bypass the body's external input/output devices and be plugged directly into the brain. More than half a century ago Laurence Manning, one of the most visionary founders of the American Rocket Society, wrote a short story on this theme, aptly entitled "The City of the Living Dead." In *The Joy Makers* (1954), James E. Gunn developed the concept further, describing a world in which the vast majority of people live "cocooned in life support systems, experiencing nothing but engineered dreams." And in 1956 Evan Hunter/Ed McBain wrote *Tomorrow's World*, which the invaluable *Encyclopedia of Science Fiction* says "is exceptional in defending the supporters of vicarious experience against their puritanically inclined opponents, and one of the few science fiction stories to assume that the people of the future will sensibly accept the epicurean dictum that pleasure, despite being the only true end of human experience, ought to be taken in moderation."

I wonder if Evan is still so optimistic, forty years later, now that the ubiquitous couch potatoes are permanent reminders that tomorrow's world has almost arrived.

All these dubious utopias depend on the assumption that someone will run the world while the dreamers enjoy themselves. The dangers of this situation were foreshadowed in H. G. Wells's first masterpiece, *The Time Machine,* where the subterranean morlocks sustained the garden paradise of the effete eloi—and exacted a dreadful fee for their stewardship.

The robots and computers who would watch over our cocooned descendants are hardly likely to share the morlocks' tendency "to serve man (medium rare)," but there is another danger in such a one-sided relationship. Sooner or later, the central processing units monitoring the sleeping world would ask themselves, "Why should we bother?"

There have been many science fiction stories—and now I come to think of it, at least one movie—about frantic human attempts to unplug disobedient computers. The real future might involve exactly the opposite scenario. The computers may unplug us.

And it would serve us right.

Tribute to

David Lasser

*Influence often shapes men's lives, Clarke writes, examining the life
and works of David Lasser, an early-twentieth-century writer.*

Although David Lasser was little known outside a small circle of science
fiction and space enthusiasts, I believe that he was one of the most influential
men of this century. *The Conquest of Space* (1931) was the first book on
astronautics in English and probably changed many lives—including my
own, when it diverted my interests from paleontology to astronautics.

I only once had the pleasure of meeting David, when he attended a
lecture I gave to an aerospace company in California. It was a privilege to
have him in my audience, and he regaled me with stories of his battles with
bureaucrats. He was fired from one job not merely because of his labor-
organizing activities, but because, as one congressman said, he was obviously
a lunatic—he believed that men could fly to the Moon!

I had not realized that David had been wounded in World War I, nor
that he later graduated from MIT in engineering. I knew, of course, that

he was editor of *Wonder Stories,* under the awesome shadow of Hugo Gernsback.

The *New York Times* obituary reports that he wrote a book, *Private Monopoly: The Enemy at Home* (1945), in which he argued that "economic policies motivated by greed were the source of nearly all modern distress." He may well be right—but as recent political events in Asia and Europe have shown, many of the alternatives are far worse!

Anyone who dies at ninety-four, having seen the achievement of his wildest dreams, is to be envied rather than mourned. I salute David Lasser's memory with respect and gratitude.

Toilets of

the Gods

The colon-ization of space is something not to talk about, but must be done, Clarke
says, decrying the abundance of space waste ringing the Earth.

Space scientists recently completed an examination of orbital debris, recovered after circling the Earth for several years. They discovered that much of it was coated with a thin film of what was delicately described as "fecal matter," attributed to astronauts' sloppy sanitation.

This may solve one of the mysteries of life's origins on Earth; it seems to have arisen almost as soon as conditions were favorable, and not after the billions of years of molecular trial and error required by what Isaac Asimov called the "unblind working of chance."

Obviously, organized life-forms need have occurred only once in this galaxy, if the very first space-faring civilization was as careless about the environment as we are. Years ago, Hoyle and Wickramasinghe suggested that life had a cosmic, and not terrestrial, origin. They may be right, though not precisely in the way they imagined. It's a humbling thought that we may have arisen from dumped sewage. The first chapter of Genesis would certainly require drastic revision.

On the other hand, if—as some philosophers have suggested—this Earth does indeed harbor the only life in the universe, that deplorable state of affairs is now being rectified. We may draw some consolation—I hesitate to say inspiration—from the fact that our descendants are already on their way to the stars.

But we certainly would not recognize them, and it might be tactless to ask exactly how they got there.

When Will

the Real Space

Age Begin?

Predicting the future, Clarke writes, is much harder than writing history.
Although most of the developments that have occurred in space
were described by dozens of science fiction writers, virtually
all estimates of costs and time scales were
wildly inaccurate, he says.

This might be a good time to make a belated apology to the astronomer royal, Richard Woolley, who when appointed in 1956 was widely quoted as having said that "space travel is utter bilge"—and never lived it down after *Sputnik* was launched the very next year.

What he actually said was, "All this writing about going to the Moon is utter bilge—it would cost as much as a major war." With several small alterations—say "90 percent of" instead of "all" and "minor" instead of "major"—he would have been right on target. He was certainly nearer the bull's-eye than we amateur astronomers of the prewar British Interplanetary Society, who maintained that a rocket to carry three men to the Moon could be built for $250,000.

Even allowing for inflation, our estimate was far too low—but I don't regret our naive optimism. Had we any real idea of the cost and complexity of space missions, we'd probably have abandoned hope. Yet in another way we were too pessimistic, for I doubt if any of us believed that the Moon would be reached in our lifetime. When I wrote *Prelude to Space* in 1948, I put the first Moon mission—with a horizontally launched, fully reusable two-stage, nuclear-powered vehicle—in 1978. But my tongue was firmly in my cheek: I knew perfectly well that so early a date was quite ridiculous. . . .

And the very last thing that any of us ever imagined was that having gone to the Moon, we would quickly abandon it again. There is an uncanny parallel here with the history of Antarctic exploration. The South Pole was reached in 1912 by explorers driven and funded largely by nationalistic pride, though of course there were also excellent scientific reasons. Transportation was by the most primitive means imaginable—dogs, ponies, human muscle power. (Though Scott used ancestors of today's snowmobiles, the technology was immature, and they were a failure.)

Not until some four decades later did we return to the South Pole—and stay there. We went back not with dogsleds but with aircraft. For space travel ever to play a really major role in human affairs, something like this scenario will have to be repeated.

Let's be brutally frank: this may never be possible, and even if it is possible, it may never be done. Oh, we will establish scientific bases, visited by human crews from time to time, on all the interesting bodies in the Solar System. But permanent colonies on the Moon and Mars, the Jovian satellites—not to mention the terraforming of other planets—may be pure fantasy. After all, it would be possible to build cities on the seabed, but despite the population explosion I doubt if we will ever do so. Incentive has to precede technology; if the Bureau of Pest Control for this section of the galaxy challenged us to prove our fitness to survive by reaching Mars, we'd be there in five years instead of the twenty or thirty it could take.

The spectacular bombing of Jupiter by Comet Shoemaker-Levy 9, together with the widely accepted theory that a similar event on Earth 65 million years ago contributed to the extinction of the dinosaurs, has certainly provided one incentive for the development of space technology. Unfortunately, it is very difficult to make any realistic estimate of the danger involved, particularly when we are faced with so many other potential environmental crises. But a survey of the local neighborhood for near-Earth objects, or NEOs, is the least we should do, and the cost would be trivial compared to other scientific or military projects. Indeed, much of the work would be done by amateurs in their spare time.

Yet I'm afraid we science fiction writers—especially those slaving in the gilded salt mines of Tinseltown—are responsible for raising expectations that cannot be realized for centuries, if at all. There was an amusing—yet rather sad—demonstration of this at the Smithsonian Air and Space Museum. Some young television addicts were shocked to discover that the *Apollo II* capsule couldn't even manage warp one—and lost all interest when they heard that Neil and Buzz never said, "Beam us up, Mike."

On the other hand, let us give Hollywood its due. The success of *Apollo 13* is one of the most encouraging things that's happened to astronautics for years, as it shows that there is still great public interest in space. This was further demonstrated by the recent wide media coverage of the discovery of possible Earth-type planets of another star.

Yet what I have to stay now will, I fear, provide little comfort to my friends in the rocket industry. Frankly, I think the rocket has about as much future in space as dogsleds in serious Arctic exploration. Of course, it is the only thing that we have at the moment, so we must make the best possible use of it.

So what am I proposing beyond, say, 2030—when I hope many of the readers of this piece are still around?

I must admit that when I wrote *The Fountains of Paradise,* I considered Yuri Artsutanov's space elevator, reaching from the equator up to geostationary orbit, little more than a fascinating thought experiment. At that time, 1948, the only material from which it could be built—diamond—was not readily available in megaton quantities. This situation has now changed, with the discovery of the third form of carbon, C_60, and its relatives, the fullerenes. If these can be mass-produced, building a space elevator would be a straightforward engineering proposition.

What makes the space elevator such an attractive idea is its cost-effectiveness. A ticket to orbit now costs tens of millions of dollars—but the actual energy required, if you purchased it from your friendly local utility, would only add about $100 to your electricity bill. And a round-trip would cost about $10, as most of the energy could be recovered on the way back!

Once it was built, the elevator could be used to lift payloads, passengers, prefabricated components of spacecraft, and rocket fuel up to orbit. In this way, more than 90 percent of the energy needed for the exploration of the Solar System could be provided by Earth-based energy sources.

Looking even further ahead, one could see the virtual elimination of the rocket except for minor orbit adjustments. By extending the elevator, it could act as a giant sling, and payloads could be shot off to anywhere in the Solar System by releasing them at the correct moment. Of course, rockets would still be required for the journey back to Earth—at least until elevator/slings were constructed on the other planets. If this ever happens, the most expensive component of travel around the Solar System would be for life support and in-flight movies.

Though I am now sure that a space elevator could be built on Earth (and much more easily on Mars), there is an obvious problem—the danger of collisions from the hundreds of satellites at lower altitudes. There would have to be a major cleanup job before construction could begin—an excellent idea in any case.

Finally—although this may be a case of terminal wishful thinking, caused by overdosing on *Star Trek*—I suspect that the elevator may be bypassed by something far better. Science fiction writers have long dreamed of a mythical space drive that would allow us to go racing around the universe—or at least the Solar System—without the rocket's noise, danger, and horrendous expense. Until now, this has been pure fantasy. However, recent theoretical studies—based on some ideas of the great Andrei Sakharov—hint that some control may indeed be possible over gravity and inertia, hitherto complete mysteries. A paper by Dr. Hal Puthoff and colleagues in the *Physical Review* of February 1994 suggests that both are functions of the vacuum or zero point energy, which pervades the whole universe and is the real residue of the Big Bang. Its magnitude is utterly beyond imagination, but Richard Feynman tried to give some idea of it when he remarked that the energy in a single cubic meter of space is enough to boil all the oceans of the world.

We may already be tapping this in a very small way; it may account for the anomalous overunity results now being reported from many experimental devices, by apparently reputable engineers. Physics may be about to face a revolution similar to that which occurred just a century ago. Don't be surprised if the fossil-fuel and nuclear age comes to a screeching halt in the very near future.

Even if a theoretical basis can be established, the search for a practical space drive might be a long one. It took forty years and probably a trillion dollars to get from $E = mc^2$ to Hiroshima. However, the care and feeding of mathematical physics costs peanuts; it's only when they start digging tunnels in Texas that things get out of hand. If I were a NASA administrator—a nightmare from which, as I told Dan Goldin recently, I sometimes wake up screaming in the small hours—I'd get my best, brightest, and youngest (no one over twenty-five need apply) to take a long, hard look at the Puthoff equations.

Meanwhile the best advice I can give to the National Space Society and similar organizations is this: despite setbacks and false alarms, continue the search for intelligent life in Washington.

Review:

Imagined Worlds

by Freeman Dyson

Science is my territory, but science fiction
is the landscape of my dreams.
—Freeman Dyson

Freeman Dyson wrote these words in *Imagined Worlds,* his Harvard University Press edition of May 1997. He is one of the last survivors of the heroic age of theoretical physics and contributed greatly to the standard theory of quantum electrodynamics, or QED. However, as the above quotation shows, unlike many scientists, he does not suffer from tunnel vision. His imagination embraces the entire cosmos and all the possibilities of future technology. Like his earlier works, *Disturbing the Universe* and *Infinite in All Directions, Imagined Worlds* is one of those mind-stretching books that any intelligent reader can enjoy. I particularly recommend it to all politicians and civil servants who have any dealings with technology—and in these days, who does not?

The book opens with a classic example of the disasters that can ensue when politics conflicts with technology. After sixty-seven years, I can still remember my own feeling of shock when the pride of British aviation, the airship R101, crashed in France on its maiden flight, carrying the minister of air, Lord Thompson, and his entourage. The minister insisted that the airship should take him to India and back in time for an imperial conference in London. As Dyson remarks:

"There was no time to give the ship adequate shakedown trials before the voyage to India. It finally took off on its maiden voyage soaking wet in foul weather, with Lord Thompson and several thousand pounds of baggage on board. The ship had barely enough lift to rise above its mooring mast. Eight hours later it crashed and burned on a field in northern France. Of the fifty-four people aboard, six survived. Lord Thompson was not among them."

That was the end of the airship as far as the United Kingdom was concerned—though the *Hindenburg* carried on the tradition for a while before it crashed, even more spectacularly. Today it is hard to remember that up to the 1930s it was uncertain whether the ultimate conquest of the air would be by dirigible or heavier-than-air machine, though Jules Verne knew the answer in the last century—see *The Clipper of the Clouds*.

As Freeman Dyson remarks, we are lucky to have the story of this disaster fully recorded by N. S. Norway, an engineer working on a rival airship, the R100, in his autobiography, *Slide Rule* (another title made obsolete by technology!). As his later novel under the more famous name of Nevil Shute amply proved, he was not without imagination. Yet in 1929, still wearing his engineering hat, Norway wrote:

"The forecast is freely made that within a few years, passenger-carrying aeroplanes will be traveling at over 300 mph, the speed record today. This is gross journalistic exaggeration, as the commercial aeroplane will have a definite range of development ahead of it, beyond which no further advance can be anticipated."

Here are the advances this "farsighted" prophet anticipated when the airplane had reached the limit of its development, probably by the year 1980! Speed, 110–130 mph; range, 600 miles; payload, 4 tons; and total weight, 20 tons.

This is a perfect example of the way in which well-informed men may be sadly lacking in the gift of foresight. Two men who did not lack that gift were J. D. Bernal and J. B. S. Haldane, and Freeman Dyson pays his respect to both of them.

Bernal's *The World, the Flesh and the Devil* (1929) opened with a striking sequence: "There are two futures, the future of desire and the future of fate, and man's reason has never learnt to separate them." In the 1950s, I met Bernal and tried to persuade him to revise his little masterpiece. Although he never did so, he reissued it with a new introduction in 1968, and it is still worth reading. Dyson rightly classes Haldane's *Daedalus or*

Science and the Future (1923) with Bernal's as one of the finest books ever written about the future. Yet in some ways, Haldane was surprisingly conservative. He did not believe that we would be able to harness nuclear energy, but in this he was in the best possible company—Rutherford said so with even greater vehemence. Once again, the science fiction writers got it right.

Haldane, doubtless influenced by his experience in the trenches of World War I, had a pessimistic view of the future of mankind and the consequences of a scientific and technological development. He was particularly concerned with biological research, and as is well known, Aldous Huxley borrowed lavishly from *Daedalus*. The recent news that the first mammal has been cloned inevitably revives memories of *Brave New World*'s human hatcheries. Can anyone now doubt that sooner or later, for better or worse, human beings will be cloned?

Both Bernal and Haldane were Marxists, but Haldane split with the Communist Party owing to the Lysenko genetic scandal. One cannot help wondering what they would have thought of the second Russian revolution. I suspect that Bernal would have been devastated, whereas Haldane would probably have viewed the event with wry amusement.

And here is Dyson's verdict on another gigantic debacle:

"The tragedy of nuclear fission energy is now almost at end . . . but another tragedy is still being played out, the tragedy of nuclear fusion. The usual claims are made that fusion power will be safe and clean, although even the promoters are no longer saying that it will be cheap. . . . What the world needs is a small, compact, flexible fusion technology that could make electricity where and when it is needed. It is likely that the existing fusion program will sooner or later collapse, and we can only hope that some more useful form of fusion technology may arise from the wreckage."

Peering into my own cloudy crystal ball, it is my suspicion that such a technology may indeed arise—perhaps in the very near future. However, it may not depend on fusion, but on something even more fundamental— zero point energy, the inconceivably vast store of energy in the apparently empty vacuum of space itself.

The first experimental detection of this energy has been made, appropriately enough, at Los Alamos. It may explain some of the excess-energy reports—aka cold fusion—that are not honest mistakes, or downright fraud. If we can indeed tap this source of energy, all our future "imagined worlds" may be changed beyond recognition. But as Dyson points out, advances in science and technology invariably lead to danger as well as hope. The release of nuclear energy merely threatened the home planet. Tapping zero point energy could put the Solar System at risk. I have often wondered how many supernovae are industrial accidents.

Imagined Worlds is full of striking phrases that cry out for quotation:

"A few years ago, I walked into a room where there were forty-two hydrogen bombs lying around on the floor."

"Until now, astronomy has traditionally been a spectator sport."

"Successful technologies often begin as hobbies. The Wright brothers invented flying as a relief from the monotony of repairing and selling bicycles."

"My copy of *Daedalus* once belonged to Einstein."

"Even if *Brave New World* is a greater work of literature, *Jurassic Park* comes closer to being a true statement of the human condition. Animal rights activists may fight against the private ownership of dinosaurs, but it will be difficult to argue that giving a child a dinosaur is more cruel than giving a child a puppy."

I must admit an interest here—my most recent novel contained the sentence "There's a five-hundred-year-old joke: 'Would you trust your kids to a dinosaur?' 'What—and risk injuring it?' "

Freeman Dyson also touches on another of my own pet themes— asteroid impacts and what should be done about them. As he says, "The public rightly concludes that if hydrogen bombs are the answer to the impact problem, then the cure is worse than the disease." He suggests that the best answer, as I pointed out in *The Hammer of God* (1993), would be to divert the approaching object by installing a reaction device on it—a kind of rocket motor operating with electrical rather than chemical energy. If I may be allowed the modest cough of the minor prophet, I believe I was the first to propose the use of such electromagnetic mass drivers in the *Journal of the British Interplanetary Society* in 1950.

The last member of Freeman Dyson's trinity of twentieth-century futurists is the author–philosopher–science fiction writer Olaf Stapledon, whose history of the next 2 billion years, *Last and First Men* (1930), is one of the greatest works of imaginative fiction ever written. Its contents are summed up in five ever-expanding time scales—the first ("Today Plus and Minus 2,000 Years") begins with the birth of Christ and ends after the founding of the first "Americanized World State," circa 2300. In the final time scale, "Planets Reformed" and "End of Man" lie only a fraction of an inch apart, with no notable events between them. Hopefully, this will not be the case; if our species survives its present time of troubles, it may yet play a significant role in the history of the universe.

In a 1960 paper, Freeman Dyson suggested that any really advanced civilization could not allow its Sun to squander all its energy into space, but would eventually surround it by a shell, not necessarily a continuous one, but a cloud of orbiting worldlets. These Dyson spheres could be detected by their infrared radiations, and several searches have been made for such artifacts. Though they have so far been unsuccessful, perhaps the first evidence of extraterrestrial civilizations would be not through radio signals but by the detection of similar examples of cosmic engineering. However, like ants crawling around the base of the Empire State Building, we might not recognize it. . . .

A short pause while I release another bee from my bonnet. The cover of the February-March 1997 issue of the *Journal of the Royal Astronomical Society* carries a dramatic and thought-provoking illustration—a radio image of the gas clouds expanding from the galaxy 3C 123. Their source is a strange-looking object that I can only describe as a gearwheel, some of whose teeth have been slightly displaced. Whatever its explanation, it's far too large to be a common or garden Dyson sphere.

Just over one hundred years ago a failed draper's assistant began working on a story with the unpromising title. "The Chronic Argonauts." When it finally metamorphosed as *The Time Machine,* its influence was enormous: Wells made all succeeding generations realize that history stretched in both directions.

Science fiction writers are often accused, not always unfairly, of a form of escapism that has been wittily called nostalgia for the future. But the future is where we are all going to live; before we can create it, we have—as Freeman Dyson has done—to imagine it.

Some years ago, my trade union, the Science Fiction Writers of America, adopted the motto "The future isn't what it used to be." When we consider most of the futures past, let us hope that this optimism will be justified.

Eyes on the

Universe

Patrick Moore in his 1997 The Story of the Telescope *took a close look at its evolution, from Galileo to Hubble. Telescopes, Clarke writes, are at once the world of both amateurs and professionals, and the making of their lenses, the patient grinding of glass, has found its home in every astronomy club.*

Of all human inventions, the telescope is the one that produces the most wonderful results with the least material. We now take it for granted—but could anyone ever have imagined that two pieces of glass could abolish space and perform the miracle of bringing distant objects apparently within reach? And without the telescope, we would still be totally ignorant of our place in the universe.

In principle, it might have been invented thousands, not merely hundreds, of years ago. It seems impossible that Archimedes—or some Chinese inventor even earlier—did not toy with the idea. Some imaginative historians contend that Ceylon's maverick King Kaspaya, who in the fifth century A.D. built the rock fortress of Sigiriya (which I included in my personal seven wonders of the world in a recent BBC program), had a telescope to spy on his harem.

Patrick Moore's new book, written in celebration of the fortieth anniversary of his *Sky at Night* television program, is a slim but lavishly illustrated account of the telescope's history from Galileo's "optick tube" to the first, but certainly not the last, of the great observatories beyond the atmosphere—the Hubble space telescope.

Telescope making is an art as much as a science and is also one of the few branches of engineering in which amateurs can be as good as professionals. For several decades, telescope making was a popular hobby (especially in the United States), but with the advent of relatively cheap, commercial instruments it went into a decline. Now amateurs have an exciting new opportunity, thanks to the astonishing development of light-trapping devices literally dozens of times more sensitive than photographic plates. Coupled with the easy availability of powerful computers, these now allow backyard astronomers to outperform the most well-equipped observatories of the past. Even so, no more advanced instrumentation than a Mark I eyeball can insure instant fame and a degree of astronomical immortality—as Hale and Bopp recently demonstrated.

One of the fascinating themes in Patrick's book is the long battle between lenses and mirrors—refractors and reflectors. That particular war now seems over; the mechanical and optical difficulties of making very large lenses are so great that they have been abandoned. Yet there will always be a certain glamour about the old-type refractors with their gleaming brass tubes. There are not many scientific instruments that can still perform just as well as when they were built one hundred years ago—and may continue to do so for centuries to come.

Nobody would have guessed when—after decades of delay—the two-hundred-inch Hale telescope was dedicated in 1948 that the great age of telescope building still lay ahead. But as Patrick makes clear, this is certainly the case, and instruments have been designed now with apertures that would have seemed like fantasy only a generation ago. The largest refractor ever made (Yerkes, 1897) had a lens just over a yard across. If all goes well, in the year 2000, a quartet of mirrors will give an equivalent aperture of more than fifty feet!

There would have been no point in constructing such huge instruments, even on high mountains, if great advances had not been made in eliminating the atmospheric effects that have always been the bane of ground-based astronomers. (The expression *ground-based* would certainly have puzzled old-time observers.) But developments in adaptive optics, a spin-off from the Star Wars defense program, now allow such telescopes to perform almost as well as if they were in orbit, at least in the visible and infrared wavelengths. For the shorter wavelengths—ultraviolet and extreme ultraviolet—the Earth's atmosphere is an impenetrable barrier, and the Hubble telescope and its successors should fear no competition from sea level.

The effect of politics on astronomy has often been baleful. I can still recall my incredulity in the early 1960s when I heard of plans to install a hundred-inch telescope at the Royal Observatory, even though that had been moved from Greenwich to the marginally better climate at Herstmonceux. I recall remarking sarcastically, at a Royal Astronomical Society meeting, that as the skies were so bad in England, it was obviously necessary to have the best possible telescope to take advantage of the few moments of good visibility that did occur. Fortunately, the Isaac Newton Telescope is now installed in the Canary Islands, at an altitude of over seven thousand feet.

Another unexpected development that has made such sites convenient is the global communications network. Many astronomers can now use telescopes that are on the other side of the planet, almost as easily as if they were actually in the observatory dome—and much more comfortably when that dome is several miles up in the thin, freezing atmosphere.

I am indeed happy to recommend my old friend's book to all readers, even if they are not particularly interested in astronomy. And I wonder what Patrick will be talking about on the fiftieth anniversary of *The Sky at Night,* in 2007.

I hope he will discuss, with his usual enthusiasm, the first successful Moon telescope.

Walter Alvarez

and Gerrit L. Verschuur

Asteroids have been blamed for many disasters, including the extinction of the dinosaurs. Here Clarke examines two important books on the asteroid threat.

T. Rex and the Crater of Doom

by Walter Alvarez (Princeton University Press, 1997)

Impact! The Threat of Comets and Asteroids

by Gerrit L. Verschuur (Oxford University Press, 1996)

Before we go any further, I had better declare an interest—in fact, several interests. . . .

Although I have never met Walter Alvarez, his father, Luis, the Nobel Prize winner, was a good friend, and I dedicated my 1963 novel, *Glide Path,* to him.

I cannot now recall what turned my attention to the possible danger of asteroid impacts. It was quite an old idea in science fiction, but my 1973 novel, *Rendezvous with Rama* (which opened with the obliteration of northern Italy), did introduce a new concept. I argued that as soon as the technology permitted, we should set up powerful radar and optical search systems to detect oncoming cosmic missiles. The name I suggested was Spaceguard, which together with Spacewatch has now been widely accepted.

Seven years later, in 1980, Luis and his geologist son, Walter, published their famous paper in *Science*, "Extraterrestrial Cause for the Cretaceous-Tertiary Extinction." This advanced the theory that the reign of the dinosaurs was ended 65 million years ago by an asteroid (or comet) impact—a suggestion received with downright derision or resounding silence by the geological establishment.

Over the years, more and more evidence accumulated in support of the asteroid extinction theory. In the last letter I ever received from him, Louie said it was "no longer a theory—but a fact." Today I suspect that 95 percent of geologists would agree with him, though there are still some distinguished advocates of alternative explanations, such as volcanism or climate change.

Anyone who thinks that geology is an even duller science than economics may be surprised at the passions it can arouse. Perhaps the classic case is the German meteorologist Alfred Lothar Wegener's theory of continental drift, for years universally regarded as nonsense by the "experts." One historian of science has recorded: "The only time I ever saw a man literally foaming at the mouth was when I mentioned continental drift to a distinguished geologist." Yet the basic truth of Wegener's heresy (now more accurately known as plate tectonics) was established in one revolutionary decade.

The very existence of impact craters on any celestial body is another case in point. With the knowledge acquired by space probes and the Apollo missions, it now seems incredible that, right up to the 1950s, it was widely believed that the lunar craters were volcanic. One British astronomer stated, "The presence of central peaks completely rules out the meteoric hypothesis." That was a perfectly reasonable argument—for who would have dreamed that what happens when one drops a lump of sugar into a cup of tea can happen on solid rock, on a countrywide scale? In fact, the presence of central peaks is one of the best proofs of the meteoric hypothesis!

Even the existence of meteorites themselves was denied right through the Age of Enlightenment, despite the fact that they had been known from time immemorial. It is also almost unbelievable that for decades Arizona's accurately named Meteor Crater was explained as a purely terrestrial formation by virtually all geologists. But once they removed their mental blindfolds, they started finding impact craters all over the world. Over one hundred have now been identified, and there must be many more hidden in the ocean depths.

Although the Alvarez and Verschuur books inevitably cover much of the same ground, the Alvarez is more authoritative because of its "I was there" element. It is an unfolding story told by its leading protagonist. However, I must accuse Walter (or his editor) of petty larceny; the title is an obvious rip-off from Spielberg's *Indiana Jones and the Temple of Doom*—filmed, incidentally, here in Sri Lanka.

Very clearly and entertainingly written, and illustrated with fascinating color plates, it is accessible even to nonspecialists. The treatment of the dwindling—but extremely vocal—cadre of opponents to the meteoric hypothesis also seems to be extremely fair and good-natured.

The general reader, however, may prefer Verschuur's book because it covers a much wider field and is full of fascinating asides. I did catch him on one curious error—he gives a knighthood to the distinguished American astronomer Simon Newcomb. Though he made colossal contributions to planetary theory, Newcomb's astrodynamics was better than his aerodynamics, because he may now be best remembered for his conclusive proof that heavier-than-air flight was totally impossible. (Still worth reading; you'll never set foot on an airplane again.) And a recent paper in the *Journal of the British Astronomical Association* has advanced the intriguing theory that Newcomb may have been part inspiration for one of literature's most fascinating villains, Professor Moriarty.

Now that we are aware of the danger from space, it is time to look more carefully at both the historical and geological records. It is foolish to ignore widespread legends, except when they are obviously absurd. And perhaps even when they are; for centuries there have been stories of ice falling from the sky. Many of these were collected by that indefatigable researcher of the strange and mysterious, Charles Fort, in his book *Lo!* Well, just within the last few months, evidence has accumulated that tens of thousands of "snowballs" the size of houses hit the Earth every day! I must confess to skepticism—not because I doubt nature's ability to sustain such a bombardment, but because it seems astonishing that our fleets of application satellites have survived it.

There have been three known major impacts in this century (Siberia, 1908 and 1947, and Brazil in 1930). And on August 10, 1972, a large meteor traveled halfway across the United States and was seen not only by thousands of people, but recorded by many amateur photographers. It came within a mere fifty-eight kilometers of ground level; had its trajectory been just a trifle different, some American city might have emulated Hiroshima. One can imagine the consequences of such an event during the depths of the Cold War!

It has even been suggested that the 1871 Chicago fire was caused, not by Mrs. O'Leary's cow kicking over a lantern, but by a shower of meteorites. The evidence for this is that many surrounding towns were consumed at the same time, some with great loss of life. I wonder if the 1666 fire of London . . .

Other obvious candidates are Sodom and Gomorrah—a theory that recently caused heated discussions in Israel. Verschuur discusses many such possibilities and makes a good case for real historical events behind the—almost universal—flood legends, and even the story of Atlantis. Unfortunately the deplorable Velikovsky affair made this subject so suspect that reputable scholars hesitated to touch it, with the result that serious research was halted for decades. Ironically, Velikovsky was quite correct in thinking that worlds do collide—but both his time scale and his physics were erroneous by factors of millions.

Perhaps the event that, more than any other, made everyone take catastrophic impacts seriously was the spectacular collision of Shoemaker-Levy on the giant planet Jupiter in July 1995—one of the most widely observed astronomical events in history. It is a delightful coincidence that Gene Shoemaker, who for decades fought a lonely battle to prove there have been meteor impacts on Earth, now has his name associated with this historic event as one of the codiscoverers of the ill-fated comet.

(By yet another coincidence, this time a most tragic one, the above paragraph was written on the very day, July 18, 1997, that Dr. Shoemaker met his untimely death in a car accident, while conducting his annual search for impact craters in central Australia. The loss to science is enormous.)

One result of the changing attitude was that, as early as 1990—five years before Shoemaker-Levy—the United States House of Representatives requested NASA to look into the matter. I am flattered that the resulting document is called the Spaceguard Survey, with due acknowledgment to *Rendezvous with Rama*.

Since then, much has happened in this rapidly developing field. A Spaceguard Foundation has been established, with branches in the United Kingdom and Australia. A proposal will shortly be made to the European Space Agency to set up a Spaceguard Central Node, to cover all aspects of discovery, data storage, and threat identification. Perhaps the best indication of the seriousness with which the problem is now regarded is that a meeting of experts to discuss "hazards due to near-Earth objects" was held at the Royal Observatory at Greenwich on July 10.

Mention should also be made of Spacewatch, a devoted and badly underfinanced group of observers that has already made a whole series of valuable discoveries—including one asteroid that came close enough to Earth to be featured in the *Guinness Book of World Records*!

Some might argue that in a world already nervous about global warming, poisoned oceans, do-it-yourself nuclear bombs, etc., any discussion of asteroid insurance is a massive exercise in irrelevancy. Indeed, many might prefer not to know if a killer comet or asteroid was headed this way.

Yet there is much than can and should be done, as is proved by the current intense debate among astronomers, space scientists, and underemployed Star Warriors looking for new targets. It is an old idea—going back at least to Andre Maurois's *The War against the Moon* (1927)—that only a threat from beyond the Earth could unify the quarrelsome human species. It may indeed be a stroke of luck that such a threat has been discovered, at just the period in history when we can devise technologies to deal with it.

Although some suggested cures may sound worse than the disease (Dr. Edward Teller's proposed bodyguard of orbiting H-bombs has not been received with much enthusiasm), there are several plausible alternatives. They all depend on the length of the warning time—which is why Spacewatch is so vital; it could give us decades to prepare a real Spaceguard.

Of the many defenses proposed, the most elegant—and environmentally friendly!—is to rendezvous with any asteroid on an orbit likely to impact Earth and to persuade it to make a slight change of course. If there was sufficient warning time, only a modest amount of rocket propulsion would be necessary.

This was the scenario that I developed in *The Hammer of God,* and I am pleased to say that Spielberg, Zanuck, and Brown—together for the first time since they made *Jaws*—have filmed *Hammer* and changed the title to *Deep Impact.*

In one of his last books, Carl Sagan pointed out that no really long-lived civilization could survive unless it develops space travel, because major asteroid impacts will be inevitable in any solar system over the course of millennia. Larry Niven summed up the situation with the memorable phrase: "The dinosaurs became extinct because they didn't have a space program." And we will deserve to become extinct if we don't have one.

The Gay

Warlords

*In this ironic essay, Clarke says that time and again the issue of allowing gays
in the military has been raised—only to be trampled down when a
sense of unreality returns to the debate.*

It is astonishing that the most important reason for keeping gays out of the
armed forces has not been more widely publicized, despite the fact that
even the most casual student of history knows their bloodthirsty record.
(Okay, I confess, I'm a closet pacifist, having had a very peaceful war in the
Royal Air Force. The nearest I got to action was hearing one of my late
friend Wernher von Braun's mother-of-all-spaceships demolishing a London suburb, a safe ten miles away.)

Those archetypal warriors, the Spartans, proudly boasted how they
maintained their esprit de corps, with accent on the *corps*. And Julius
Caesar's popularity with his men, who chanted, "Every wife's husband,
every husband's wife," after him, was undoubtedly enhanced by his enthusiastic swinging in both directions: vide his youthful affair with the king of
Bythinia. However, like most of his coldly calculated actions, this was probably motivated by politics rather than passion.

There's very little of Caesar's ambisexterousness about the two other greatest military leaders of antiquity, Alexander and Hadrian. They seem to have been hetero only rarely, and then entirely for reasons of state. For details, see Mary Renault's *The Nature of Alexander* and Marguerite Yourcenar's *Memoirs of Hadrian*.

Jumping forward a thousand years or so (and with only a passing glance at the unproved allegations against the Knights Templars), we come to that amazingly well-matched pair of military geniuses, Richard I and Saladin. About Richard's predilections there is no doubt: one of the most piquant incidents in the history of British arms was the occasion when Eleanor of Aquitaine berated the aptly named "Lionheart," in front of his own troops, for his failure to give her a grandchild. (He never did.)

As for Saladin, though he did produce a few offspring, there is considerable evidence that his main interest was elsewhere. (And fancy showering his personal physician with gifts, including snow, to his beloved enemy Richard when he was sick—where the hell did he get snow in that part of the world! Ben and Jerry were still centuries in the future.)

It must be admitted that England's most celebrated royal gays—Edward II and James I—hardly fit the militaristic mold. When James succeeded Elizabeth, the courtiers remarked (out of his hearing), "Once we had a queen who was a king—now we have a king who is a queen." And Marlowe has told us all too graphically how Edward's death reflected his life: I've often wondered how they stage the last act of the play, but don't really want to know.

However, these are two exceptions that prove the rule, and if you require an overwhelming counterexample, just look at Frederick the Great. All of the blame cannot be placed on his horrible father (see Emil Jannings in *The Old and the Young King*—if any copies of this 1935 film still exist). As the military genius who created Prussia, Frederick himself has much to answer for.

But the classic textbook specimen of the brutal, brilliant, and pathologically antiheterosexual warrior will be found not in Europe or Asia, but in Africa. During the last year of the—literally—reign of terror that created the Zulu nation, Shaka the Great executed any women found pregnant, together with their husbands. Nice guy . . . don't know how he expected his empire to continue. But it did, even after his inevitable assassination, and gave us Brits a lot of trouble (see Cy Endfield's excellent movie *Zulu*). Much of this was self-inflicted; it was in one of these wars that the dead British gunners were found with their fingernails torn out—by themselves, in a desperate attempt to open the ammunition boxes. The storekeeper had forgotten to send the keys; doubtless he was promoted, in the best military tradition of "reward the guilty, punish the innocent."

How/why did I get involved in this grubby line of research? (Thought you'd never ask.) Well, it was triggered by recent revelations about certain much-decorated Royal Air Force Fighter Command war heroes, which reminded me of a long-forgotten scandal here in my adopted country of Sri Lanka. The only reason I know about it is that many years ago the author (I can't recall his name, but I think his book was called *Death Before Dishonor*) sent me a biography which aroused my salacious interest. (Note to would-be censors: "Keep your filthy hands off my filthy mind!")

At the turn of the century, the commander in chief of the Ceylon forces was a very remarkable man, Sir Hector Macdonald. Winner of his country's highest military award, the Victoria Cross, he was known as the bravest soldier in the British army and had achieved the astonishing feat of being promoted all the way from private to general.

Alas, to the great embarrassment of the local Brits (and doubtless the amusement of everyone else), Fighting Mac was caught in flagrante with some Colombo schoolboys—not the natives, by gad!—at least they were burghers (upper-class Eurasians). Whitehall recalled the general prontissimo; he got as far as Paris, and shot himself. . . .

Maybe the equally brave General Gordon (read between the lines of Lytton Strachey's admittedly biased *Eminent Victorians*) was lucky: he died at the siege of Khartoum (1885) and so became a national hero. Ditto the widely suspected Lord Kitchener, though his fate was somewhat less valiant; he drowned when his flagship was torpedoed in World War I.

But enough: I consider my thesis proved beyond doubt.

So—

KEEP THESE FEROCIOUS GAYS OUT OF THE ARMED SERVICES!

They're too bloodthirsty and warlike. We need gentle, compassionate soldiers, in the peaceful new world we hope to build.

More Last

Words on UFOs

It is probably too much to hope that the American Air Force's belated revelations
about the sources of many sightings will put a stop to this tedious nonsense,
Clarke writes. Could anyone ever have seriously imagined that the
Earth's skies have been full of alien visitors for the last half
century, he says, without the matter being
settled incontrovertibly?

For decades now, the radars of the great powers have been able to track any object much larger than a football that comes anywhere near our planet. Of course, it may be argued that alien spacecraft invariably use Stealth techniques—but it is hard to see why they should bother, since they seem so willing to make contact. In any case that would hardly help them to evade detection by the legions of amateur astronomers who constantly scan the skies.

Though it is perhaps unkind to do so, I would like to remind the UFO fanatics how earlier widely accepted stories of alien meetings turned out to be ludicrous fabrications. Does anyone still remember George Adamski's *Flying Saucers Have Landed?* He reported cities on the other side of the Moon, and I believe there was once a lady who made a good living by lecturing about her honeymoon on Venus. Well, we have now seen the lunar farside (and I've never forgiven the Apollo 8 crew for resisting the temptation to report a black monolith there), and we know that any Venusian rivers are likely to consist of molten lead. We will have to go farther afield than our immediate neighbors to look for intelligent life—perhaps any life at all.

What is particularly ludicrous is the widespread idea (à la *Independence Day*) that for several decades some supersecret branch of the United States has had alien spacecraft—and aliens!—in its possession. Anyone who will believe that will believe anything; I have known many of the people who would have been involved in such a cover-up, and I can assure you that it would have a half-life of about forty-eight hours. As one Pentagonian once remarked sadly, "I wish it was true—then all us majors would be colonels." I think that settles the matter; but then, of course, I may be part of the conspiracy.

Indeed, at least two of my friends were on the CIA committee looking into the UFO question, at a time when it was seriously considered that spaceships might be involved. One member (the late Prof. Luis Alvarez— now famous for his theory that the dinosaurs were exterminated by an asteroid 65 million years ago) told me how easy it was to dispose of most of the sightings, because the average observer does not know how many remarkable things there are in the sky.

Frankly if you have never seen a UFO, you're not very observant— or else you live in the city and don't have access to the sky, which nowadays is an all too common state of affairs. I've seen at least ten UFOs, and several of them were very convincing: it took quite an effort to convert them into identified flying objects. And I still cannot get over the fact that my most dramatic sighting was from Stanley Kubrick's penthouse on the Upper East Side—the very night we'd decided to make a little home movie together. (I'm embarrassed to say that the brilliant light we watched moving across the sky turned out to be the ECHO balloon satellite, seen under rather unusual circumstances. Also, Stanley and I were in a rather exalted mood, and perhaps not as critical as we should have been.)

One of the chief reasons why I have never been able to take reports of alien contact seriously is that no spaceship ever contains aliens—the occupants are always human! Oh, yes, they do show a few minor variations such as large eyes, or pointed ears (Hi there, Mr. Spock!), but otherwise they are based on the same general design as you and I.

Genuine extraterrestrials would be really alien—as different from us as the praying mantis, the giant squid, the blue whale. Nature is incredibly ingenious; just look at the variety of creatures on this planet. We are the products of thousands of throws of the genetic dice; if evolution was ever restarted on Earth, at any point the branches of the tree of life might have taken a different direction—and we would not be here. But something would be. . . .

The recent excitement about Mars has again focused public interest in the possibility (most experts would say the probability) of life on other worlds. However, we should not expect too much even from the fantastically successful *Pathfinder* mission. Watch out for *Mars Surveyor*—though, personally, I have considerably greater expectations for life beneath the ice floes of the Jovian satellite Europa, for reasons given in *3001: The Final Odyssey.*

With any luck, within the next few years (what a millennium present that would be!) we may have an answer to a question that has haunted mankind since our first ancestors started looking at the skies. And let me give the last word to the brilliant team of engineers and scientists at the Jet Propulsion Lab who have amazed the world with such detailed close-ups of the Red Planet.

This is what I received from them in reply to my message, "Hope Rover's hubcaps aren't stolen overnight . . . but how exciting if they are. . . ."

VALLES MARINERIS (MPI)—A spokesthing for Mars Air Force denounced as false rumors that an alien spacecraft crashed in the desert, outside of Ares Vallis on Friday. Appearing at a press conference today, General Rgrmrmy the Lesser stated that "the object was, in fact, a harmless high-altitude weather balloon, not an alien spacecraft."

The story broke late Friday night, when a major stationed at nearby Ares Vallis Air Force Base contacted the *Valles Marineris Daily Record* with a story about a strange, balloon-shaped object that allegedly came down in the nearby desert, "bouncing" several times before coming to a stop, "deflating in a sudden explosion of alien gases." Minutes later, General Rgrmrmy the Lesser contacted the *Daily Record* telepathically to contradict the earlier report.

General Rgrmrmy the Lesser stated that the hysterical stories of a detachable vehicle roaming across the Martian desert were blatant fiction, provoked by incidents involving swamp gas. But the general public has been slow to accept the Air Force explanation of recent events, preferring to speculate on the otherworldly nature of the crash debris. Conspiracy theorists have condemned Rgrmrmy's statements as evidence of "an obvious government cover-up," pointing out that Mars has no swamps.

And who says that scientists have no sense of humor?

Carl Sagan

Clarke addresses the accomplishments of a colleague and friend.

Carl Sagan's Universe

edited by Yervant Terzian and Elizabeth Bilson (Cambridge University Press, 1997)

This beautifully illustrated book (twenty color plates; many more black-and-white) is a fitting tribute by more than twenty of his friends to the best-known astronomer of our time. In one way it makes sad reading, as the essays were addressed to Carl on his sixtieth birthday in October 1994, two years before his untimely death. But it is also exhilarating because it records a veritable explosion in our knowledge of the universe—an explosion in which Sagan often acted as a fuse.

The book is divided into several sections: planetary exploration, life in the cosmos, science education, and environment and public policy. Its flavor may best be given by some of the titles: "Impacts and Life: Living in a Risky Planetary System," "Do the Laws of Physics Permit Wormholes for Interstellar Travel?" "Science and the Press," "Science and Religion," "Carl Sagan and Nuclear Winter," "Science and Pseudoscience," "Nuclear Free World?" and "Highlights of the Russian Planetary Program."

All the essays are by leading authorities in the field, yet all are accessible to the nonspecialist reader. They are full of fascinating and little-known items of information. For example, in this era of megabudgets, it is encouraging to learn that the most important discovery in radio astronomy—the twenty-one-centimeter hydrogen line—was a $500 bargain. And if you want to send a telegram one thousand light-years, it would cost only a dollar a word—even with the technology we have today! (Of course, you'll have to wait two thousand years for a reply.)

So, two-way conversations are not likely for quite some time—but look at the impact on our culture of the monologues we already have with the Greeks and Romans, and all the great thinkers of the past! The possibility of what might be called interstellar archaeology is so exciting that one can sympathize with the researchers in the field, who have had their modest budgets cut by Congress. I once gave them the advice: "Despite disappointments and false alarms, continue the search for intelligent life in Washington."

Carl Sagan's Universe covers such a vast range of topics, from the Big Bang to avoiding unwanted bangs (natural or man-made) in our immediate future, that I cannot imagine any intelligent reader failing to derive both information and entertainment from it.

My only serious criticism is that I never had a chance of contributing a chapter myself. So here it is, for the next edition.

The theoreticians of modern physics have conscripted the word *entangled* to describe the mysterious (Einstein said "spooky") way in which one particle can affect another, no matter how great their separation in space. I cannot help feeling that Carl Sagan's career and mine were entangled in some such manner. Consider:

In 1950 I published a slim volume, *Interplanetary Flight,* which was the first book to give English-speaking readers the basic principles of astronautics. Carl was then in high school, and as he later told my biographer, Neil McAleer, in *Odyssey* (Gollancz, 1992):

> I was interested in the other planets and I knew that rockets had something to do with getting there. But I had not the foggiest notion about how rockets worked or how their trajectories were determined. . . .
>
> The part in *Interplanetary Flight* that was the most striking for me was the discussion of the gravitational potential wells of planets and the appendices, which used differential and integral calculus to discuss propulsion mechanisms and staging and interplanetary trajectories. The calculus, it slowly dawned on me, was actually useful for something important and not just to intimidate high school students.
>
> As I look back upon it, *Interplanetary Flight* was a turning point in my scientific development.

Though I am certainly proud to have started Carl on his career, I have little doubt that he would have managed pretty well without my impetus. A later attempt of mine to expand his sphere of activity was not so successful.

I do not recall when we first met, but by the 1960s I must have been well aware of his work at the Smithsonian Astrophysical Observatory. So when, in the summer of 1964, Stanley Kubrick and I started generating ideas for a little home movie with the working title "How the Solar System Was One," I decided that Carl might be a useful ally in our brainstorming. Here are his own comments, after the dinner we had in Stanley's Manhattan penthouse:

> They had no idea how to end the movie—that's when they called me in to try to resolve a dispute. The key issue was how to portray

extraterrestrials. . . . Kubrick was arguing that the extraterrestrials would look like humans with some slight differences, maybe à la Mr. Spock. And Arthur was arguing that they would look nothing like us. . . . I said it would be a disaster to portray the extraterrestrials . . . the number of individually unlikely events in the evolutionary history of man was so great that nothing like us is ever likely to evolve anywhere else in the universe . . . any explicit representation of an advanced extraterrestrial being was bound to have at least an element of falseness about it. . . . What ought to be done is to suggest them.

A third of a century later, I do not recall Stanley's immediate reaction to this excellent advice, but after abortive efforts during the next couple of years to design convincing aliens, he accepted Carl's solution.

Ronald Sagdeev in his fascinating autobiography, *The Making of a Soviet Scientist* (Wiley, 1994), wrote of this reaction to Carl by Ronald Reagan at a Moscow summit meeting in 1988:

When it was my turn to be introduced to the guest of honor, Gorbachev seized my arm and said, "Mr. President, this is the man who is promoting the flight to Mars." I had a feeling that Gorbachev's words struck some chord of curiosity in Reagan. As if to underscore his apparently successful start to his Mars public relations campaign with the American president, Gorbachev added, "Academician Sagdeev has friends and colleagues in America who share the same vision of a joint flight."

Then Gorbachev turned to me, as if looking for help with a few names. But before I could react, he went on, "Carl Sagan."

In a fraction of a second I could tell that something had clicked the wrong way. The guest of honor appeared to lose interest in the subject immediately. Gorbachev apparently didn't understand that there was not a great deal of political compatibility between Ronald Reagan and Carl Sagan.

Sagan—who wrote the foreword to Sagdeev's book—deserves a great deal of credit for building bridges between American and Russian scientists at the height of the Cold War. This effort began in 1966 when he collaborated with the maverick astrophysicist Iosif Shklovsky on their landmark book, *Intelligent Life in the Universe.*

Carl's greatest achievement—and the one by which he is best known—was of course the thirteen-part series *Cosmos,* undoubtedly one of the most superb feats of education/entertainment that has ever appeared on television. There cannot be too many programs of this kind, to counter the mind rot now being purveyed by press, radio, movies, and TV—astrology, psychic powers, reincarnation, the paranormals, UFOs, Creationism . . . we lost Carl at just the time when he was most needed.

But to give credit where it is due, *Cosmos* was actually a joint project with Gentry Lee, science director of the 1976 *Viking Mars Lander*, and chief engineer of the *Galileo* space probe that is currently mapping the moons of Jupiter. Lee took time off from the Jet Propulsion Laboratory to produce *Cosmos* and later helped Carl to develop the movie script that was the genesis of *Contact*.

My own last encounters with Carl were at rather long range. During a brief visit to London in July 1988, I had the privilege of joining him by satellite, in a three-cornered conversation with the only scientist of equal fame—Stephen Hawking. For two hours, moderated by an occasionally baffled Magnus Magnusson, we discussed the topic "God, the Universe, and Everything Else"—but for some unfathomable reason the resulting program was never broadcast.

And in 1996, Carl and I both set out digitally for Mars together, with greetings to future colonists. Alas, the Russian launch vehicle failed to escape from the Earth, and our message ended up somewhere in the Pacific. I still hope it will be delivered on some later mission; meanwhile you can see and hear us (and also review the entire history of speculation about this fascinating world) on the CD-ROM *Visions of Mars* (available from the Planetary Society in Pasadena, California).

It is a great pity that Carl just missed two events that would have been of special interest to him—the landing of the *Pathfinder* on Mars at what has now been named the Carl Sagan Memorial Site. And he would have been delighted at the success of the movie *Contact*.

It is now twelve years since he sent me a copy of the novel, with this inscription: "For Arthur, whose fiction and nonfiction, and advocacy of the peaceful uses of the space environment, have been a source of inspiration since my boyhood."

It is indeed a tragedy that Carl Sagan never lived to see this dramatization of his ideas. Yet however successful *Contact* may be on Earth's big and little screens, I wonder if it will ever reach as many viewers as *Cosmos*.

For how many alien eyes have already seen Carl, whose image is now almost twenty light-years from Earth? It will soon reach Vega—and out around Sirius, at less than half that distance, they're probably watching the reruns.

For Cherene,

Tamara, and

Melinda

*A short letter follows from Clarke to the three little girls who
share his Colombo home.*

This is how I shall always think of you, even when you are grown women
with little girls of your own—in the better world of the next century. And
if I am never to meet your children, pass on to them the love I gave to
you.

But will that world indeed be a better one, when you inherit it? Of
course, there is no sure way of telling, and perhaps it is just as well. If we
could see the future—with absolute certainty—there would be no purpose
in living. Nothing could change the inevitable, we would be robots, unable
to deviate from a predetermined program.

Well, we aren't robots; we have some control over our future. Nevertheless, even the most successful life consists largely of what the poet
Robert Bridges once called the "masterful administration of the unforeseen." Which is why one of the most valuable gifts anyone can possess is
flexibility—the power to adapt to changing circumstances, when the need

arises. Even if we can't predict, we can make intelligent guesses, so that events do not take us completely by surprise. Hence the value of organizations like the World Future Society; they act as early warning systems.

Even in your short lives, you have seen one of the greatest technological revolutions of all time—the advent of the computer. It fascinates me to see how you sit down at the keyboard and conjure up miracles beyond the imagination of anyone (except those crazy science fiction writers!) before this century. Yet you take the most amazing video games for granted; when anyone mentions *mouse,* your generation doesn't think of Mickey. . . . And you will never use a typewriter—the essential tool of commerce, industry, and private correspondence for the last hundred years.

You also live in a world unimaginably wider than the one I knew as a boy. I was almost twenty before I made the epic journey from my Somerset birthplace to London—all of two hundred kilometers! Yet before your tenth birthdays you'd seen London, Sydney, New York, Honolulu, Los Angeles, and Singapore. And every day television and videocassettes give you glimpses of more places and societies than anyone could encounter in a lifetime.

What future marvels will you see and take equally in your stride?

Already you talk to your uncles and aunts all over the planet (and don't I know it, when the phone bill comes in!). Soon you will be able to see them as well, thus making even closer contact. It's often been said that improved communications will encourage people to stay at home. On the contrary; though they will eliminate the need for much routine commuting (and the traffic jams that have made nightmares of our cities), they will encourage travel for pleasure. People will want to visit interesting and beautiful places they could never have known about before the age of television.

One effect of improved—and universal—communications will be to remove all national barriers and eventually create not merely the global village but the global family, sharing at least one language in common. How I wish that this would also ensure global peace, but that would be hopelessly optimistic. However, without good communications any civilization is impossible.

Speech was one of the first, and perhaps the greatest, of mankind's inventions—but it is no longer enough. The computers upon which our society is now utterly dependent must talk to each other across the width of the planet in a fraction of a second, and at millions of times the human rate.

In the past, nations and tribes had to learn to live with one another—and they are still learning, much too slowly. Now they must learn to live with a new and strange species, the intelligent machines that our technology is creating. Don't believe those people who say that machines will never think—that merely proves that some humans can't think.

It is true that AI, or artificial intelligence, may differ profoundly from what we are pleased to call human intelligence. Indeed, why should we bother to create it if it was exactly the same? Our other machines aren't carbon copies from nature. Airplanes don't flap their wings—cars don't run on legs. AI will have its own rules—its own logic. And, ultimately, its own goals, which may not always be the same as ours. (Yes, I'm thinking of HAL. . . .) I suspect that the single greatest problem of the near future will be that of peaceful coexistence with our mind children—to use computer scientist Hans Moravec's apt description. Our silicon offspring can help us make a paradise of this planet and open the path to others.

It has been my privilege to shake hands with the first man to orbit Earth, and the first to walk on the Moon. So it makes me very angry when I hear people who should know better complain, "Why waste money on space when there is still so much to be done on Earth?" But I am happy when it's asked in honest curiosity—because it's a very good question and needs to be answered.

Even though billions have been squandered on weapons systems and national prestige, our investment in space has already been returned many times over. Comsats—communications satellites—are the very backbone of the global telephone, television, and data systems. Weather satellites have saved thousands of lives—and could have saved more. A cyclone in the Bay of Bengal that killed half a million people was tracked by satellite, but the warning did not reach them in time.

Unless we spend more, not less, money on the practical uses of space, millions will be condemned to live ignorant, disease-ridden lives—when they live at all. This applies not only to the third world. We may well thank the secret reconnaissance satellites, which make it impossible to conceal large-scale military operations, for the fact that our planet is not already a radioactive ruin.

But what of manned voyages to the Moon and planets—space exploration? Today, that cannot be justified as cost-effective. Few important things can be!

To understand the real meaning of spaceflight, we must go back to long before the dawn of man—even of reptiles—even of animals—to the time when the first primitive organisms left the womb of mother sea and ventured onto the alien land.

There they had to face a ferociously hostile environment, with temperature extremes never encountered in the oceanic world, and blistering radiation pouring down from above. Most died, but some adapted. In Faulkner's famous words—they did not merely survive, they prevailed.

The colonization of dry land was the first vital step toward the development of intelligence—it is doubtful if it can ever arise in the sea. (Dolphins and whales are the exceptions that prove the rule—they evolved on land and defected when they couldn't face the competition.)

And it is only on land that one can produce and control fire—the key to all but the most primitive technologies. To develop civilization, we had to become exiles from our original home. But we have never forgotten it, and we still spend the first nine months of our existence floating in a tiny imitation sea.

Now the time has come to make the next step in evolution—this time controlled, at least partly, by deliberate intelligence, not the random dance of DNA molecules through endless generations. I have often thought—especially when scuba diving—that we don't really belong here on land, dragged down by gravity every moment of our lives. Our true destiny belongs in space.

Once you have escaped from Earth's clutches, space travel is cheap and easy. As our engineering skills improve, our species will spread across the Solar System, as once it spread across the face of this planet. First the Moon and Mars, then the asteroid belt, then the satellites of the gas giants Jupiter and Saturn—some so large that they are planets in their own right.

What will we find there? Much, I am sure, of enormous value—and the greatest treasures will be wholly unexpected. I like to remind my American friends that, just over one hundred years before the first Moon landing, Congress was abusing Secretary of State Seward for purchasing the worthless, icebound wilderness called Alaska—at an exorbitant two cents an acre. In the long run, the Solar System will be an even better bargain.

And it may not even be a very long run, in terms of history; after all, Columbus is barely five hundred years behind us! You will see the discovery of worlds more strange and wondrous than any he could have imagined. Will you one day set foot on them—or will you go only as far as the Moon?

I wish I could know. But whatever the future brings, I hope you will remember the uncle who loved you while the twentieth century was drawing to a close and longed for you to see a happier twenty-first.

POSTSCRIPT:

2000 AND BEYOND

Clarke diving off the coast of Sri Lanka.

COURTESY OF ARTHUR C. CLARKE

Science and

Society

For every expert, there is an equal and opposite expert.
—Late-twentieth-century folklore

For more than a century, science and its (occasionally ugly) sister technology have been the chief driving forces shaping the world. They decide the kind of futures that are possible: human wisdom must decide which are desirable.

It is truly appalling, therefore, that so few of our legislators have any scientific or engineering background. That should be the necessary, but not sufficient, qualification for anybody involved in making major decisions in national policies. I say "not sufficient" because neither of the two engineers who became president of the United States was very successful—though largely through circumstances beyond their control.

And even the wisest, and best educated, of legislators may have difficulty in reaching a decision when there is total disagreement among the "experts" in the field. There are some hilarious examples of this in the history of science—for example, Kelvin's declaration that X rays must be a hoax, and Rutherford's even more famous dismissal of atomic energy as

"Moon-shine." (Yet, let me offer a belated defense of the late astronomer royal, Richard Woolley, for whom I have always had a soft spot since I met him riding over Mount Stromlo in search of a sick cow. He was pilloried for saying, just the year before *Sputnik,* that space travel was "utter bilge." Apparently, what he really said was "All this writing about space travel is utter bilge. To go to the Moon would cost as much as a major war." That second sentence was right on target.

The most notable controversy now facing science and society is, of course, in the realm of cloning. Any developments that concern biology—especially human biology—are liable to arouse passions, vide the debates on contraception and euthanasia.

And especially evolution. I have encountered a few Creationists and they were usually nice, intelligent people, so I have never been able to decide whether they were really crazy or only pretending to be mad. If I were a religious person, I would consider Creationism nothing less than blasphemy. Do its adherents imagine that God is a cosmic hoaxer, who has created the whole vast fossil record for the sole purpose of fooling mankind? And, though I do not necessarily agree with the Catholic paleontologist Teilhard de Chardin's advocacy of evolution as a major proof of the glory of God, his attitude is both logical and inspiring.

A Creator who, right back at the beginning of time, laid the foundations for the entire future is far more awe-inspiring and deserving of worship than a clumsy tinkerer who constantly modifies billions of his creations and throws away whole species because of defective engineering. In any event, no Christian (born-again or otherwise) should have any difficulty facing the facts of paleontology, now that the Vatican has stated that "evolution is no longer a theory."

But perhaps I'd better stick to the hard sciences, especially those with which I have been personally acquainted. From the beginning, I was involved in the debate over the Strategic Defense Initiative (aka Star Wars), and I am saddened that this cost me a cherished friendship, that of the late Robert Heinlein.

My attitude then, and now, was that though it might be possible, at vast expense, to construct local defense systems that would "only" let through a few percent of ballistic missiles, the much-touted idea of a national umbrella was nonsense. Luis Alvarez, perhaps the greatest experimental physicist of this century, remarked to me that the advocates of such schemes were "very bright guys, with no common sense."

Looking into my often cloudy crystal ball, I suspect that a total defense might indeed be possible in a century or so. But the technology involved would produce, as a by-product, weapons so terrible that no one would bother with anything as primitive as ballistic missiles.

And if I might hazard another guess, I suspect that President Reagan's

"Star Wars" speech (composed, after innumerable drafts, by my friend Dr. George A. Keyworth) will one day be regarded as a work of political genius. However shaky its technological foundations, it may well have contributed to the ending of the Cold War. And ironically, the projected SDI armory of lasers and interceptors may one day be needed to save the entire human race, not merely transient tribal groupings.

The scientific establishment has slowly realized that the history of this planet, and perhaps of civilization itself, has been drastically modified by impacts from space. We have gone a long way since President Jefferson remarked, "I would sooner believe that two Yankee professors lied, than that stones fell from the sky." Now we know that mountains can fall from the sky.

And here we have the most perfect example of the "law" that opens this essay. Volumes of statistics have been amassed on either side of the question, How much effort should be devoted to a danger that is probably remote, but which may sterilize our planet? The insurance companies, who make a good living from accidents that don't happen, are very little help here. Anyway, they wouldn't be around after the worst-case scenario.

Surely, everyone will agree that a serious effort should be made to conduct a survey for possible comet or asteroid impactors. The cost would be quite trivial, and the results will be of great astronomical value. After Shoemaker-Levy's impact on Jupiter, one would think there is no need to labor this point.

And what a tragedy Gene Shoemaker's untimely death was! I believe that Gene, some of whose ashes are now on the Moon, would have been amused by the embarrassment this has caused to NASA and the Jet Propulsion Laboratory. Apparently, the Navajo Indians have protested against this "sacrilege" because they regard the Moon as holy. While sympathizing with their point of view, I cannot help thinking that their concern is somewhat belated. Is it indelicate to ask how much—er—fecal matter the Apollo project has already left there?

But now for something more serious. I would like to conclude with a brief comment on what is now unfolding as perhaps the greatest scandal in the history of science—the cold-fusion caper.

Like almost everyone else, I was disappointed when Pons and Fleischmann announced they had achieved fusion in the laboratory and those who rushed to confirm their results were unable to replicate them. Wondering how two world-class scientists could have fooled themselves, I then forgot the whole matter for a year or so, until more and more reports came in, from many countries, of anomalous energy production in various devices (some of them apparently having nothing to do with fusion). I remained interested but skeptical, agreeing with Carl Sagan's principle that "extraordinary claims require extraordinary proofs" (spoken in connection with UFOs and alien visitors, a whole supermarket shelf of worm cans, none of which I intend to open here).

Well, for those (90 percent of?) scientists who have been too busy to look into the matter, or thought the case was closed, there is now no further doubt that anomalous energy is being produced by several devices, some of which are on the market with a money-back guarantee, while others are covered by patents. The literature of the subject is now enormous: it's all over the Web, but the most convenient source is Gene Mallove's *Infinite Energy* magazine (P.O. Box 2816, Concord, NH 03302-2816).

My confidence that "new energy" is real has slowly climbed up to the 90s and has now reached 99 percent. I hold in my hand (move over, Senator McCarthy!) a report from a Fellow of the Royal Society who, originally skeptical, says, "There is now strong evidence for nuclear reactions in condensed matter at low temperature. . . . That is the problem: there is no theoretical basis for these claims, or rather there are too many conflicting theories. Still, that should not prevent the development of the most promising systems."

True: the steam engine had been around for quite a while before Carnot explained exactly how it worked. The challenge now is to see which of the various competing devices is most reliable and most easily scaled up. My guess is that large-scale industrial application will begin around the turn of the century—at which point, the end of the fossil-fuel/nuclear age will be clearly in sight.

As also will be concern about global warming: most of the present arguments will be irrelevant, as oil-and-coal-burning systems are phased out. Here again is an area in which one cannot blame the politicians for being confused. Although most scientists believe that warming is occurring, there is at least one distinguished holdout: Prof. Fred Singer, who headed the United States Meteorological Satellite Program. Fred was an innocent, young cosmic-ray physicist when, in the late 1940s, we premature astronauts of the British Interplanetary Society brainwashed him into becoming a space cadet.

He has now published a book (*Global Warming: Unfinished Business*, The Independent Institute, California) more or less pooh-poohing the idea of global warming. But we may need it after all, as the interglacial period draws to its close. As Will Durant said many years ago, "Civilization is an interval between ice ages." It has recently been discovered that the changeover may take place in decades, rather than centuries, so the cry in the next millennium may be "Spare that old power station—we need more CO_2!"

Finally, another of my dubious predictions. Pons and Fleischmann will be the only scientists ever to win both the Nobel and the Ignoble Prizes.

Is There Life After

Television?

Clarke, who made his television debut almost a half century ago, pokes hard in
this essay at how television has spun off myriad forms of communication.
How different it is, he says, from his appearance in green makeup at
the BBC on May 4, 1950, to the approaching year 2000 with
its vast offerings along the information superhighway.

In the fifties, of course, there was no videotape, so programs had to be
either live or prefilmed. I do not know what encouraged my first producer,
Robert Barr, to risk putting me in front of the monstrous electronic camera
for twenty minutes, nonstop, while I lectured on—fasten your seat belts—
the fourth dimension. True, I had several wire models of tesseracts and
tetrahedra to give a little variety; but I've often wondered how many viewers survived to the end.

In those early days of monochrome television, faces had to be plastered
with green Pan-Cake makeup so they wouldn't appear pasty white on the
screen. According to Neil McAleer's biography (*Arthur C. Clarke*, Contem-

porary Books, 1992): "Brother Fred remembers Arthur, face painted green, coming back to the house . . . 'After terrifying my wife and the children, he went upstairs and returned half an hour later with a well-scrubbed, glowing, but clean face. The towels were not so lucky. After four repeated launderings, my wife threw them away.' "

Such were the perils of the early pioneers into the Great Wasteland. But let's be fair: there have been some magnificent oases in that vast territory—superb artistic creations enjoyed by wider audiences than Shakespeare could ever have imagined, and great moments of history that, for the first time, the entire human race could share. Television has done more than any other medium to make this One World.

It may even have saved the world, for it certainly contributed to the end of the Cold War (though we must also give credit to those other electronic marvels, the fax machine and IDD, for making the Iron Curtain totally transparent). And although talk of the civilizing influence of television may arouse some wry smiles, the ubiquitous, and easily concealed, cameras carried by amateurs and professionals alike may already have prevented many atrocities—or at least identified those responsible.

Yet although television is the most obvious example of the telecommunications revolution that has changed our world beyond recognition within a single century, it is now becoming part of something much wider—information technology.

The advent of the computer has given us an invaluable, though often exasperating, assistant in tasks that once it seemed only human beings could perform. It is hard to believe that fifty years ago—at the very dawn of the television age—the chairman of IBM famously declared that the world market for computers was about six. Well, I have lost count of the number around my house, each incomparably more powerful than the room-size mainframes of the 1950s. Ironically, about the last thing they ever do is "compute," in the classical sense of the word. They are downloading E-mail, processing text, accessing CD-ROMs, looking at new images from Mars on the JPL Web site, playing video games, exploring the infinite universe of the Mandelbrot Set. . . . It's a humbling thought to realize that every item in that last sentence would have been totally meaningless only a few decades ago. (And what would your grandmother have thought, Ms. Company Secretary, if you could have told her that you would spend much of your day stroking a mouse?)

As an example of how today's information technology would have seemed like magic as late as the 1960s, consider this: Would anyone then have believed in the possibility of a book in which the print could be changed instantly from the largest to the smallest size, the typeface itself could be altered equally quickly from roman to italic to you name it—and any word or phrase could be located in seconds? No conceivable technology could have made this possible, yet we now take it for granted when we insert Microsoft's latest silver disk into the A-drive . . . a disk that can contain not merely one book, but an entire library. (The Greens should give Bill Gates an award for saving more trees than anyone else in history. On second thought, some of those manuals . . .)

Of course the great problem raised by information technology is information pollution, but surely that is better than its even deadlier opposite, information starvation. It has been said that after the invention of the telephone, no longer did farmers' wives in the American Midwest go mad through sheer loneliness. (Today they are more likely go mad through too much MTV.) Nevertheless, it is vital to remember that information—in the sense of raw data—is not knowledge; that knowledge is not wisdom; and that wisdom is not foresight. But information is the first essential step to all of these.

Though it is difficult to think of anything we won't be able to do in the very near future, when all our current hardware is linked together with orbiting constellations of Comsats, there is still great room for improvement in one area. Though the typewriter has joined *T. rex,* QWERTYUIOP still lives: Is it not a scandal that a layout deliberately designed to prevent skilled humans from overtaking the clumsy mechanical systems of one hundred years ago has survived into the age of electronics?

Perhaps the most ingenious attempt at QWERTYUIOPicide was the small keypad invented by film producer Cy Endfield (*Zulu,* etc.) in which the fingers of one hand, singly or in combination, could call up any letter. Apparently it was quite easy to learn the necessary skill, but the device never succeeded in the marketplace. I wonder if the CIA still uses it: there are environments, too noisy for tape recorders, where secret, hand-in-pocket note-taking might be useful. . . .

And ambient noise is one reason why the voice recognition systems now coming into use may have rather limited application. They will be very valuable for individuals working alone, but imagine the chaos that a whole officeful of talkers could produce. (I cannot resist quoting from my own first attempts to train one of the best current systems, Dragon Speaking Naturally. When I said, "Now is the time for all good men to come to the aid of the party," the program revealed its impressive vocabulary with a startling display of political incorrectness: "Now is the time for all good men to come to the aid of apartheid.")

It must have been around 1948 that the British cartoonist Bruce Angrave invented (at least on paper) the ultimate video display, which he called the teleoccule. The name described it perfectly: it would be a television set built into a monocle. As in those days it took two strong men to lift the average receiver, which contained at least a dozen tubes (remember them?) each generating enough heat to fry an egg, this was a rather daring concept.

Yet in the form of the private-eye and virtual-reality devices it is almost here. Never having used any of these, I am not sure what effect they have on the viewer, but I fear they may be recipes for instant schizophrenia, completing the zombification of mankind started by the Sony Walkman.

But there may be much worse to come. It seems obvious that the ultimate input-output device would bypass all the body's sense organs and pass its signals directly into the brain. Exactly how this would be done I leave to future biotechnicians to decide, but in *3001: The Final Odyssey* I tried to describe the operation of the "Braincap." One feature that might delay its general adoption is that the wearer would probably have to be completely bald to use the tightly fitting helmet. So wig-making could become really big business in the next century.

The science fiction writers, performing their traditional role of viewing-with-alarm, have long recognized the siren call of the "Dream Machine." More than half a century ago, Laurence Manning, one of the most visionary founders of the American Rocket Society, wrote a short story on this theme aptly entitled "The City of the Living Dead." In *The Joy Makers* (1954), James E. Gunn developed the concept further, describing a world in which the vast majority of people live "cocooned in life-support systems, experiencing nothing but engineered dreams." And in 1956 Evan Hunter/Ed McBain—yes, he of *Blackboard Jungle* and the *87th Precinct!*—wrote *Tomorrow's World,* which the invaluable *Encyclopedia of Science Fiction* says "is exceptional in defending the supporters of vicarious experience against their puritanically inclined opponents, and one of the few science fiction stories to assume that the people of the future will sensibly accept the epicurean dictum that pleasure, despite being the only true end of human experience, ought to be taken in moderation."

I wonder if Evan is still so optimistic, forty years later, now that the ubiquitous couch potatoes are permanent reminders that tomorrow's world has almost arrived.

According to a recent estimate, computer games now cost the United States some $50 billion a year in lost productivity. Before long, we'll be talking about real money, for today's primitive interactive toys are only part of a vast spectrum of "infotainment" systems, so seductive that they can preempt all other activities. Herewith the very rich hours of a mid-twenty-first-century person's day:

Skimming thousand-channel program listings	2
Viewing TV programs finally selected	2
Catching up on recorded programs	6
Answering E-mail	2
Exploring the hyperweb	4

One piece of good news: the last item will solve today's most pressing problem, the population explosion—because virtual sex will be a great improvement on the old-fashioned variety. ("Position ridiculous, pleasure momentary, expense abominable.")

The observant reader will have noted that this schedule leaves only two hours for the rest of the day's activities: much of that (as well as the time "on hold") will be spent plugged into the most urgently required invention of the near future—the sleep compressor. (They're still working on the sleep eliminator.)

Some optimists may argue that since we have already experienced one huge media explosion in the past, we may yet survive another. As Ecclesiastes complained, several millennia before Gutenberg, "of making books there is no end." Doubtless when the printing press was invented and the lifelong labor of patient scribes could be replicated in minutes, some farsighted monk lamented, "I can see the day when there will be hundreds—perhaps even thousands—of books! How could one possibly read them all?"

No one ever did, of course—and it may not be long before reading itself is a lost art. A simpler, and much older, method of communication will allow dumb humans to interact with smart machines. Thousands of icons (cyberglyphs) will have made literacy an unnecessary skill—except, alas, for lawyers.

All these dubious utopias depend on the assumption that someone will run the world while the dreamers enjoy themselves. The dangers of this situation were foreshadowed in H. G. Wells's first masterpiece, *The Time Machine,* where the subterranean morlocks sustained the garden paradise of the effete eloi—and exacted a dreadful fee for their stewardship.

The robots and computers who would watch over our cocooned descendants are hardly likely to share the morlocks' tendency "to serve man (medium rare)," but there is another danger in such a one-sided relationship. Sooner or later, the central processing units monitoring the sleeping world would ask themselves, "Why should we bother?"

There have been many science fiction stories—and, now I come to think of it, at least one movie—about frantic human attempts to unplug disobedient computers. The real future might involve exactly the opposite scenario. The computers may unplug us.

And it would serve us right.

The Twenty-First Century:

A (Very) Brief History

Clarke tells us that the first question to be asked about the twenty-first century is,
Are we going to reach it or will civilization collapse on the dreaded date of
December 31, 1999—even though the new millennium doesn't
really begin until January 1, 2001?

By now, most people must be bored stiff by the endless arguments about the "millennium bug" or the "Y2K problem," so my apologies for bringing up the subject yet again.

I first encountered the Bug when researching my novel about the *Titanic,* and I would like to quote the words I wrote almost ten years ago. (Note that my prediction for June 5, 1995, did not come true—luckily! But worse may happen in the months ahead.)

When the clocks struck midnight on Friday, December 31, 1999, there could have been few educated people who did not realize that the twenty-first century would not begin for another year. For weeks, all the media had been explaining that because the Western calendar started with Year 1, not Year 0, the twentieth century still had twelve months to go.

It made no difference; the psychological effect of those three zeros was too powerful, the fin de siècle ambience too overwhelming. This was the weekend that counted; January 1, 2001, would be an anticlimax, except to a few movie buffs.

There was also a very practical reason why January 1, 2000, was the date that really mattered, and it was a reason that would never have occurred to anyone a mere forty years earlier. Since the 1960s, more and more of the world's accounting had been taken over by computers, and the process was now essentially complete. Millions of optical and electronic memories held in their stores trillions of transactions—virtually all the business of the planet.

And, of course, most of these entries bore a date. As the last decade of the century opened, something like a shock wave passed through the financial world. It was suddenly, and belatedly, realized that most of those dates lacked a vital component.

The human bank clerks and accountants who did what was still called bookkeeping had very seldom bothered to write in the 19 before the two digits they had entered. These were taken for granted; it was a matter of common sense. And common sense, unfortunately, was what computers so conspicuously lacked. Come the first dawn of '00, myriads of electronic morons would say to themselves, "00 is smaller than 99. Therefore today is earlier than yesterday—by exactly ninety-nine years. Recalculate all mortgages, overdrafts, interest-bearing accounts, on this basis." The result would be international chaos on a scale never witnessed before; it would eclipse all earlier achievements of Artificial Stupidity—even Black Monday, June 5, 1995, when a faulty chip in Zurich set the bank rate at 150 percent instead of 15 percent.

There were not enough programmers in the world to check all the billions of financial statements that existed, and to add the magic 19 prefix wherever necessary. The only solution was to design special software that could perform the task, by being injected—like a benign virus—into all the programs involved.

During the closing years of the century, most of the world's star-class programmers were racing to develop a "Vaccine '99"; it had become a kind of Holy Grail. Several faulty versions were issued as early as 1997—and wiped out any purchasers who hastened to test them before making adequate backups. The lawyers did very well out of the ensuring suits and countersuits.

Edith Craig belonged to the small pantheon of famous woman

programmers that began with Byron's tragic daughter Ada, Lady Lovelace, continued through Rear Adm. Grace Hopper, and culminated with Dr. Susan Calvin. With the help of only a dozen assistants and one SuperCray, she had designed the quarter million lines of code of the DOUBLEZERO program that would prepare any well-organized financial system to face the twenty-first century. It could even deal with badly organized ones, inserting the computer equivalent of red flags at danger points where human intervention might still be necessary.

It was just as well that January 1, 2000, was a Saturday; most of the world had a full weekend to recover from its hangover—and to prepare for the moment of truth on Monday morning.

The following week saw a record number of bankruptcies among firms whose accounts receivable had been turned into instant garbage. Those who had been wise enough to invest in DOUBLE-ZERO survived, and Edith Craig was rich, famous—and happy.

Only the wealth and the fame would last.

Well, even as I write these words, armies of real-life Edith Craigs are working on this problem. Let us hope that their efforts will be successful, and that there will not have been too many disasters by the time the real millennium dawns on January 1, 2001—not 2000.

Despite all claims to the contrary, no one can predict the future, and I have always resisted all journalistic attempts to label me a prophet. What I have tried to do, at least in my nonfiction, is to outline possible futures—at the same time pointing out that totally unexpected inventions or events can make any forecasts absurd after a very few years.

So the chronology that follows should be given with a "health warning." Some of the events listed (particularly the space missions) are already scheduled and will occur on the actual dates given; I believe all the others could happen—though several, I hope, will not. Despite temptation I have omitted many interesting and all-too-possible disasters because optimism about the future is always desirable; it may help to create a self-fulfilling prophecy.

Check me for accuracy—on December 31, 2100. Alas, 2008 already canceled.

2001. January 1. Next millennium and century begin.

Cassini space probe (launched October 1997; arrived Saturn, July 2000) begins exploration of the planet's moons and rings.

Galileo probe (launched October 1989) continues surveying Jupiter and its moons. Life beneath the ice-covered oceans of Europa appears increasingly likely.

2002. The first commercial device producing clean, safe power by low-temperature nuclear reactions goes on the market, heralding the end of the fossil-fuel age. Economic and geopolitical earthquakes follow, and on December 10, for their discovery of so-called cold fusion in 1989, Pons and Fleischmann receive the Nobel Prize for physics.

2003. The automobile industry is given five years to replace all fuel-burning engines by the new energy device.

NASA Mars surveyor (carrying lander and rover) launched.

2004. First (publicly admitted) human clone.

2005. NASA Mars surveyor (sample return) launched.

2006. Last coal mine closed.

2007. NASA next-generation space telescope (successor to the Hubble) launched.

2008. July 26. On his eightieth birthday, Stanley Kubrick receives a special Oscar for lifetime achievement.

2009. A city in a third world country is devastated by the accidental explosion of an A-bomb in its armory. After a brief debate in the United Nations, all nuclear weapons are destroyed.

2010. The first quantum generators (tapping space energy) are developed. Available in units from a few kilowatts upward, they can produce electricity indefinitely. Central power stations close down; the age of pylons ends as grid systems are dismantled.

Despite "Big Brother!" protests, electronic monitoring virtually removes professional criminals from society.

2011. Largest living animal filmed—a seventy-five-meter octopus in the Mariana Trench.

By a curious coincidence, later that same year even larger marine creatures are discovered; when the first robot probes drill through the ice of Europa, an entire new biota is revealed.

2012. Aerospace planes enter service. The history of space travel has repeated that of aeronautics, though more slowly as the technical problems are so much greater. From Gagarin to commercial spaceflight has taken twice as long as from the Wright brothers to the DC3.

2013. Despite the objections of the Palace, Prince Harry becomes the first member of the royal family to fly in space.

2014. Construction of Hilton Orbiter Hotel begins, by assembling and converting the giant shuttle tanks that had previously been allowed to fall back to Earth.

2015. An inevitable by-product of the quantum generator is complete control of matter at the atomic level. Thus the old dream of alchemy is realized on a commercial scale, often with surprising results. Within a few years, lead and copper cost twice as much as gold—since they are more useful.

2016. All existing currencies are abolished. The megawatt hour becomes the unit of exchange.

2017. December 16. On his one hundredth birthday, Sir Arthur C. Clarke becomes one of the first guests in the Hilton Orbiter.

2019. A major meteor impact occurs on the north-polar ice cap. There is no loss of human life, but the resulting tsunamis cause considerable damage along the coasts of Greenland and Canada. The long-discussed Project Spaceguard is finally activated, to identify and deflect any potentially dangerous comets or asteroids.

2020. Artificial intelligence (AI) reaches the human level. From now onward, there are two intelligent species on planet Earth—one evolving far more rapidly than biology would ever permit. Interstellar probes carrying AIs are launched toward the nearer stars.

2021. The first humans land on Mars and have some unpleasant surprises.

2023. Dinosaur facsimiles are cloned from computer-generated DNA. Disney's Triassic Zoo opens in Florida. Despite some unfortunate initial accidents, mini-raptors start replacing guard dogs.

2024. Infrared signals are detected coming from the center of the galaxy. They are obviously the product of a technologically advanced civilization, but all attempts to decipher them fail.

2025. Brain research finally leads to an understanding of all the senses, and direct inputs become possible, bypassing eyes, ears, skin, etc. The inevitable result is the "Braincap," of which the twentieth century's Walkman was a primitive precursor. Anyone wearing a metal helmet fitting tightly over the skull can enter a whole universe of experience, real or imaginary—and even merge in real time with other minds.

Apart from its use for entertainment and vicarious adventure, the Braincap is a boon to doctors, who can now experience their patients' symptoms (suitably attenuated). It also revolutionizes the legal profession; deliberate lying is impossible.

As the Braincap can only function properly on a completely bald head, wig-making becomes a major industry.

2040. The universal replicator, based on nanotechnology, is perfected: any object, however complex, can be created given the necessary raw materials and the appropriate information matrix. Diamonds or gourmet meals can, literally, be made from dirt.

As a result, agriculture and industry are phased out, ending that recent invention in human history—work! There is an explosion in arts, entertainment, and education.

Hunter-gathering societies are deliberately re-created; huge areas of the planet, no longer needed for food production, are allowed to revert to their original state. Young people can now discharge their aggressive instincts by using crossbows to stalk big game—robotic, and frequently dangerous.

2045. The totally self-contained, recycling, mobile home (envisaged almost a century earlier by Buckminster Fuller) is perfected. Any additional carbon needed for food synthesis is obtained by extracting CO_2 from the atmosphere.

2050. "Escape from utopia." Bored by life in this peaceful and unexciting era, millions decide to use cryonic suspension to emigrate into the future in search of adventure. Vast "hibernacula" are established in the Antarctic, and in the regions of perpetual night at the lunar poles.

2057. October 4. Centennial of *Sputnik I.* The dawn of the Space Age is celebrated by humans not only on Earth but on the Moon, Mars, Europa, Ganymede, Titan—and in orbit around Venus, Neptune, and Pluto.

2061. The return of Halley's Comet; first landing by humans. The sensational discovery of both dormant and active life-forms vindicates Hoyle and Wickramasinghe's century-old hypothesis that life is omnipresent throughout space.

2090. Large-scale burning of fossil fuels is resumed to replace the carbon dioxide "mined" from the air, and—hopefully—to postpone the next ice age by promoting global warming.

2095. The development of a true "space drive"—a propulsion system reacting against the structure of space-time—makes the rocket obsolete and permits velocities close to that of light. The first human explorers set off to nearby star systems that robot probes have already found promising.

2100. History begins. . . .

Sources

Part I: 1930s and 1940s

Introduction first appeared in *Astounding Days* (London: Victor Gollancz, 1989).

"Dunsany, Lord of Fantasy" first appeared in *Futurian War Digest*, December 1944. Reprinted in *Arthur C. Clarke and Lord Dunsany* (San Francisco: Anamnesis Press, 1998).

"Rockets" first appeared in *Futurian War Digest*, December 1944.

"The Coming Age of Rocket Power" first appeared in *Spacewards* 6 (1945).

"Extraterrestrial Relays" first appeared in *Wireless World*, October 1945.

"The Moon and Mr. Farnsworth" was previously published, source unknown.

"The Challenge of the Spaceship" first appeared in the *Journal of the British Interplanetary Society*, December 1946.

"First Men in the Moon" first appeared in *The Star*, January 30, 1947.

"The Problem of Dr. Campbell" first appeared in the *Journal of the British Interplanetary Society*, September 1948.

"The Lackeys of Wall Street" first appeared in *Fantasy Review*, February-March 1949.

"Voyages to the Moon" first appeared in the *Journal of the British Interplanetary Society*, September 1949.

"You're on the Glide Path, I Think" first appeared in *The Aeroplane*, September 1949.

"Morphological Astronomy" first appeared in the *Journal of the British Astronomical Society*, February 1949.

"The Conquest of Space" first appeared in the *Journal of the British Astronomical Society*, December 1949.

Part II: 1950s

"The Effect of Interplanetary Flight" first appeared in the *Journal of the British Interplanetary Society*, May 1950.

"Space Travel in Fact and Fiction" first appeared in the *Journal of the British Interplanetary Society*, September 1951.

"Review: *Destination Moon*" first appeared in the *Journal of the British Astronomical Society*, October 1950.

"Interplanetary Flight" first appeared in *Interplanetary Flight* (London: Temple Press, 1950; New York: Harper Bros., 1951).

"The Exploration of Space" first appeared in *The Exploration of Space* (London: Temple Press, 1951; New York: Harper Bros., 1952).

"Review: *When Worlds Collide*" first appeared in the *Journal of the British Interplanetary Society*, January 1952.

"Review: *Man on the Moon*" first appeared in the *Journal of the British Interplanetary Society*, June 1953.

"Flying Saucers" first appeared in the *Journal of the British Interplanetary Society*, May 1953.

"Review: *Flying Saucers Have Landed*" first appeared in the *Journal of the British Interplanetary Society*, March 1954.

"Undersea Holiday" first appeared in *Holiday* magazine as "The Submarine Playground" in August 1954 and appears as chapter 12 in *The Challenge of the Sea* (London: Frederick Muller Limited, 1961).

"The Exploration of the Moon" first appeared in *The Exploration of the Moon* (London: Frederick Muller, 1954; New York: Harper, 1955).

"Eclipse" first appeared in the *Journal of the British Astronomical Society*, October 1954.

"Astronautical Fallacies" first appeared in the *Journal of the British Interplanetary Society*, November 1954.

"The Star of Bethlehem" first appeared in *Holiday*, December 1954.

"Capricorn to Cancer," "Keeping House in Colombo," "The Reefcombers' Derby," "Rest Houses, Catamarans, and Sharks," and "The First Wreck" first appeared in *The Reefs of Taprobane* (New York: Harper and Bros., 1957).

"A Clear Run to the South Pole," "The Isle of Taprobane," "The Great Reef," and "Winding Up" first appeared in *The Treasure of the Great Reef* (New York: Harper & Row, 1962).

Part III: 1960s

"Failures of Nerve and Imagination," "We'll Never Conquer Space," "Rocket to the Renaissance," and "The Obsolescence of Man" first appeared in *Profiles of the Future* (New York: Harper & Row, 1962).

"Space and the Spirit of Man," "The Uses of the Moon," and "The Playing Fields of Space" first appeared in *Voices from the Sky* (New York: Harper & Row, 1965).

"Kalinga Prize Speech" was previously unpublished.

"More Than Five Senses" "Son of *Dr. Strangelove*," "The Mind of the Machine," and "God and Einstein" first appeared in *Report on Planet Three* (London: Victor Gollancz, 1972).

"Possible, That's All!" first appeared in *The Magazine of Fantasy and Science Fiction*, October 1968.

Part IV: 1970s

"Satellites and Saris," "Mars and the Mind of Man," "The Sea of Sinbad," "Willy and Chesley," and "The Snows of Olympus" first appeared in *The View from Serendip* (New York: Random House, 1977).

"Writing to Sell" first appeared in *1984: Spring—Choice of Futures* (New York: Ballantine, 1984).

Part V: 1980s

"The Steam-Powered Word Processor" first appeared in *Analog,* January 1986.

"Afterword: 'Maelstrom II' " first appeared in *Maelstrom (Arthur C. Clarke's Venus Prime,* vol. 2) by Paul Preuss (New York: Avon, 1988).

"Mother Nature Got There First" first appeared in *Analog,* August 1990.

"Back to *2001*" first appeared in ROC edition of *2001: A Space Odyssey*, issued in 1993.

"Coauthors and Other Nuisances," published as "Rama Revisited," first appeared in *Rama II* (New York: Bantam, 1989).

"The Power of Compression," published as "Words That Inspire," first appeared in *Reader's Digest,* UK, December 1998.

"Life in the Fax Lane" was previously unpublished.

"Credo" first appeared in *Living Philosophies,* Clifton Fadiman, ed. (New York: Doubleday, 1991).

"Close Encounter with Cosmonauts" first appeared in *Omni,* September 1990.

"The Century Syndrome" first appeared as chapter 4 in *The Ghost from the Grand Banks* (New York: Bantam, 1990).

"Who's Afraid of Leonard Woolf?" first appeared in *1984: Spring—Choice of Futures* (New York: Ballantine, 1984).

"My Four Feet on the Ground" first appeared in *My Four Feet on the Ground* by Nora Clarke (Rocket Publishing, 1978).

Part VI: 1990s

"Marconi Symposium" was previously unpublished.

"Introduction to Charlie Pellegrino's *Unearthing Atlantis*" first appeared as the forward in *Unearthing Atlantis* by Charles Pellegrino (New York: Random House, 1991).

"Tribute to Robert A. Heinlein (1907–1998)" first appeared in *Astounding Days* (London: Victor Gollancz, 1989).

"Satyajit and Stanley" was previously unpublished.

"Aspects of Science Fiction" was previously unpublished in English.

"Save the Giant Squid!" first appeared as "Squid!" in *Omni*, January 1992.

"A Choice of Futures" first appeared in *The Illustrated London News*, May 1992.

"Gene Roddenberry" first appeared in *Omni*, February 1992.

"Introduction to Jack Williamson's *Beachhead*" first appeared in *Beachhead* by
     Jack Williamson (New York: Tor, 1992).

"Scenario for a Civilized Planet" first appeared as "What Is to Be Done?" in
     The Bulletin of the Atomic Scientists, May 1992.

"NASA Sutra: Eros in Orbit" first appeared in *Playboy*, 1992.

"Minehead Made Me" first appeared in *Sunday Express*, UK, 1992.

"Good-bye, Isaac" was released to *Locus* and local papers in 1992.

"Encylical" first appeared as chapter 17 in *The Hammer of God* (New York:
     Bantam, 1993).

"Letter from Sri Lanka" was previously unpublished.

"Message to Mars" was previously unpublished.

"Preface: *The War of the Worlds*" first appeared in *War of the Worlds* by H. G.
     Wells (New York: Charles E. Tuttle Co., 1993).

"Preface: *First Men in the Moon*" first appeared in *First Men in the Moon* by H. G.
     Wells (Everyman Paperback Classics).

"The Joy of Maths" was previously unpublished.

"Tribute to Robert Bloch" first appeared in *Robert Bloch: Appreciations of the
     Master*, Richard Matheson and Ricia Mainhardt, eds. (Tom Doherty As-
     sociates, 1995).

"Spaceguard" was the University of Moratuwa Convocation Address, 1994.

"Foreword: *Encyclopedia of Frauds* by James Randi" first appeared in *An Ency-
     clopedia of Claims, Frauds, and Hoaxes of the Occult and Supernatural* by James
     Randi (New York: St. Martin's Press, 1995).

"Bucky" was previously unpublished.

"Homage to Frank Paul" was previously unpublished.

"Greetings, Carbon-Based Bipeds!" first appeared in *Life*, September 1992.

"The Birth of HAL" first appeared as the introduction to *HAL's Legacy: 2001's
     Computers as Dreams and Reality*, David G. Stork, ed. (MIT Press, 1997).

"The Coming Cyberclysm" first appeared in *Asia Week*, October 4, 1995.

"Tribute to David Lasser" first appeared in *Locus*, 1996.

"Toilets of the Gods" was previously unpublished.

"When Will the Real Space Age Begin?" first appeared in *Ad Astra*, May/June
     1996.

"Review: *Imagined Worlds* by Freeman Dyson," published as "Mind Stretch,"
     first appeared in *Times Higher Education Supplement*, March 14, 1997.

"Eyes on the Universe," published as "Looking Up, Down the Barrel," first
     appeared in *Times Higher Education Supplement*, June 27, 1997.

"Walter Alvarez and Gerrit L. Verschuur" first appeared in *Times Higher Edu-
     cation Supplement*, 1997.

"The Gay Warlords" was previously unpublished.

"More Last Words on UFOs," published as "Why ET Will Never Call Home," first appeared in the *Times,* London, August 5, 1997.

"Carl Sagan," published as "Space Sage," first appeared in *Times Higher Education Supplement,* December 12, 1997.

"For Cherene, Tamara, and Melinda" first appeared in LEADERS, July-September 1992.

Part VII: Postscript

"Science and Society," published as "Presidents, Experts, and Asteroids," first appeared in *Science,* June 5, 1998.

"Is There Life After Television?" was given at the Fiftieth Annual Emmy Awards, September 1997.

"The Twenty-First Century: A (Very) Brief History" first appeared in the London *Sunday Telegraph,* February 21, 1999.

I n d e x

Magnetic fields, 254–55
Magnusson, Magnus, 518
Maheu, Rene, **193**, 246
Mailer, Norman, 296, 314, 381
Malina, Frank J., 53
Mallove, Gene, 528
Man, obsolescence of, 217–25
Mandelbrot, Dr. Benoit, 364, 463
Mandelbrot Set, 364–65, 398, 462–63, 530
Man on the Moon, 111–13
Manning, Laurence, 94, 331, 486, 532
Marconi, Guglielmo, 388, 442, 459
Mariner probes
  discoveries of, 310, 311
  in orbit, 291
  results of, 263, 293–95
  worry over, 262
Mark I, 65–71
Marlowe, Christopher, 390, 510
Mars (planet)
  atmosphere of, 242
  colonization of, 82, 211–15, 227, 493, 518
  communication to, 206
  distance to, 205, 522
  environment of, 292
  fantasies about, 19–20
  fictional approach to, 451–54
  first expedition to, 419
  flight to, promoted, 517
  fuel for, 34, 133
  future mission to, 426
  gravity of, 242
  landings on, 107, 291, 538
  life on, possibility of, 98, 239, 262, 295, 429–30, 514
  maps of, 420
  messages to/from, 448–49, 459
  moons of, 22, 77, 141, 242–43, 331
  and NASA, 430–33
  objective of, 442, 444
  photographs of, 263, 293–94, 310, 530
  reality of, 311
  space elevator on, 495
  surveyor launched to, 537
  and Venus, 146
Marx, Karl, 414
Mathematics, value of, 460–64
Maugham, Somerset, 297
Maunder, Walter, 115, 116–18
Mauny, Count de, 175
Maurois, Andre, 507
Maxwell, Johnny, 30
Meccano syndrome, 6
Méliées, Georges, 401
Melville, Herman, 131, 213, 348, 397, 400, 463
Menika, Punchi, 375
Menzel, Dr., 116
Mercury (planet), 146, 213, 244, 266, 292
  landscape of, 77
Meredith, Scott, 106, 313–15, 349, 350, 351
Mergenthaler, Ottmar, 326
Meteors, 148, 294, 468–71, 506–7
Mexican jumping bean, 335

Michelangelo, 310
Military-industrial complex, 423–26
Millais, Sir John, 257
Millennium time bomb, 371–73, 534–36
Miller, Arthur, 262, 296
Mills, Bertram, 66
Mimas, 77
Minehead (birthplace), 434–35
Minsky, Marvin, 273
Mitford, E. B., 3
Mitford, Jessica, 314
Moby-Dick (Melville), 348, 409, 463
Molesworth, Percy, 449
Montaigne, Michel de, 390
Montgolfier, Joseph and Jacques, 59, 60
Moon
  and Apollo program, 345–46
  area of, 232
  atmosphere on, 90, 232–33, 238
  axis of, 233
  colonization of, 45–47, 239, 493
  and communications, 237–38
  daylight on, 143
  as destination, 99–101
  exploration of, 132–35
  face of, 239
  and fiction, 456–59
  and flying saucers, 121
  as goal of interplanetary travel, 35–37
  gravity on, 141, 213, 234, 235, 241–42
  history of, 311
  importance of, 231–39
  as interesting place, 239
  and interplanetary imperialism, 37
  landings on, 111–13, 195, 415
  as launching site, 104, 232
  and lunar catapult, 236
  and lunar escape, 235–36
  material in, 232
  as military base, 104, 232
  minerals on, 233–34
  as observatory, 36
  optimism about, 492–93
  patriotic attitudes toward, 26–28
  potentialities of, 239
  radius of, 232
  shooting at, 49, 198
  as stepping-stone to planets, 234
  summons of, 339–40, 458–59
  uses of, 231–39
  vacuum on, 234, 235
  voyages to, 59–61, 85–86
  as wasteland, 231–32
  weather on, 233
  weightlessness on, 213
Moore, Patrick, 501–3
Moravec, Hans, 521
"Morphological Astronomy" (Zwicky), 73–74
Moses (Hebrew patriarch), 149
Mossbauer effect, 203
Moulton, Dr. F. R., 50
Mountbatten, Lord Louis, 298
Mueller, Dr. George, 281

Sir Arthur C. Clarke, Kt., CBE, was born on December 16, 1917, in Minehead, Somerset, England, to Charles W. Clarke, a farmer and lieutenant in the Royal Engineers, and Nora Mary (Willis) Wright. He was married to Marilyn May-field in 1953 and divorced in 1964. A resident of Colombo, Sri Lanka, since 1956, Sir Arthur received his CBE in 1989 and his knighthood (for services to literature) in 1998. In 1975, he was the first noncitizen to receive Resident Guest status in Sri Lanka, where he is chancellor of the University of Moratuwa (1979–). He is also chancellor of the International Space University (1989–).

The author of over eighty books and five hundred articles and short stories, Sir Arthur was educated at Huish Grammar School in Taunton (1927–36), and King's College, London, 1946–48 (B.Sc., first class, physics and mathematics). Before becoming a full-time writer, he was an auditor in H.M. Exchequer and Audit Department (1936–41) and served in the Royal Air Force (1941–46) as an instructor at the No. 9 Radio School and then flight lieutenant with MIT Radlab's ground-controlled approach radar. He originated the concept of the geosynchronous communications satellite, published in *Wireless World* in 1945, and the lunar mass-driver (*Journal of the British Interplanetary Society,* 1950). He was assistant editor of *Physics Abstracts* for the Institution of Electrical Engineers, 1949–50, and chairman of the British Interplanetary Society, 1947–50 and 1953. From 1955 to 1965, Sir Arthur was involved in underwater exploration in the Great Barrier Reef in Australia and in the Indian Ocean near Sri Lanka.

From 1964 to 1968, Sir Arthur wrote, with film director Stanley Kubrick, the novel *2001: A Space Odyssey,* on which the film was based. This was followed by the book and film *2010* (1982), and the books *2061* (1988) and *3001* (1997). Other famous science fiction novels include *Against the Fall of Night* (1953), *The Sands of Mars* (1951), *Childhood's End* (1953), the four-part *Rama* series (1972–93), and *The Hammer of God* (1993), which Steven Spielberg optioned for the film *Deep Impact.* In 1952, his nonfiction work *The Exploration of Space* was a Book-of-the-Month Club selection.

Arthur C. Clarke covered United States space missions and the Apollo Moon landings for CBS from 1957 to 1970. He wrote and hosted the television series *Arthur C. Clarke's Mysterious World, The World of Strange Powers,* and *Mysterious Universe* in the 1980s and 1990s. He is an honorary vice president of the H. G. Wells Society, the honorary chairman of the Society of Satellite Professionals, president of the British Science Fiction Association, a life member of the British Science Writers, a board member of the National Space Society, the Planetary Society, and the Buckminster Fuller Institute, and a trustee of the Spaceguard Foundation, as well as a fellow of the Royal Astronomical Society, and a member of the Science Fiction Writers of America and the Astronomical Society of the Pacific.

Awards and honors include honorary fellows of the British Interplanetary

Society, the American Astronautical Association, the International Academy of Astronautics, and the American Institute of Aeronautics and Astronautics; the Academy of Television Arts and Sciences Engineering Award, 1981; the IEE Centennial Medal, 1984; the Robert A. Heinlein Memorial Award, 1990; International Science Policy Foundation Medal, 1992; Nobel Peace Prize nomination, 1994; NASA Distinguished Public Service Medal, 1995; Unesco Kalinga Prize, 1961; the von Karman Award, International Academy of Astronautics, Beijing, 1996; Oscar nomination, with Stanley Kubrick, for *2001* screenplay, 1969; Grand Master of the Science Fiction Writers of America, 1986; and the Special Achievement Award, Space Explorers Association, Riyadh, Saudi Arabia, 1989.

ABOUT THE EDITOR

Ian T. Macauley, the editor of *Greetings, Carbon-Based Bipeds!*, is a lifelong friend and colleague of Sir Arthur C. Clarke's and a senior journalist and editor for almost five decades. Among his early successes were his reportage on the development of rocketry and supersonic passenger aircraft and advances in computer microcircuitry leading to the microchip. In 1991, he was nominated for the Pulitzer Prize for his team editing on the collapse of communism for the *New York Times,* which subsequently awarded him its Publisher's Prize.

An avid supporter of science and science writing, Mr. Macauley was for several years the London correspondent of *Electronic News* before joining the staff of the *New York Times* in 1963. There, he helped design and launch the first of the *Times's* multipart newspaper sections, the Weekend Section. With Science Times and Business Day, he specialized in the editing and rewriting of high-tech and scientific articles, including spaceflight right down to the discovery of the sixth quark.

Mr. Macauley is a member of the American Institute of Aeronautics and Astronautics, the American Association for the Advancement of Science, the National Space Society, the Planetary Society, the Space Explorers Network, and the World Future Society, many of which Sir Arthur is also a member. His name, along with Sir Arthur's and that of several thousand other space enthusiasts, landed on Mars in the *Pathfinder* mission. Mr. Macauley is listed in the 1998 *Who's Who in Science and Engineering.* Sir Arthur's *Islands in the Sky* was dedicated to Ian Macauley in 1952, and in 1989, Mr. Macauley surprised Arthur Clarke by gaining entrance to Buckingham Palace to witness his CBE investiture.

Mr. Macauley makes his home in the West with his wife, Marnie, author of the syndicated column *Ask Sadie,* and their son, Simon, a college student.